AMERICAN MATHEMATICAL SOCIETY TRANSLATIONS

Series 2

Volume 93

NINE PAPERS ON FUNCTIONAL ANALYSIS

by

Ju. M. Berezanskiĭ
M. M. Dragilev
A. R. Esajan
A. G. Kostjučenko
M. G. Kreĭn
B. S. Mitjagin
D. A. Raĭkov
V. Ja. Stecenko
B. E. Veĭc

Published by the
AMERICAN MATHEMATICAL SOCIETY
Providence, Rhode Island
1970

Library of Congress Catalog Number A51-5559
International Standard Book Number 0-8218-1793-0

Printed in the United States of America

TABLE OF CONTENTS

Page

Amer. Math. Soc. Transl.
(2) Vol. 93, 1970

POSITIVE-DEFINITE FUNCTIONALS ON NUCLEAR SPACES*

A. G. KOSTJUČENKO AND B. S. MITJAGIN[1)]

CONTENTS

Introduction

The structure problem for positive-definite functionals on normed rings has been completely solved. The most powerful method used to give the solution in this case is based on an application of the Gel'fand theory of normed rings. Recently, in connection with the stormy development of the theory of linear topological spaces and of their applications, the problem arose of representing positive-definite functionals defined on various linear topological spaces. The simplest theorem of this type is the Bochner-Schwartz theorem [1]. There is also the work of I. M. Gel'fand and Shah Tao-shing [2] and N. Ja. Vilenkin [3] in which the general form of a positive-definite functional on the space of even functions in $S_{1/2}^{1/2}$ is found. The chief new fact uncovered in the consideration of positive-definite functionals on topological spaces is the possibility that the representing measure may not be unique.

In this article we present a general approach, essentially a synthesis of of the idea of M. G. Kreĭn, [4] contained in his method of direction functionals and of the general theorem [5] on the expansion in eigenfunctions of arbitrary selfadjoint operators. The idea of such a synthesis is not new [6] but the problem of the uniqueness of the measure has not been studied. Moreover, up to now only one operator has been considered, not allowing the

---

* Translation of Trudy Moskov. Mat. Obšč. 9 (1960), 283–316.

1) The basic results of this paper were reported to the October 27, 1959, session of the Moscow Mathematical Society.

possibility of obtaining multi-dimensional theorems. In our presentation the nature of the possibility that the measure is not unique will be clearly seen; namely, it is related to the question of whether certain symmetric operators that arise naturally are selfadjoint or not. We note that in the case of normed rings these symmetric operators are always bounded. In this paper we obtain a series of both new and already known multi-dimensional theorems.

The results of this paper are partially contained in [31,32]. We make several remarks on the notation. In this article we consider only complete locally convex linear topological spaces, and this will not be mentioned henceforth. For the definition of the notion of nuclearity and of the tensor see [12].

§ 1. Some problems in spectral theory

Theorem 1.1. *Let the Hilbert space H be the completion of the nuclear space Φ with respect to some scalar product (ϕ, ψ). Let $E_{\lambda\mu} = E(\Delta^{\lambda}_{-\infty}, \Delta^{\mu}_{-\infty})$ be a spectral family of projections on H, let e be a fixed vector, and let $\sigma(\lambda, \mu) = (E_{\lambda\mu}e, e)$.*

Then the derivative

$$\frac{\overline{dE_{\lambda\mu}}}{d\sigma(\lambda, \mu)} = \chi_{\lambda\mu},$$

extsts almost everywhere with respect to the measure $\sigma_{\lambda\mu}$ and is a continuous linear functional on Φ acting according to the formula

$$\chi_{\lambda\mu}(\varphi) = \frac{\overline{d(E_{\lambda\mu}e, \varphi)}}{d\sigma(\lambda, \mu)}. \tag{1.1}$$

For the proof we must repeat with insignificiant changes §2 of Chapter IV, Example 2° on Russian page 201 and page 210 of [7]. The following theorem is proved as in §4 of Chapter IV [7].

Theorem 1.2. *The set of functionals $\chi_{\lambda\mu}$ constructed in Theorem 1.1 is orthogonal and complete in the sense that for each $\phi \in \Phi$*

$$P_e\varphi = \iint \chi_{\lambda\mu}(\varphi)\chi_{\lambda\mu}\, d\sigma(\lambda, \mu) \in \Phi',$$

$$|P_e\varphi|^2 = \iint |\chi_{\lambda\mu}(\varphi)|^2\, d\sigma(\lambda, \mu).$$

In particular, if $\chi_{\lambda\mu}(\phi) = 0$ for all (λ, μ) then $P_e\phi = 0$.[1)]

---

1) H_e is the space spanned by the vectors $\{E_{\lambda\mu}e\}$, and P_e is the operator on H orthogonal to the projection onto H_e.

Two unbounded selfadjoint operators are called *commuting* if their spectral families commute. Let A_1 and A_2 be commuting continuous operators on Φ and symmetric on H. We will call the functional $\chi_{\lambda_1\lambda_2} = \chi_\lambda$ a common eigenfunctional of A_1 and A_2 with eigenvalue $\lambda = (\lambda_1, \lambda_2)$ if

$$A_1'\chi_\lambda = \lambda_1\chi_\lambda, \quad A_2'\chi_\lambda = \lambda_2\chi_\lambda.$$

Theorem 1.3. *If symmetric linear operators A_1 and A_2 acting on the nuclear space Φ and commuting in Φ can be extended to selfadjoint commuting operators on H, then they admit a complete system of common eigenfunctionals χ_λ in Φ, i.e.*

$$\varphi = \sum_\alpha \int\int \chi_\lambda^\alpha(\varphi)\, \chi_\lambda^\alpha\, d\sigma^\alpha(\lambda) \in \Phi', \quad \lambda = (\lambda_1, \lambda_2)$$

or

$$(\varphi, \psi) = \sum_\alpha \int\int \chi_\lambda^\alpha(\varphi)\, \overline{\chi_\lambda^\alpha(\psi)}\, d\sigma^\alpha(\lambda). \tag{1.2}$$

For the proof we set $E_{\lambda_1\lambda_2} = E_{\lambda_1}^{A_1} E_{\lambda_2}^{A_2}$ and use Theorem 1.2. The fact that χ_λ are common eigenfunctionals is proved just as for a single operator ([7], Russian p. 216).

Remark 1.1. For simplicity the theorems of this section are stated for two operators. It is clear that analogous propositions are true for systems of any finite[1)] number of operators, and we will use this in the sequel.

If we knew beforehand that for any $\lambda = (\lambda_1, \lambda_2)$ there exists not more than one common eigenfunctional (up to a factor) χ_λ, then as in [7, Russian page 223], formula (1.2) is simplified; namely,

$$(\varphi, \psi) = \int \chi_\lambda(\varphi)\, \overline{\chi_\lambda(\psi)}\, d\sigma(\lambda). \tag{1.3}$$

The problem of whether the measure σ in (1.3) is unique is rather important. We will now give a criterion for uniqueness.

Theorem 1.4. *Let the closures of the operators A_1 and A_2 on H be selfadjoint commuting operators. Then the measure σ is unique.*

Proof. For simplicity we will assume that A_1 and A_2 have simple

1) The theorem remains valid in the case of a countable system of mutually commuting selfadjoint operators $\{A_i,\ i = 1, 2, \cdots\}$. In this case the corresponding measure σ is concentrated on R^∞, the direct product of a countable number of lines [30, Russian page 315].

spectrum. By Theorem 1.3 we have the representation (1.3). Suppose there is a representation with another measure

$$(\varphi, \psi) = \iint \chi_\lambda(\varphi)\overline{\chi_\lambda(\psi)}\, d\sigma_1(\lambda). \tag{1.4}$$

As is obvious from this equation, the closure L of the set L_0 of functions $\tilde{\phi}(\lambda) = \chi_\lambda(\phi)$, $\phi \in \Phi$, is a subspace of $L_2(\sigma_1)$. We now remark that L_0 is closed under multiplication by λ_1 and λ_2, since the χ_λ are common eigenfunctionals of A_1 and A_2. Equations (1.3) and (1.4) show that there is an isometric correspondence between the operators of multiplication by λ_1 and λ_2 on $L \subset L_2(\sigma_1)$ and on $L_2(\sigma)$, and so they must also be selfadjoint operators on L. By the relation that exists between the spectral families of the isometric operators,

$$\int_{-\infty}^{\lambda_1}\int_{-\infty}^{\lambda_2} \chi_\mu(\varphi)\overline{\chi_\mu(\psi)}\, d\sigma(\mu) = \int_{-\infty}^{\lambda_1}\int_{-\infty}^{\lambda_2} \chi_\mu(\varphi)\overline{\chi_\mu(\psi)}\, d\sigma_1(\mu)$$

for all λ_1 and λ_2, whence $\sigma(\lambda) = \sigma_1(\lambda)$.

Remark 1.2. In the one-dimensional case it is easy to show that when A is not selfadjoint all the measures σ giving the representation (1.3) are obtained by all possible maximal extensions of A.

We note that in some sense there is a converse to the assertion of Theorem 1.4: if the measure is unique then there do not exist two different commuting selfadjoint extensions of A_1 and A_2 (the proof is analogous to the proof of Lemma 1.1 below).

In this account the assumption that the extensions of A_1 and A_2 commute played an essential role. In each concrete case this condition is verified with great difficulty [9,10]. Therefore general criteria that the closures and extensions of A_1 and A_2 commute are of interest. We will give one such criterion below for a wide class of spaces. It will play a fundamental role in the last three sections of this paper, devoted to various concrete examples.

Let A be a continuous operator on a nuclear space Φ such that the equation $A'\chi = \lambda\chi$ for all λ in Φ' has really only one (up to a constant factor) solution χ_λ. We will also assume that for any λ_0 there exists a vector $\phi \in \Phi$ such that $\tilde{\phi}(\lambda) = \chi_\lambda(\phi) \neq 0$ in a neighborhood of λ_0. If the closure of A on the Hilbert space H, the completion of Φ with respect to some scalar product, has selfadjoint extensions, then

$$(\varphi, \psi) = \int \chi_\lambda(\varphi)\, \chi_\lambda(\psi)\, d\sigma_0(\lambda), \tag{1.3'}$$

where σ_0 is some measure (we will assume that the eigenfunctionals χ_λ are normalized).

Lemma 1.1. *If* (1.3') *is possible with only one measure* $\sigma_0(\lambda)$, *then the closure of* A *on* H *is selfadjoint.*

Proof. The formula

$$E_\Delta \varphi = \int_\Delta \chi_\lambda(\varphi)\, \overline{\chi_\lambda}\, d\sigma_0(\lambda) \text{ in } \Phi' \tag{1.4}$$

for $\phi \in \Phi$ is valid for the spectral family E_λ of any selfadjoint extension of A to H, i.e. for any $\phi, \psi \in \Phi$ we have

$$(E_\Delta \varphi, \psi) = \int_\Delta \chi_\lambda(\varphi)\, \overline{\chi_\lambda(\psi)}\, d\sigma_0(\lambda). \tag{1.4'}$$

Indeed, let g be a spanning vector in H (for simplicity we may speak of one spanning vector—see the paragraph before Theorem 1.4). Then

$$\frac{dE_\lambda g}{d(E_\lambda g, g)} = a(\lambda)\, \chi_\lambda \text{ in } \Phi'.$$

But in Φ'

$$\varphi = \int_{-\infty}^{\infty} \chi_\lambda(\varphi)\, \overline{\chi_\lambda}\, |a(\lambda)|^2\, d(E_\lambda g, g)$$

and

$$\varphi = \int_{-\infty}^{\infty} \chi_\lambda(\varphi)\, \overline{a(\lambda)}\, dE_\lambda g.$$

(Here we use the Gel'fand-Kostjučenko theorem ([7], Russian pp. 209–217).) We can also write

$$(E_\Delta \varphi, \psi) = \int_\Delta \chi_\lambda(\varphi)\, \overline{\chi_\lambda(\psi)}\, |a(\lambda)^2|\, d(E_\lambda g, g).$$

$$d\sigma_0(\lambda) = |a(\lambda)|^2\, d(E_\lambda g, g). \tag{1.5}$$

Therefore (1.5) leads to (1.4).

But this implies that the operators E_λ^1 and E_λ^2 coincide on H, i.e. the selfadjoint extension of A is unique. Indeed, by (1.4), for all $\phi \in \Phi$ the vectors $E_\lambda^1\phi$ and $E_\lambda^2\phi$ coincide as elements of Φ'; but since both vectors belong to H and the scalar product is nondegenerate, $E_\lambda^1\phi = E_\lambda^2\phi$ in H. Thus, since Φ is dense in H and E_λ^1 and E_λ^2 are bounded on H, $E_\lambda^1 = E_\lambda^2$. The lemma is proved.

Let Φ_1 and Φ_2 be nuclear spaces and $\Phi = \Phi_1 \hat{\otimes} \Phi_2$ their projective tensor product. If A_1 and A_2 are continuous operators on Φ_1 and Φ_2, respectively, then we can define the operators $A_1 \otimes 1_2$ and $1_1 \otimes A_2$ on $\Phi_1 \otimes \Phi_2$ by the formulas

$$(A_1 \otimes 1_2)(\varphi_1 \otimes \varphi_2) = (A_1\varphi_1) \otimes \varphi_2, \quad (1_1 \otimes A_2)(\varphi_1 \otimes \varphi_2) = \varphi_1 \otimes (A_2\varphi_2).$$

These operators are continuous in the topology of Φ and are continued by continuity to all of Φ. As before, H denotes the completion of Φ with respect to some scalar product (ϕ, ψ).

Theorem 1.5. *Let the following conditions be satisfied:*

a) *The closure of $A_1 \otimes 1_2$ on H is a symmetric operator with selfadjoint extensions.*

b) *The closure of A_1 on the Hilbert space H, the completion of Φ_1 with respect to the scalar product*

$$(\varphi_1, \psi_1)_\xi = (\varphi_1 \otimes \xi, \psi_1 \otimes \xi), \tag{1.6}$$

is a selfadjoint operator for all $\xi \in \Phi_2$.

Let the same conditions be satisfied for A_2.

Then the closures of $A_1 \otimes 1_2$ and $1_1 \otimes A_2$ on H are mutually commuting selfadjoint operators.

Proof. Let $\xi \in \Phi_2$, and let H_ξ be the closure of the manifold $\Phi_1 \otimes \xi$ in H. $A_1 \otimes 1_2$ has dense domain of definition $\Phi_1 \otimes \xi$ in H_ξ which is closed under $A_1 \otimes 1_2$. We denote the closure of A_1 in $H_{\xi 1}$ by $\bar{A}_1$. By b) this is selfadjoint. There is a correspondence

$$\varphi_1 \longleftrightarrow \varphi_1 \otimes \xi, \quad A_1\varphi_1 \longleftrightarrow (A_1 \otimes 1_2)(\varphi_1 \otimes \xi),$$

between $H_{\xi 1}$ and H_ξ and the operators $\bar{A}_1$ and $\overline{A_1 \otimes 1_2}$ on these spaces, and by (1.6) this correspondance is isometric. This implies that the operator $\overline{A_1 \otimes 1_2}$, acting on H_ξ, must be selfadjoint. We denote its spectral family by E_λ^0.

We now remark that the spectral family E_λ of any selfadjoint extension of $\overline{A_1 \otimes 1_2}$ to H (they exist by a)) coincides with E^0_λ on the subspace H_ξ, i.e.

$$E_\lambda h = E^0_\lambda h, \quad h \in H_\xi.$$

This follows from the fact that their resolvents coincide on $H_\xi = \overline{\Phi_1 \otimes \xi}$ and

$$E_\lambda h = \lim_{\tau \to 0} \frac{1}{2\pi i} \int_{-\infty}^{\lambda} (R_{\varrho+i\tau} - R_{\varrho-i\tau})\, h\, d\varrho. \tag{1.7}$$

Since the linear hull of the elements $\phi_1 \otimes \xi$ is dense in Φ and in H, $\overline{A_1 \otimes 1_2}$ is selfadjoint on H.

We can repeat the same argument for $\overline{1_1 \otimes A_2}$.

Thus $B_1 = \overline{A_1 \otimes 1_2}$ and $B_2 = \overline{1_1 \otimes A_2}$ are selfadjoint operators on H. Since they are the closures of symmetric operators from the set $\Phi_1 \otimes \Phi_2$, for any λ_1 and λ_2 with nonzero imaginary part not only are $(B_1 - \lambda_1)D_{B_1}$ and $(B_2 - \lambda_2)D_{B_2}$ dense in H, but so are the manifolds

$$(B_1 - \lambda_1)(\Phi_1 \otimes \Phi_2) \quad \text{and} \quad (B_2 - \lambda_2)(\Phi_1 \otimes \Phi_2).$$

For any $\phi_i \in \Phi_i$ $(i = 1, 2)$ we have

$$B_1(\varphi_1 \otimes \varphi_2) = (A_1 \otimes 1_2)(\varphi_1 \otimes \varphi_2) = A_1\varphi_1 \otimes \varphi_2;$$

an analogous equation holds for A_2.

By the same arguments as above we see that the closure of $1_1 \otimes A_2$ from the manifold $(A_1 - \lambda_1)\Phi_1 \otimes \Phi_2$ is a selfadjoint operator on H. Therefore

$$(A_1 \otimes 1_2 - \lambda_1)(1_1 \otimes A_2 - \lambda_2)\Phi = (1_1 \otimes A_2 - \lambda_2)(A_1 \otimes 1_2 - \lambda_1)\Phi$$

is dense in H. On Φ the operators B_1 and B_2 commute in the most ordinary sense. Thus for $\phi \in \Phi_1 \otimes \Phi_2$

$$(B_1 - \lambda_1)(B_2 - \lambda_2)\varphi = (B_2 - \lambda_2)(B_1 - \lambda_1)\varphi = f.$$

This implies that $R^1_{\lambda 1} R^2_{\lambda 2} f = R^2_{\lambda 2} R^1_{\lambda 1} f$. Since $R^1_{\lambda 1}$ and $R^2_{\lambda 2}$ are bounded operators and the set of vectors of the form f is dense in H,

$$R^1_{\lambda_1} R^2_{\lambda_2} = R^2_{\lambda_2} R^1_{\lambda_1}.$$

Formula (1.7) thus implies that the spectral families $E^1_{\lambda_1}$ and $E^2_{\lambda_2}$ commute. The theorem is completely proved.

In concluding this section we note that results similar to Theorems 1.3 and 1.4 have been stated in a communication of C. Foiaş [8].

§2. Representation of positive-definite functionals. General scheme

Definition 2.1. A complete linear topological space Φ is called a *nuclear algebra with involution* if Φ is a nuclear space with two operations that are continuous in its topology: a commutative multiplication $\phi \otimes \psi$ and an involution $^*: \phi \to \phi^*$ with the usual properties.

An element ϕ of the algebra is called *Hermitian* if $\phi^* = \phi$.

Definition 2.2. A continuous operator A on Φ is called *characteristic*[1)] if for any $\phi, \psi \in \Phi$

$$A\varphi \circ \psi^* = \varphi \circ (A\psi)^*. \tag{2.1}$$

Definition 2.3. A continuous linear functional T on Φ is called *positive-definite* if $T(\phi \circ \phi^*) \geq 0$ for all $\phi \in \Phi$.

In the nuclear algebras that are interesting for applications there is, as a rule, no identity. But let there be an identity sequence δ_n in Φ, i.e. $\delta_n \in \Phi$ and $\delta_n \circ x \to x$ $(n \to \infty)$ in the topology of Φ for all $x \in \Phi$. Without loss of generality we may assume that the elements of this sequence are Hermitian (otherwise we may replace δ_n by $\delta_{1n} = \frac{1}{2}(\delta_n + \delta^*_n)$).

Suppose further that there is at least one strongly positive-definite functional T_0 on Φ (i.e. $T_0(\phi \circ \phi^*) > 0$ for all $\phi \in \Phi$, $\phi \neq 0$), with representation known to us. Then in the study of positive-definite functionals we may restrict our consideration to strongly positive-definite functionals: instead of a positive-definite functional T it suffices to study the strongly positive-definite functional $T + T_0$.

An operator A on an algebra with involution is called real if $(Ax)^* = Ax^*$ for all $x \in \Phi$.

The relation

$$A(x \circ y) = Ax \circ y = x \circ Ay. \tag{2.2}$$

1) If Φ is a ring of functions on some Lie group G then the real Laplace operators on the group are characteristic. The Lie operators of the group will be the fundamental characteristic operators.

is valid for characteristic operators. Indeed, (2.1) implies

$$A(x \circ y) \circ z^* = x \circ y \circ (Az)^* = x \circ Ay \circ z^* = Ax \circ y \circ z^*.$$

Setting $z^* = \delta_n$ and passing to the limit as $n \to \infty$, we obtain (2.2).

Every characteristic operator is real. Indeed,

$$Ax \circ y^* = x \circ (Ay)^*,$$

but by (2.2)

$$Ax \circ y^* = x \circ Ay^* = x \circ (Ay)^*.$$

Using the identity sequence we deduce from the last equation that A is real.

The characteristic operators on the algebra Φ form a subring $\mathfrak{A}$ of the ring of all continuous operators on Φ with the topology of strong convergence.

The ring $\mathfrak{A}$ is commutative.

Indeed,

$$AB(\delta_n \circ y) = A(B\delta_n \circ y) = B\delta_n \circ Ay = \delta_n \circ BAy.$$

Passing to the limit we obtain $AB = BA$ for all $A, B \in \mathfrak{A}$.

We may consider each Hermitian element $x \in \Phi$ as the operator X of multiplication by x,

$$X\varphi = x\varphi.$$

Since the algebra Φ is commutative, such an operator X will be characteristic. In this sense we may consider the subalgebra Φ_* of all Hermitian elements as being embedded in $\mathfrak{A}$.

Φ_* is dense in $\mathfrak{A}$. This follows from the equation

$$Ax = \lim_{n\to\infty} A\delta_n \circ x$$

and the fact that A is real.

Any positive-definite functional is real, i.e. $T(\phi^*) = T(\phi)$.

Proof. Each element of the algebra Φ can be represented in the form $x = x_1 + ix_2$, where x_1 and x_2 are Hermitian. Therefore it suffices to show that $T(x)$ is real for all Hermitian elements.

Since T is a positive-definite functional and $\delta_n^2 = \delta_n \circ \delta_n^*$, we have $T(\delta_n^2) \geq 0$. Then

$$T((\delta_n + x) \circ (\delta_n + x)^*) \geqslant 0 \quad \text{and} \quad T(\delta_n^2) + 2T(\delta_n \circ x) + T(x \circ x^*) \geqslant 0,$$

which implies that $T(\delta_n \circ x)$ is real for all n and therefore

$T(x) = \lim_{n\to\infty} T(\delta_n \circ x)$ is real.

Let A be a characteristic operator on Φ such that the subspace $\Phi'_\lambda = \{\chi: A'\chi = \lambda\chi_\lambda\}$ of the adjoint Φ' is one-dimensional for any (complex) λ, and let χ_λ be normalized so that

$$\lim_{n\to\infty} \chi_\lambda(\delta_n) = 1. \tag{2.3}$$

Lemma 2.1. *Under these hypotheses*

a) $\chi_\lambda(\phi \circ \psi) = \chi_\lambda(\phi)\chi_\lambda(\psi)$, *i.e.* χ_λ *are multiplicative functionals on* Φ;

b) $\chi_\lambda(\phi^*) = \overline{\chi_{\bar\lambda}(\phi)}$;

c) $\chi_\lambda(\phi \circ \psi^*) = \chi_\lambda(\phi)\overline{\chi_{\bar\lambda}(\psi^*)}$; *in particular, when* λ *is real* χ_λ *is a positive-definite functional.*

Proof. We remark that for fixed ψ the functional f defined by $f(\phi) = \chi_\lambda(\phi \circ \psi)$, belongs to Φ'_λ. Indeed,

$$(A'f)(\varphi) = f(A\varphi) = \chi_\lambda(A\varphi \circ \psi) = \chi_\lambda(A(\varphi \circ \psi)) = A'\chi_\lambda(\varphi \circ \psi)$$
$$= \lambda\chi_\lambda(\varphi \circ \psi) = \lambda f(\varphi).$$

Since Φ'_λ is one-dimensional, $\chi_\lambda(\phi \circ \psi) = g(\psi)\chi_\lambda(\phi)$. Clearly $g(\psi)$ is a linear functional, and by the same reasoning $g(\psi) = a(\lambda)\chi_\lambda(\psi)$, i.e. $\chi_\lambda(\phi \circ \psi) = a(\lambda)\chi_\lambda(\phi)\chi_\lambda(\psi)$. The normalization condition (2.3) gives $a(\lambda) = 1$. Thus a) is proved.

Assertion b) is established analogously.

After this assertion c) is obvious. Lemma 2.1 is completely proved.

In a natural way, any positive-definite functional T gives a scalar product $(\phi, \psi) = T(\phi \circ \psi^*)$ on Φ. In view of (2.1) each characteristic operator A is symmetric in this scalar product. We complete Φ with respect to this scalar product to a Hilbert space H_T. We note that

$$(x^*, y^*)_T = \overline{(x, y)_T}, \quad (x \circ y, z)_T = (x, y^* \circ z)_T \tag{2.4}$$

always holds and, in particular,

$$(x, x)_T = (x^*, x^*)_T.$$

Theorem 2.1. *The closure* $\bar{A}$ *on* H_T *of any characteristic operator* A *is a symmetric operator with the same defect indices.*

Proof. By (2.4) the involution extends to all of H_T. Therefore it suffices to show that $A'\chi = \lambda\chi$ implies $A'\chi^* = \bar{\lambda}\chi^*$. Let $x \in \Phi$. Since A is real, (2.4) implies

$$(\chi^*, Ax) = \overline{(\chi, (Ax)^*)} = \overline{(A\chi, x^*)} = \bar{\lambda}(\chi^*, x),$$

which is what we were required to prove.

Suppose further that there exists a system $A_1, \cdots, A_n$ of characteristic operators such that

1. For each $\lambda = (\lambda_1, \cdots, \lambda_n)$ there exists not more than one (or m) common eigenfunctional χ_λ, and for all of them

$$\chi_\lambda = \lim_{n\to\infty} \chi_\lambda(\delta_n) = 1. \tag{2.3'}$$

2. The closures $\bar{A}_k$ of the operators A_k on H_T

a) are selfadjoint pairwise commuting operators on H_T or

b) admit selfadjoint pairwise commuting extensions.

Then by Theorem (1.3)

$$T(x \circ y^*) = (x, y)_T = \int \chi_\lambda(x)\overline{\chi_\lambda(y)}\, d\sigma(\lambda) \tag{2.5}$$

and by the normalization (2.3′) the measure $\sigma(\lambda)$ is, in general, infinite. Setting $y = \delta_n$ in (2.5) and passing to the limit as $n \to \infty$, we obtain

$$T(x) = \int \chi_\lambda(x)\, d\sigma(\lambda). \tag{2.6}$$

Passing to the limit on the left is possible since T is a continuous functional, while we will assume that it is valid on the right. Thus we obtain a representation[1)] for any positive-definite functional. Indeed, if T_0 is a strongly positive-definite functional, then

$$T_0(x) = \int \chi_\lambda(x)\, d\sigma_0(\lambda) \text{ and } (T_0 + T)(x) = \int \chi_\lambda(x)\, d\sigma_1(\lambda)$$

with certain measures. Therefore

$$T(x) = \int \chi_\lambda(x)\, d[\sigma_1(\lambda) - \sigma_0(\lambda)]$$

---

1) The representation of a positive-definite functional has a simpler form than (2.6) when the system $\{A_i, i = 1, 2, \cdots\}$ has a simple spectrum in common. The physical significance of such a system consists of the fact that it is a complete collection of operators of the physical quantities uniquely defining the stationary state of a quantum-mechanical system.

and the fact that T is positive implies that $\sigma(\lambda) = \sigma_1(\lambda) - \sigma_0(\lambda)$ is a measure.

Theorem 2.2. *Under the above hypotheses a positive-definite functional* T *admits the representation* (2.6), *and when condition* 2a) *is fulfilled the measure* $\sigma(\lambda)$ *is unique.*

The first assertion is proved above, and the second is a corollary of Theorem 1.4.

Remark 2.1. In Lemma 2.1 the assumption that Φ'_λ is one-dimensional is not essential. Indeed, let Φ'_λ be m-dimensional, let $\chi^1_\lambda, \cdots, \chi^m_\lambda$ be a basis for Φ'_λ, and let

$$\chi^k_\lambda(\delta_n) \longrightarrow 1 \tag{2.3''}$$

for all k $(k = 1, 2, \cdots, m)$. Then by the same arguments as in Lemma 2.1 we obtain

$$\chi^k_\lambda(\varphi \circ \psi) = \sum_{i,\, j=1}^{m} a^k_{ij}(\lambda)\, \chi^i_\lambda(\varphi)\, \chi^i_\lambda(\psi) \qquad (k = 1, \ldots, m),$$

and since $\{\chi^k_\lambda;\ k = 1, \cdots, m\}$ is a basis for Φ'_λ and (2.3″) is satisfied, all the assertions of Lemma 2.1 are verified for χ^k_λ $(k = 1, \cdots, m)$ without difficulty.

The picture is the same in the case of a system of operators.

We note a useful proposition about characteristic operators.

Lemma 2.2. *Let* Φ_1 *and* Φ_2 *be algebras with involutions* $*_1$ *and* $*_2$, *respectively. We introduce the involution* $* = *_1 \otimes *_2$ *on their projective tensor product* Φ. *If* A_1 *is a characteristic operator on* Φ_1 *then* $A_1 \otimes 1_2$ *will be a characteristic operator on* Φ.

Indeed,

$$\begin{aligned}(A_1 \otimes 1_2)(\varphi_1 \otimes \varphi_2) \circ (\psi_1 \otimes \psi_2)^* &= (A_1\varphi_1 \otimes \varphi_2) \circ (\psi_1^{*1} \otimes \psi_2^{*2}) \\ &= (A_1\varphi_1 \circ \psi_1^{*1}) \otimes (\varphi_2 \circ \psi_2^{*2}) = (\varphi_1 \circ (A_1\psi_1)^*) \otimes (\varphi_2 \circ \varphi_2^{*2}) \\ &= (\varphi_1 \otimes \varphi_2) \circ (A_1\psi_1 \otimes \psi_2)^* = (\varphi_1 \otimes \psi_1) \circ ((A_1 \otimes 1_2) \circ (\varphi_2 \otimes \psi_2)).\end{aligned}$$

A positive-definite functional f_1 is called *subordinate* to a positive-definite functional f if for some c the inequality $cf(x) \geq f_1(x)$ is satisfied for all elements x of the form $x = \phi \circ \phi^*$. A positive-definite functional f is called *indecomposable* if each positive-definite functional f_1 subordinate to f is divisible by f.

Theorem 2.3. *For real* λ, *the* χ_λ *are positive-definite functionals. If*

(2.6) *holds for all positive-definite functionals on* Φ *then the* χ_λ *are indecomposable positive-definite functionals.*

Proof. The first assertion is contained in Lemma 2.1 c).

Let g be subordinate to χ_{λ_0}. By the hypotheses of the theorem

$$g(\varphi \circ \psi^*) = \int \chi_\lambda(\varphi)\chi_\lambda(\psi^*)\,d\sigma(\lambda), \qquad g(\varphi \circ \varphi^*) = \int |\chi_\lambda(\varphi)|^2\,d\sigma(\lambda).$$

Since g is subordinate, if $\chi_{\lambda_0}(\phi) = 0$ then $g(\phi \circ \phi^*) = 0$. Since for any λ and λ_0 there exists ϕ such that $\chi_\lambda(\phi) = 0$ in some neighborhood of λ_0 and $\chi_{\lambda_0}(\phi) = 0$ (this is equivalent to the fact that on Φ, $\lambda_1 \neq \lambda_2$ implies $\chi_{\lambda_1} \neq \chi_{\lambda_2}$), we have $\sigma(\{\lambda_0\}') = 0$. Thus σ is concentrated at λ_0 and therefore $g(\phi) = \int \chi_\lambda(\phi)d\sigma(\lambda) = \alpha\chi_{\lambda_0}(\phi)$, which is what we were required to prove.

§ 3. Positive-definite functionals on S_α^β spaces

The space S_α^β, where $\alpha = (\alpha_1, \ldots, \alpha_n)$ and $\beta = (\beta_1, \ldots, \beta_n)$, consists of the infinitely differentiable functions $\phi(t)$ on the n-dimensional space R^n such that

$$\sup_t \left| t_1^{k_1} \ldots t_n^{k_n} \frac{\partial^{q_1 + \ldots + q_n}}{\partial t_1^{q_1} \ldots \partial t_n^{q_n}} \varphi(t_1, \ldots, t_n) \right| \leqslant C A^k B^q k^{k\alpha} q^{q\beta}$$

for some C, $A = (A_1, \ldots, A_n)$ and $B = (B_1, \ldots, B_n)$ depending, in general, on ϕ. For the introduction of a topology on S_α^β, its completeness and a whole series of properties see([11], Chapter 4) and [12], by one of the authors. We note two propositions [12] necessary for the inclusion of the examples considered below in the general scheme.

Lemma 3.1. S_α^β *is nuclear for all* α *and* β.

Lemma 3.2. *We have the representation*

$$S_\alpha^\beta = S_{\alpha_1}^{\beta_1} \hat{\otimes} \ldots \hat{\otimes} S_{\alpha_n}^{\beta_n}.$$

We introduce multiplication and involution into S_α^β by the following formulas:

$$\left.\begin{aligned} (\varphi \circ \psi)(t) &= \int_{R^n} \varphi(\xi)\psi(t - \xi)\,d^n\xi, \\ \varphi^*(t) &= \overline{\varphi(-t)}. \end{aligned}\right\} \tag{3.1}$$

On the space $S_{\alpha_k}^{\beta_k}$ we let A_k denote the operator $A_k = id/dt_k$. It is clear that the multiplication and involution formulas (3.1) are obtained by Lemma 3.2 from the corresponding one-dimensional multiplications and involutions as in Lemma 2.2. The operators A_k are characteristic: $A_k\phi \circ \psi^* = \phi \circ (A_k\psi)^*$, the well-known rule for differentiation of a convolution. For the identity sequence in S_α^β we may take, for example $\delta_\nu(t) = \nu^n\omega(\nu t)$, where $\omega(t) \in S_\alpha^\beta$, is real on the real space R^n and $\int_{R^n}\omega(t)d^nt = 1$.

Lemma 3.3. *Only one (up to a constant factor) functional* $\chi_\lambda = e^{i(\lambda,t)}$ *can be a solution of the system*

$$A_k'\chi = \lambda_k\chi \qquad (k = 1, 2, \ldots, n) \tag{3.2}$$

in $(S_\alpha^\beta)'$, *i.e.*

$$\chi_\lambda(\varphi) = \int_{R^n} \varphi(t)\, e^{i\sum_1^n \lambda_k t_k}\, d^nt. \tag{3.3}$$

Proof. If we pass from (3.2) to the corresponding system in Fourier transforms, we have

$$(-S_k - \lambda_k)\tilde{\chi} = 0 \qquad (k = 1, 2, \ldots, n) \tag{3.4}$$

in (S_β^α). This means that on S_β^α, $\tilde{\chi}$ is zero at those and only those functions $\psi(s)$ that are zero at the point $-\lambda$, i.e. $\psi(-\lambda_1, \cdots, -\lambda_n) = 0$.

These functions form a subspace of index one in S_β^α; therefore the general solution of (3.4) contains one arbitrary constant. Clearly, $\delta(s + \lambda)$ is the solution of (3.4). Reverting to the Fourier preimage, we obtain (3.3).

In this discussion everything is true for the case when S_β^α consists of entire functions, i.e. when $\alpha_k < 1$ $(k = 1, \cdots, n)$ ([11], Russian p. 210). If some $\alpha_k \geq 1$ then for $\operatorname{Im}\lambda_k \neq 0$ systems (3.4) have only the zero solution. The lemma is proved.

Let us now consider S_α^β on the one-dimensional space R^1. By Theorem 1.3 and §2, a positive-definite functional on S_α^β admits the representation

$$T(\varphi \circ \psi^*) = \int \chi_\lambda(\varphi)\, \overline{\chi_\lambda(\psi)}\, d\sigma(\lambda), \tag{3.5}$$

where $\chi_\lambda(\phi)$ is nothing other than the usual Fourier transform of the function $\phi(t)$. In the sequel we will denote it by $\tilde{\phi}(\lambda)$. As ϕ runs over S_α^β, $\tilde{\phi}(\lambda)$ runs over all of S_β^α ([11], Russian p. 235). Since (3.5) is defined for all ϕ, $\psi \in S_\alpha^\beta$, we have

$$T(\varphi\circ\varphi^*)=\int|\tilde{\varphi}(\lambda)|^2\,d\sigma(\lambda)<\infty$$

for all $\tilde{\phi}\in S^{\alpha}_{\beta}$. The growth of the functions in S^{α}_{β} at infinity is bounded,

$$|\tilde{\varphi}(\lambda)|\leqslant Ce^{-b|\lambda|^{1/\beta}} \tag{3.6}$$

([11], Russian p. 210), where $C, b>0$ depend on $\tilde{\phi}$ and with a change in $\tilde{\phi}$ may be anything. Hence[1)] the measure $\sigma(\lambda)$ satisfies

---

1) We will make this more precise. It follows from the following proposition.

Lemma 3.4. *For any* $\epsilon>0$ *there is a function* $\phi_0(t)$ *in* S^{β}_{α}, $n=1$, *such that* $|\phi_0(t)|\geq Me^{-\epsilon|t|^{1/\alpha}}$ *for real* t.

Proof. If $\beta>1$ we may put

$$\varphi_0(t)=e^{-\varepsilon|t|^{1/\alpha}}(1-g(t)),$$

where $g(t)$ is a function of Gevrey's class β and

$$g(t)=\begin{cases}1 & \text{for } |t|<1,\\ 0 & \text{for } |t|>2.\end{cases}$$

If $\beta<1$ we use the function

$$\psi(t)=\prod_1^{\infty}\left(1-\frac{1}{n^{\varrho}}\right),\quad \varrho>1.$$

It is shown in ([11], Russian pp. 268–275) that $\psi(t)$ is an entire function of order $\leq 1/\rho$, and, moreover,

$$\ln|\psi(t)|=\frac{\pi}{\sin\frac{\pi}{\varrho}}\cos\frac{\vartheta-\pi}{\varrho}\cdot r^{\frac{1}{\varrho}}+O(\ln r),\quad 0<\vartheta<\pi.$$

Then for $\rho<2$ the rotation of the function $\psi(t^2)$ by the angle γ, $0<\gamma<\pi(1-\rho/2)$, is the function $\psi(e^{i\gamma}t^2)=g(t)$, which belongs to S_{α}, $\alpha=\rho/2$, $\beta=1-\rho/2$, and

$$|g(t)|\geqslant Me^{-L|t|^{1/\alpha}}\quad\text{for real } t,$$

where M and L are positive constants. Hence for any $\epsilon>0$ there is a number $a>0$ such that

$$|g(at)|\geqslant Me^{-\varepsilon|t|^{1/\alpha}}\quad\text{for real } t$$

and $\phi_0(t)=g(at)$ satisfies the conditions of the lemma.

Since $\rho<2$ in our considerations, the required function has been constructed for $\alpha<1/2$. If $\alpha>1/2$ then $1>\alpha'=n\alpha>1/2$ for some integer n. Then, if $\phi_0(t)$ is the function constructed above for the space $S^{\beta'}_{\alpha'}$, $\beta'=1-n(1-\beta)$, we have $\phi_0(t^n)\in S^{\beta}_{\alpha}$ and

$$\int \exp(-b|\lambda|^{1/\beta})\, d\sigma(\lambda) < \infty \quad \text{for any } b > 0. \tag{3.7}$$

We remark that

$$\widetilde{\delta}_\nu(s) = \widetilde{\nu\omega(\nu t)} = \widetilde{\omega}\left(\frac{s}{\nu}\right), \quad \widetilde{\omega}(0) = 1,$$

and

$$|\widetilde{\delta}_\nu(s)| \leqslant \Omega_0 \tag{3.8}$$

uniformly in s and ν. Set $\psi^* = \delta_n$ in (3.5). Then

$$T(\varphi \circ \delta_n) = \int \chi_\lambda(\varphi)\overline{\chi_\lambda(\delta_n)}\, d\sigma(\lambda), \tag{3.9}$$

and $\widetilde{\delta}_n(\lambda) \to 1$ uniformly on any finite segment. Since $\phi \in S^\alpha_\beta$, (3.6) and (3.8) guarantee the possibility of passing to the limit under the integral sign in the right-hand side of (3.9). Thus we arrive at the formula

$$T(\varphi) = \int \chi_\lambda(\varphi)\, d\sigma(\lambda). \tag{3.10}$$

We will show that such a representation is possible with only one measure $\sigma(\lambda)$. Suppose that

$$T(\varphi \circ \psi^*) = \int \widetilde{\varphi}(\lambda)\overline{\widetilde{\psi}(\lambda)}\, d\sigma(\lambda) = \int \varphi(\lambda)\overline{\widetilde{\varphi}(\lambda)}\, d\sigma_1(\lambda) \tag{3.11}$$

for all $\widetilde{\phi}, \widetilde{\psi} \in S^\alpha_\beta$. Then

$$\int |\widetilde{\varphi}(\lambda)|^2\, d\sigma(\lambda) = \int |\widetilde{\varphi}(\lambda)|^2\, d\sigma_1(\lambda) < \infty \tag{3.12}$$

for any $\widetilde{\phi} \in S^\alpha_\beta$, and the measures

$$d\sigma^\varphi = |\widetilde{\varphi}(\lambda)|^2\, d\sigma \quad \text{and} \quad d\sigma_1^\varphi = |\widetilde{\varphi}(\lambda)|^2\, d\sigma_1$$

$$|\varphi_0(t^n)| \geqslant Me^{-\varepsilon|t^n|^{1/\alpha'}} = Me^{-\varepsilon|t|^{1/\alpha}}$$

For $S^\beta_{1/2}$ the function $e^{-\epsilon t^2}$ serves as the necessary example.

We remark that the following lemma is proved analogously.

Lemma 3.5. *For any $K > 0$ there is a function $\psi_0(t)$ in the space ${}^+S^\beta_\alpha$ such that*

$$|\psi_0(i\tau)| \geqslant M_1 e^{k|\tau|^{1/(1-\beta)}} \quad \textit{for real } \tau.$$

This lemma is used in §4.

Propositions analogous to Lemmas 3.4 and 3.5 are true in the multidimensional case: for the function $\phi_0(t_1, \cdots, t_n)$ we may take the product of the corresponding one-dimensional functions.

are bounded. $e^{is\lambda}$ is a multiplier on $\tilde{S}_{\tilde{\beta}}^{\alpha}$ for any real s ([11], Russian p. 227); therefore, together with $\tilde{\phi}$ we have $\tilde{\psi}_s(\lambda) = e^{is\lambda}\tilde{\phi}(\lambda) \in S_{\beta}^{\alpha}$, and, by (3.11),

$$\int e^{is\lambda} |\tilde{\varphi}(\lambda)|^2 d\sigma = \int e^{is\lambda} |\tilde{\varphi}(\lambda)|^2 d\sigma_1, \quad \int e^{is\lambda} d\sigma^{\varphi} = \int e^{is\lambda} d\sigma_1^{\varphi}$$

for all s. By (3.12) and the uniqueness of the Fourier transform of a bounded measure we have $d\sigma^{\phi} = d\sigma_1^{\phi}$, and since ϕ is an arbitrary element of S_{α}^{β} we have $d\sigma = d\sigma_1$. We have thus proved the following lemma.

Lemma 3.6. *A positive-definite functional on* S_{α}^{β} $(n = 1)$ *admits the representation*

$$T = \int e^{it\lambda} d\sigma(\lambda), \tag{3.13}$$

where the measure $\sigma(\lambda)$ *is unique and satisfies* (3.7).

The uniqueness of the measure in (3.5) and (3.10), by Lemma 1.1, implies that the operator $A = id/dt$ is selfadjoint on any H_T generated by a strongly positive-definite functional T. By this remark and Lemmas 3.2 and 2.2, the operators A_k $(k = 1, 2, \cdots, n)$ on S_{α}^{β}, where $\alpha = (\alpha_1, \cdots, \alpha_n)$ and $\beta = (\beta_1, \cdots, \beta_n)$, satisfy the hypotheses of Theorem 1.5 on the Hilbert space H_T generated by a strongly positive-definite functional T on S_{α}^{β}.

Theorem 3.1. *Any continuous positive-definite functional* T *on* S_{α}^{β} *admits the representation*

$$T = \int_{R^n} e^{i(\lambda, t)} d\sigma(\lambda), \quad (\lambda, t) = \sum_1^n \lambda_k t_k, \tag{3.14}$$

where the measure σ *is unique and satisfies the following condition for any* $b_k > 0$ $(k = 1, 2, \cdots, n)$:

$$\int_{R^n} \exp\left(-\sum_1^n b_k |\lambda_k|^{\frac{1}{\beta_k}}\right) d\sigma(\lambda) < \infty. \tag{3.15}$$

T *may be extended to a continuous positive-definite functional on* S^{β}.

Conversely, each measure σ *satisfying* (3.15) *defines a positive-definite functional by* (3.14) *that is continuous on* S^{β} *and* S_{α}^{β} *for any* α.

Proof. By the remarks following Lemma 3.6, Theorem 1.5 and Lemma 3.3 lead to the representation

$$T(\varphi \circ \psi^*) = \int_{R^n} \tilde{\varphi}(\lambda) \overline{\tilde{\psi}(\lambda)} d\sigma(\lambda),$$

where the measure $\sigma(\lambda)$ is unique.

We recall that the inequality

$$|\tilde{\varphi}(\lambda)| \leqslant C \exp\left(-\sum_1^n b_k |\lambda_k|^{\frac{1}{\beta_k}}\right)$$

is satisfied for functions in S_β^α ([11], Russian p. 282), where C, $b_k > 0$ depend on ϕ. As in the one-dimensional case, we obtain (3.15) and justify the passage to the limit in the scheme of Theorem 2.2. Thus (3.14) is proved. The remaining assertions of the theorem are obvious.

A functional T is called *spherically symmetric* if $T(\phi(t)) = T(\phi(ut))$ for any orthogonal transformation u on the n-dimensional space R^n.

Theorem 3.2. *A spherically symmetric positive-definite functional T on S_α^β admits the representation*

$$T = \int_0^\infty \Omega_n(tr)\, d\sigma_0(r), \tag{3.16}$$

i.e.

$$T(\varphi) = \int_0^\infty \left(\int_{R^n} \Omega_n(tr)\, \varphi(t)\, d^n t\right) d\sigma_0(r),$$

where

$$\Omega_n(\xi) = \int_{|\lambda|=1} e^{i\lambda\xi}\, d^{n-1}S$$

and $\sigma_0(r)$ is the unique measure on the half-lines and

$$\int_0^\infty e^{-br^{1/\beta_0}}\, d\sigma_0(r) < \infty, \quad \beta_0 = \max_{1\leqslant k\leqslant n} \beta_k \tag{3.17}$$

for any $b > 0$.

T can be extended to a continuous spherically symmetric positive-definite functional on $S^{\beta_0,\dots,\beta_0}$.

We will not present a detailed proof. We note only that (3.16) and the uniqueness of the measure are obtained from Theorem 3.1, and that condition (3.17) on the measure is obtained from the fact that for $b_k > 0$ and $\gamma_k > 0$ the asymptotics of the integral

$$F(r) = \int_{|x|=r} \exp\left\{-\sum_1^n b_k |x_k|^{\gamma_k}\right\} d^{n-1}S$$

are determined by γ_0, the smallest of the numbers γ_k.

The representation (3.16) with a bounded measure for a spherically symmetric positive-definite function was first noted by Schoenberg [13].

We can introduce an involution in S_α^β different from (3.1), namely,

$$\varphi^*(t) = \overline{\varphi(\overline{t})}.$$

The functionals that are positive-definite for such an involution will be called positive-definite functionals in the sense of Bernstein, as opposed to the positive-definite functionals in the sense of Bochner considered in Theorem 3.1. We must take $B_k = d/dt_k$ $(k = 1, \cdots, n)$ as the fundamental characteristic operators in this case.

Theorem 3.3. *A functional on S_α^β that is positive-definite in the sense of Bernstein ($\alpha_k < 1$ for $k = 1, \cdots, m$; $\alpha_k \geq 1$ for $k = m+1, \cdots, n$; $0 \leq m \leq n$) admits the representation*

$$T(\varphi) = \int\limits_{R^m} \int\limits_{R^{n-m}} \widetilde{x}(\lambda_1, \ldots, \lambda_m; t_{m+1}, \ldots, t_n)\, d^{n-m}t\, d\sigma(\lambda), \tag{3.18}$$

where

$$\widetilde{x}(\lambda, t) = \int\limits_{R^m} \exp\left\{\sum_1^m \lambda_k t_k\right\} x(t_1, \ldots, t_n)\, dt_1 \ldots dt_m$$

and $\sigma(\lambda)$ is a measure that is uniquely determined and satisfies

$$\int\limits_{R^m} \exp\left\{\sum_1^m a_k |\lambda_k|^{\frac{1}{1-\alpha_k}}\right\} d\sigma(\lambda) < \infty \tag{3.19}$$

for any $a_k > 0$ $(k = 1, \cdots, m)$. A positive-definite functional T can be extended to a continuous functional that is positive-definite on S_α in the sense of Bernstein.

The converse is also true, i.e. a measure satisfying (3.19) gives, by (3.18), a continuous positive-definite functional in the sense of Bernstein on S_α and S_α^β for any β.

The proof essentially repeats the proof of Theorem 3.1. We note only two points. The uniqueness of the measure in the one-dimensional case follows from the uniqueness of the two-sided Laplace transform by the same scheme as in the proof of Lemma 3.6.

By Lemma 3.3 the common eigenfunctionals of the operators B_k can only be $\chi_\lambda = e^{(t,\lambda)}$, i.e.

$$\chi_\lambda(\varphi)=\int_{R^n}\exp\Big(\sum_1^n \lambda_k t_k\Big)\varphi(t)\,d^n t,$$

but for real λ_k different from zero this functional will not be continuous on S_α^β if $\alpha_k \geq 1$ ([11], Russian pp. 206 and 281). Therefore in (3.18) only those χ_λ with $\lambda_k = 0$ for $k = m+1, \cdots, n$ are involved.

We can give a mixed involution on S_α^β as follows:

$$\varphi^*(t_1, \ldots, t_m, t_{m+1}, \ldots, t_n)=\varphi(-\overline{t_1}, \ldots, \overline{t_m}, \overline{t_{m+1}}, \ldots, \overline{t_n}). \tag{3.20}$$

For simplicity suppose $\alpha_k < 1$ $(k = m+1, \cdots, n)$.

Theorem 3.4. *A positive-definite functional on S_α^β, in the sense of the involution* (3.20), *can be represented in the form*

$$T=\int_{R^m}\exp\Big(i\sum_{j=1}^m t_j\lambda_j+\sum_{k=m+1}^n t_k\lambda_k\Big)\,d\sigma(\lambda), \tag{3.21}$$

where $\sigma(\lambda)$ is a measure that is uniquely determined and satisfies

$$\int_{R^n}\exp\Big(-\sum_{j=1}^m b_j\,|\lambda_j|^{\frac{1}{\beta_j}}+\sum_{k=m+1}^n C_k\,|\lambda_k|^{\frac{1}{1-\alpha_k}}\Big)\,d\sigma(\lambda)<\infty. \tag{3.22}$$

The functional T admits a continuous extension to a positive-definite functional on $S^{\beta_1,\cdots,\beta_m,\infty}_{\infty,\alpha_{m+1},\cdots,\alpha_n}$.

The converse is also true, i.e. each measure satisfying (3.22) *gives, by* (3.21), *a continuous positive-definite functional on $S^{\beta_1,\cdots,\beta_m,\infty}_{\infty,\alpha_{m+1},\cdots,\alpha_n}$.*

In our scheme we can obtain various theorems of the S. N. Bernstein type on the representation of an absolutely monotone function on the half-line and in the corresponding multi-dimensional cases, in particular, the representation of a multi-dimensional continuous positive mixed positive-definite function (results obtained by A. Devinatz [14]).

We present in detail an example of this type.

Let K be a convex cone in the n-dimensional space R^n and $K' = \{\lambda = (\lambda, t) \geq 0,\ t \in K\}$ its dual cone. Let $C_0^\infty(K)$ denote the space of infinitely differentiable functions on K that, together with all their derivatives, decrease at infinity faster than any power of $|t|^{-1}$ and equal zero on the boundary of K. This space is nuclear in the corresponding topology. We introduce multiplication and involution in $C_0^\infty(K)$ by the formulas

$$(\varphi\circ\psi)(t)=\int_{Q_t}\varphi(\xi)\,\psi(t-\xi)\,d^n\xi;\quad \varphi^*(t)=\overline{\varphi(t)}, \tag{3.23}$$

where $Q_t = - K + t$.

Theorem 3.5. *A continuous positive-definite functional on* $C_0^\infty(K)$ *admits the representation*

$$T = \int_{K'} e^{-(t,\lambda)}\, d\rho(\lambda), \tag{3.24}$$

where $\rho(\lambda)$ *is a uniquely determined measure on* K'.

Proof. We call the set $Q = \{t\colon (t, e_i) \geq 0,\ 1 \leq i \leq n\}$ an affine octant in R^n, where $\{e_i,\ i = 1, \cdots, n\}$ are linearly independent vectors in R^n. For the case when the cone K is an affine octant the proof is obtained by the general scheme if for the fundamental characteristic operators on $C_0^\infty(K)$ we take differentiation in the directions e_i $(i = 1, \cdots, n)$ and represent $C_0^\infty(K)$ as the tensor product of n corresponding spaces of functions on half-lines.

Let $K \supset K_1 \cup K_2$ and $K_1 \cap K_2 \supset K_3$, where K_1, K_2 and K_3 are affine octants. Clearly, $K' \subset K_1' \cap K_2'$ and $K_1' \cup K_2' \subset K_3'$. If $Q \supset S$ we can extend a function from S to Q by setting it equal to zero on $Q \backslash S$, and we may consider $C_0^\infty(S)$ as being naturally embedded in $C_0^\infty(Q)$. Here the involution and multiplication (3.23) in $C_0^\infty(Q)$ induce the corresponding involution and multiplication in $C_0^\infty(S)$. Therefore T will also be a positive-definite functional on $C_0^\infty(K_i)$ $(i = 1, 2, 3)$, and by the theorem for the case of affine octants we have

$$T = \int_{K_i'} e^{-(t,\lambda)}\, d\rho_i(\lambda) \quad \text{on} \quad C_0^\infty(K_i) \qquad (i = 1,\ 2,\ 3), \tag{3.24}$$

in each of which the measure is defined in a natural way.

Let us consider the measure μ_1 on K_3', coinciding with ρ_1 on K_1' and equal to zero on $K_3' \backslash K_1'$. Since $C_0^\infty(K_1) \supset C_0^\infty(K_3)$, formula (3.24) for $i = 1$ implies the possibility of representing T on $C_0^\infty(K_3)$ by μ_1:

$$T = \int_{K_3'} e^{-(t,\lambda)}\, d\mu_1(\lambda).$$

But the uniqueness of the measure in (3.24) for $i = 3$ implies that $\rho_3 = \rho_1$ on K_1' and equals zero on $K_3' \backslash K_1'$. An analogous discussion for the measure ρ_2 leads to the representation

$$T = \int_{K_1' \cap K_2'} e^{-(\lambda, t)} \, d\rho(\lambda)$$

on $C_0^\infty(K_1 \cup K_2)$.

The following is clear from the above. Suppose we are given a family of affine octants $\{K_\gamma\}$ with the property that any two octants K_γ and K_δ can be connected by a finite chain of octants in the family $K_\gamma = K_0$, $K_1, \cdots, K_m = K_\delta$ such that for all $i = 0, \cdots, m-1$ the set $K_i \cap K_{i+1}$ contains an affine octant. Then a positive-definite functional on $C_0^\infty(\bigcup K_\gamma)$ admits the representation

$$T = \int_{\cap K_\gamma'} e^{-(\lambda, t)} \, d\rho(\lambda)$$

with a unique measure $\rho(\lambda)$. Since a convex cone is the union of the affine octants it contains, the theorem is completely proved.

The scheme of passing from affine octants to the cone is borrowed from S. Bochner's book [15] and evidently belongs to W. Gilbert [16].

Remark 3.1. We distinctly emphasize the fact that the measure $\rho(\lambda)$ in (3.24) turns out to be concentrated on the dual cone K' because for $\lambda \notin K'$ the functional $e^{-(\lambda, t)}$ will not be continuous on $C_0^\infty(K)$. If we consider not $C_0^\infty(K)$ but the space of functions decreasing at infinity, for example, exponentially with order greater than one, then for a positive-definite functional T we will have a representation of type (3.24) with a measure $\rho(\lambda)$ on all of R^n.

We can show that for some l the measure $\rho(\lambda)$ in (3.24) satisfies

$$\int_{K'} \frac{d\rho(\lambda)}{1 + |\lambda|^l} < \infty.$$

§ 4. Positive-definite functionals on ${}^+S_\alpha^\beta$ spaces of even functions

A representation for positive-definite functionals on the space ${}^+S_{1/2}^{1/2}$ of even functions in $S_{1/2, \cdots, 1/2}^{1/2, \cdots, 1/2}$ was found by I. M. Gel'fand and Shah Tao-shing [2] in the one-dimensional case, and by N. Ja. Vilenkin [3] in the multi-dimensional case. Below we give a complete description of the picture for various α and β in the one-dimensional case and for a wide class of indices in the multi-dimensional case.

Also, we will denote the subspace of even functions in a space Φ by ${}^+\Phi$ and the odd functions by ${}^-\Phi$.

On ${}^{+}S_{\alpha}^{\beta}$ the Bochner and Bernstein involutions coincide, since $\phi(t) = \phi(-t)$. We will take $C_k = d^2/dt_k^2$ as the fundamental characteristic operators.

We first consider the one-dimensional case, since even that contains many difficulties.

For $\alpha < 1$ the solutions of the equation $C'\chi = \lambda\chi$ in $({}^{+}S_{\alpha}^{\beta})'$ will be

$$\chi_{\lambda} = \begin{cases} \operatorname{ch} \sqrt{\lambda}\, t, & \lambda > 0, \\ \cos \sqrt{-\lambda}\, t, & \lambda < 0, \end{cases}$$

and for $\alpha > 1$

$$\chi_{\lambda} = \cos \sqrt{-\lambda}\, t, \quad \text{when } \lambda \leqslant 0,$$

while for $\lambda > 0$ there are no nontrivial solutions in this case. These assertions follow from Lemma 3.3. As a closed subspace of the nuclear space S_{α}^{β}, ${}^{+}S_{\alpha}^{\beta}$ is a nuclear space. Therefore the general scheme of §2 is applicable. The passage to the limit is justified just as in §3. Thus a positive-definite functional T on ${}^{+}S_{\alpha}^{\beta}$ admits the representation

$$\left.\begin{aligned} T(\varphi) &= \int_0^{\infty} \left(\int_{-\infty}^{\infty} \varphi(t) \cos \sqrt{\mu}\, t\, dt \right) d\sigma_1(\mu) + \\ &\quad + \int_0^{\infty} \left(\int_{-\infty}^{\infty} \varphi(t) \operatorname{ch} \sqrt{\lambda}\, t\, dt \right) d\sigma_2(\lambda) \quad \text{for } \alpha < 1, \\ T(\varphi) &= \int_0^{\infty} \left(\int_{-\infty}^{\infty} \varphi(t) \cos \sqrt{\mu}\, t\, dt \right) d\sigma(\mu) \quad \text{for } \alpha > 1. \end{aligned}\right\} \tag{4.1}$$

We will use the following fact in solving the uniqueness problem of the measure.

Lemma 4.1. *The Cauchy problem for the equation*

$$\frac{\partial u}{\partial \tau} = \frac{\partial^2 u}{\partial t^2}, \quad u(0, t) = \varphi(t) \in S_{\alpha}^{1-\alpha}, \tag{4.2}$$

in the space $(S_{\alpha}^{1-\alpha})'$ *has a solution for* $1 > \alpha \geq 1/2$ *which, moreover, is unique. The same assertion is true for the spaces* S_{α}^{0}, $\alpha > 1$.

Lemma 4.1 is a special case of Theorem 1 in ([7], Russian p. 52).

We will now show that the following lemma is valid.

Lemma 4.2. *The closure of the operator* $C = d^2/dt^2$ *on the Hilbert space* H_T *generated by a strongly positive-definite functional* T *on* ${}^{+}S_{\alpha}^{1-\alpha}$,

$1 > \alpha \geq \frac{1}{2}$ *(or* S_α^0, $\alpha > 1$*), is a selfadjoint operator.*

Proof. Since C is a characteristic operator it has equal defect indices. Let E_λ^1 and E_λ^2 be the spectral families of two of its selfadjoint extensions. If $\phi \in S_\alpha^{1-\alpha}$, $1 > \alpha \geq \frac{1}{2}$, then the series

$$e^{\tau \frac{d^2}{dt^2}} \varphi = \sum_{j=0}^{\infty} \frac{\tau^j}{j!} \frac{d^{2j}\varphi}{dt^{2j}}$$

converges in the topology of ${}^+S_\alpha^{1-\alpha}$ ([11], Russian p. 232). Therefore any ϕ in ${}^+S_\alpha^{1-\alpha}$ belongs to the domain of definition of the operator $e^{\tau C_1}$ in H_T, where C_1 is any selfadjoint extension of C. But then

$$u_1(\tau, t) = \int e^{\lambda\tau} dE_\lambda^1 \varphi,$$

and also

$$u_2(\tau, t) = \int e^{\lambda\tau} dE_\lambda^2 \varphi$$

will be solutions to the Cauchy problem (4.2). By uniqueness (Lemma 4.1),

$$\int e^{\lambda\tau} d(E_\lambda^1 \varphi, \psi) = \int e^{\lambda\tau} d(E_\lambda^2 \varphi, \psi)$$

for all $\tau > 0$ and $\phi, \psi \in {}^+S_\alpha^\beta$. We conclude from the uniqueness of the two-sided Laplace transform that

$$(E_\lambda^1 \varphi, \psi) = (E_\lambda^2 \varphi, \psi), \quad \varphi, \psi \in {}^+S_\alpha^\beta,$$

and $E_\lambda^1 = E_\lambda^2$, i.e. the closure of C on H^T is selfadjoint.

Lemma 4.2 and Theorem 1.4 imply the uniqueness of the measure in (4.1) for $\alpha \geq \frac{1}{2}$, i.e. for those pairs of indices for which (α, β) lies in G_2 (see the figure).

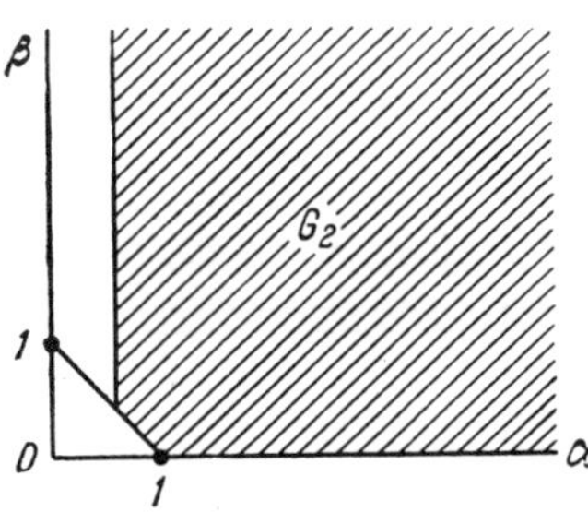

It is shown by example in [2] that the measure in the representation of a positive-definite functional on S_0^β can be not unique. The same example (which we give below) shows that the measure is also not unique in the representation of a positive-definite functional on S_α^β if (α, β) lies in G_1, i.e. $\alpha < \frac{1}{2}$.

Lemma 4.3 [2]. *If* $1 > \epsilon > 0$, *then, for an entire function* $\psi(s)$ *of order*

$p < 1 + \epsilon$,

$$\int_0^\infty \exp\left\{-x^{1+\varepsilon}\cos\frac{\pi(1+\varepsilon)}{4}\right\}\sin\left[x^{1+\varepsilon}\sin\frac{\pi(1+\varepsilon)}{4}\right]\psi(x)\,dx$$
$$= \int_0^\infty \exp\left\{-y^{1+\varepsilon}\cos\frac{\pi(1+\varepsilon)}{4}\right\}\cos\left[y^{1+\varepsilon}\sin\frac{\pi(1+\varepsilon)}{4}\right]\psi(iy)\,dy. \tag{4.3}$$

The Fourier transform of a function in S_α^β is an entire function of order $p \le \max\{1/\beta, 1/(1-\alpha)\}$. Thus, for all (α, β) in G_1, $p < 2$ and there exists an ϵ such that $1 > \epsilon > 0$, $1 + \epsilon > p$ and (4.3) is satisfied for the Fourier transform $\psi(s)$ of a function in S_α^β.

Now set

$$d\sigma_1(\lambda) = \max\left(0, \quad \exp\left\{-\lambda^{1+\varepsilon}\cos\frac{\pi(1+\varepsilon)}{4}\right\}\cos\left[\lambda^{1+\varepsilon}\sin\frac{\pi(1+\varepsilon)}{4}\right]\right)d\lambda,$$
$$d\sigma_2(\lambda) = \max\left(0, -\exp\left\{-\lambda^{1+\varepsilon}\cos\frac{\pi(1+\varepsilon)}{4}\right\}\cos\left[\lambda^{1+\varepsilon}\sin\frac{\pi(1+\varepsilon)}{4}\right]\right)d\lambda,$$
$$d\varrho_1(\lambda) = \max\left(0, \quad \exp\left\{-\lambda^{1+\varepsilon}\cos\frac{\pi(1+\varepsilon)}{4}\right\}\sin\left[\lambda^{1+\varepsilon}\sin\frac{\pi(1+\varepsilon)}{4}\right]\right)d\lambda,$$
$$d\varrho_2(\lambda) = \max\left(0, -\exp\left\{-\lambda^{1+\varepsilon}\cos\frac{\pi(1+\varepsilon)}{4}\right\}\sin\left[\lambda^{1+\varepsilon}\sin\frac{\pi(1+\varepsilon)}{4}\right]\right)d\lambda.$$

Then for the positive-definite functional

$$T(\varphi) = \int_0^\infty \tilde{\varphi}(i\lambda)\,d\sigma_1(\lambda) + \int_0^\infty \tilde{\varphi}(\lambda)\,d\sigma_2(\lambda)$$
$$= \int_0^\infty \tilde{\varphi}(i\lambda)\,d\varrho_1(\lambda) + \int_0^\infty \tilde{\varphi}(\lambda)\,d\varrho_2(\lambda),$$

on ${}^+S_\alpha^\beta$, $(\alpha, \beta) \subset G_1$, the measure is not uniquely determined.

In (4.1), by making the substitution $\lambda \to \lambda^2$ we arrive at the simpler form

$$T(\varphi) = \int \tilde{\varphi}(\mu)\,d\sigma_1^1(\mu) + \int \tilde{\varphi}(i\lambda)\,d\sigma_2^1(\lambda),$$

where $\tilde{\phi}(\mu)$ is the usual Fourier transform and σ_1^1 and σ_2^1 are even measures defined by

$$d\sigma_i^1(\mu) = \begin{cases} \frac{1}{2}\,d\sigma_i(\mu^2), & \mu > 0, \\ -\frac{1}{2}\,d\sigma_i(\mu^2), & \mu < 0, \end{cases} \qquad (i = 1, 2).$$

The following theorem is thus proved.

Theorem 4.1. *Each continuous positive-definite functional* T *on* ${}^{+}S_{\alpha}^{\beta}$, $n = 1$, *admits the representation*

$$T = \int_{+} e^{it\lambda}\, d\sigma(\lambda), \tag{4.4}$$

where $\sigma(\lambda)$ *is an even measure concentrated on the cross* $\operatorname{Re}\lambda \cdot \operatorname{Im}\lambda = 0$ *for* $\alpha < 1$ *and on the real line for* $\alpha \geq 1$ *and satisfies*

$$\int_{+} \exp\{-c\,|\operatorname{Re}\lambda|^{\frac{1}{\beta}} + d\,|\operatorname{Im}\lambda|^{\frac{1}{1-\alpha}}\}\, d\sigma(\lambda) < \infty \tag{4.5$'$}$$

for all $c, d > 0$ *if* $\alpha < 1$, *and*

$$\int_{-\infty}^{\infty} \exp\{-c\,|\operatorname{Re}\lambda|^{\frac{1}{\beta}}\}\, d\sigma(\lambda) < \infty \tag{4.5}$$

for all $c < 0$ *if* $\alpha \geq 1$.

For $\alpha \geq 1/2$ *the measure is uniquely determined, while for* $\alpha < 1/2$ *it is, in general, not uniquely determined.*

The converse is also true, i.e. a measure staisfying (4.5) *determines a continuous positive-definite functional on* ${}^{+}S_{\alpha}^{\beta}$ *by* (4.4).

We note that we could have obtained the uniqueness of the measure for $\alpha \geq 1/2$ by using a result of B. M. Levitan and N. N. Meĭman [17].

A result definitive in some sense for the uniqueness problem of the measure was obtained by E. Vul [18]. We consider the space ${}^{+}Z_{\Omega}^{\Omega}$ of even entire functions $\phi(t)$ for which the estimate

$$|\varphi(x+iy)| \leqslant c\exp\{-\Omega(ax) + \Omega(by)\}$$

holds for $c, a, b > 0$ depending, in general, on ϕ, where $\Omega(\xi)$ is some positive convex function ([11], Russian p. 294). Using the precise approximation theorems of S. N. Mergeljan, Vul arrived at the following result.

Lemma 4.4. *Let a positive-definite functional* T *on* ${}^{+}Z_{\Omega}^{\Omega}$ *admit the representation*

$$T = \int_{+} e^{it\lambda}\, d\sigma(\lambda). \tag{4.4$'$}$$

For each such T *the measure in* (4.4$'$) *is unique if and only if*

$$\int^{\infty} \frac{d\xi}{\Omega(\sqrt{\xi})} = \infty. \tag{4.6}$$

The representation (4.4′) holds, as follows from §2, if the space Z_{Ω}^{Ω} is nuclear. For a wide class of functions $\Omega(\xi)$ the nuclearity of Z_{Ω}^{Ω} is virtually proved in [12]. Because of the relation between the spaces of type S studied in [12] and the spaces Z_{Ω}^{Ω} ([7], Chapter 1) we will not go into further detail.

The following theorem holds in the multi-dimensional case.

Theorem 4.2. *Each continuous positive-definite functional T on* ${}^{+}S_{\alpha_1,\ldots,\alpha_n}^{\beta_1,\ldots,\beta_n}$, $\alpha_k > 1/2$, *admits the representation*

$$T = \int_{\Gamma} e^{it\lambda}\, d\sigma(\lambda), \tag{4.7}$$

where $\sigma(\lambda)$ is an even measure concentrated in the n-dimensional complex space on the direct product Γ of the sets Γ_k,

$$\Gamma_k = \begin{cases} \operatorname{Re}\lambda_k \cdot \operatorname{Im}\lambda_k = 0, & \text{if } \alpha_k < 1, \\ \operatorname{Im}\lambda_k = 0, & \text{if } \alpha_k \geqslant 1, \end{cases}$$

is unique and for all c_k, $d_k > 0$ $(k = 1, \ldots, n)$ satisfies

$$\int_{\Gamma} \exp\left\{ -\sum_{k=1}^{n} c_k |\operatorname{Re}\lambda_k|^{\frac{1}{\beta_k}} + \sum_{k=1}^{n} d_k |\operatorname{Im}\lambda_k|^{\frac{1}{1-\alpha_k}} \right\} d\sigma(\lambda) < \infty. \tag{4.8}$$

The converse is also true, i.e. a measure satisfying (4.8) *determines a continuous positive-definite functional on* ${}^{+}S_{\alpha}^{\beta}$ *by* (4.7).

Proof. The proof is basically the same as that of Theorem 3.1. However, in this case there is an essential singularity. The manifold of solutions of the system

$$C_k'\chi = \lambda_k\chi \qquad (k = 1, \ldots, n) \tag{4.9}$$

in $({}^{+}S_{\alpha}^{\beta})'$ is not one-dimensional as in all the examples considered above, but 2^{n-1}-dimensional. For simplicity we will restrict ourselves to the case $n = 2$. In this case the linearly independent solutions of (4.9) will be

$$\chi_\lambda^1 = \cos\sqrt{\lambda_1}t_1 \cos\sqrt{\lambda_2}t_2, \quad \chi_\lambda^2 = \sin\sqrt{\lambda_1}t_1 \sin\sqrt{\lambda_2}t_2.$$

For each strongly positive-definite functional T on ${}^{+}S_\alpha^\beta$ we have the representation

$$T(\varphi \circ \psi^*) = \sum_\gamma \int\int \chi_\lambda^\gamma(\varphi)\overline{\chi_\lambda^\gamma(\psi)}\, d(E_\lambda g^\gamma, g^\gamma) \tag{4.10}$$

by Theorem 1.3, where $\{g^\gamma\}$ is a collection of vectors generating the entire space H_T.

It is not difficult to show that

$${}^{+}S_{\alpha_1\alpha_2}^{\beta_1\beta_2} = {}^{+}S_{\alpha_1}^{\beta_1} \widehat{\otimes}\, {}^{+}S_{\alpha_2}^{\beta_2} \dot{+}\, {}^{-}S_{\alpha_1}^{\beta_1} \widehat{\otimes}\, S_{\alpha_2}^{\beta_2} = \Phi_1 \dot{+} \Phi_2.$$

We may take the system $\{g^\gamma\}$ as consisting of two subsystems $\{g^{\gamma 1}\}$ and $\{g^{\gamma 2}\}$ such that $g^{\gamma 1} \in \Phi_1$ and $g^{\gamma 2} \in \Phi_2$. Then it is clear that

$$\chi_\lambda^{\gamma_i} = b_{\gamma_i}(\lambda)\, \chi_\lambda^i \qquad (i = 1, 2).$$

Substituting these formulas in (4.10) and summing over the indices γ_1 and γ_2 we arrive at the representation

$$T(\varphi \circ \psi^*) = \sum_{k=1}^{2} \int\int \chi_\lambda^k(\varphi)\overline{\chi_\lambda^k(\psi)}\, d\sigma_k(\lambda).$$

The uniqueness of the measures σ_1 and σ_2 in this representation is established just as in Theorem 1.4, only in this case H_T must be realized as a space of functions with values in R^2 and scalar product

$$(g, f) = \sum_1^2 \int\int g_k \bar{f}_k\, d\sigma_k.$$

As for the rest, the proof of Theorem 4.2 repeats the proof of Theorem 3.1.

Having Lemma 4.4 in mind, a result analogous to Theorem 4.2 can be stated for the multi-dimensional spaces ${}^{+}Z_\Omega^\Omega = {}^{+}Z_{\Omega_1,\dots,\Omega_n}^{\Omega_1,\dots,\Omega_n}$ if (4.6) is satisfied for all Ω_k.

We consider one more example without giving detailed proofs. Let Φ_M be the space of infinitely differentiable functions on the line such that

$$|\varphi^{(k)}(t)|\, M_p(t) \leqslant C_{kp}(\varphi).$$

$M_p(t)$ are positive even functions, $M_p(t) \geq 1$, $M_p(t) \leq M_{p+1}(t)$. Under certain conditions on $M_p(t)$, Φ_M is a nuclear space ([7], Russian p. 204). In addition, assume that for any p there exist p' and C such that

$$\int_{|t-s|}^{t+s} M_p(\tau)\,d\tau \leq C M_{p'}(t)\, M_{p'}(s). \tag{4.11}$$

Let ${}^{-}\Phi_M$ be the space of odd functions in Φ_M. We introduce a commutative multiplication ([19], Russian p. 248) and involution by

$$(\varphi\cdot\psi)(t) = \frac{1}{2}\int_{-\infty}^{\infty}\int_{|t-s|}^{|t+s|} \varphi(\tau)\,\psi(s)\,ds\,d\tau, \quad \varphi^*(\tau) = \overline{\varphi(t)}. \tag{4.12}$$

By (4.11) this multiplication is continuous on ${}^{-}\Phi_M$. Any δ'-shaped sequence of functions in ${}^{-}\Phi_M$ may be taken as an identity sequence. $C = d^2/dt^2$ will be a fundamental characteristic operator. The solution to $C'\chi = \lambda\chi$ in $({}^{-}\Phi_M)'$ can only be

$$\chi_\lambda = \begin{cases} \dfrac{\operatorname{sh}\sqrt{\lambda}\,t}{\sqrt{\lambda}}, & \lambda > 0, \\[2ex] \dfrac{\sin\sqrt{-\lambda}\,t}{\sqrt{-\lambda}}, & \lambda < 0. \end{cases}$$

$M_p(t)$ grows not slower than a power, since for $\lambda < 0$, the χ_λ are continuous functionals on ${}^{-}\Phi_M$. For $\lambda > 0$, χ_λ will be a continuous functional on ${}^{-}\Phi_M$, in general, not for all λ but for a) $0 \leq \lambda \leq \lambda_0$ or b) $0 \leq \lambda < \lambda_0$ $(0 \leq \lambda_0 \leq \infty)$, where the value of λ_0 and even which of these cases occurs depend on the growth of the function $M_p(t)$. For example, if $M_p(t) = \operatorname{sh} 2t$ ([19], Russian p. 248) then $\lambda_0 = 2$ and a) occurs.

Theorem 4.3. *A continuous positive-definite functional on* ${}^{-}\Phi_M$ *for the operators* (4.12) *admits the representation*

$$T(\varphi) = \int_0^\infty \left[\int_0^\infty \varphi(t)\frac{\sin\lambda t}{\lambda}\,dt\right] d\sigma(\lambda) + \int_0^{\lambda_0}\left[\int_0^\infty \varphi(t)\frac{\operatorname{sh}\mu t}{\mu}\,dt\right] d\rho(\mu).$$

The behavior of σ at infinity and of ρ in a neighborhood of λ_0 is determined by the order of growth of the function M_p.

The uniqueness of the measures σ and ρ in the representation can be guaranteed if, for example, $M_p(t) \leq C_p e^{rt^2}$.

We can also consider this example in the multi-dimensional case analogous to Theorem 4.2.

§ 5. The moment problem

The algebras studied in §§3 and 4 consisted of functions defined on the n-dimensional space R^n. In this section we consider the algebra $K = K(a_{np})$ of sequences $\phi = \{\phi_n, n = 0, 1, \cdots\}$ with the multiplication

$$(\varphi\psi)_n = \sum_0^n \varphi_{n-i}\,\psi_i \qquad (n = 0,\ 1,\ \ldots), \tag{5.1}$$

the involution $(\phi^*)_n = \bar{\phi}_n$ and the topology of the countably normed space with the system of norms

$$|\varphi|_p = \sum_0^\infty |\varphi_n|\, a_{np},$$

where $\{a_{np}\}$ is a countable collection of sequences such that $a_{np} \le a_{n,p+1}$.

We present several facts about such algebras.

Lemma 5.1. *In order that the multiplication* (5.1) *be continuous in the topology of* $K(a_{np})$ *it is necessary and sufficient that for any* p *there exist* C *and* p' *such that*

$$a_{n+m,\,p} \leqslant C a_{n,\,p'},\quad a_{mp'} \tag{5.2}$$

for all n *and* m.

This lemma is a natural generalization of the corresponding proposition about normed rings of power series [20], and is proved in exactly the same way.

Lemma 5.2. *The dual space of* K *is the space of sequences* $f = \{f_n, n = 0, 1, \cdots\}$ *with the inductive limit topology of the Banach spaces with norms* $|f|_p = \sup_n |f_n|/a_{np}$, *where* $f(\phi) = \Sigma_0^\infty f_n\phi_n$.

Lemma 5.3. K *is a nuclear space if and only if for any* p *there exists* p' *such that*

$$\sum_0^\infty \frac{a_{np}}{a_{np'}} < \infty. \tag{5.3}$$

The proofs of these lemmas are not complicated, and we omit them.

Suppose that (5.2) and (5.3) hold. We may consider K (see [20]) as the algebra of formal power series

$$\varphi = \sum_0^\infty \varphi_n X^n.$$

The multiplication (5.1) expresses the multiplying out of power series. In this algebra $e = X^0$ is the identity. Therefore each characteristic operator A in K is the operator of multiplication by some Hermitian element, i.e. a power series with real coefficients. Indeed, $A(\phi\psi) = A\phi \cdot \psi$, and putting $\phi = e$ we obtain $A\psi = Ae \cdot \psi = x_0\psi$.

The simplest characteristic operator in K is multiplication by X,

$$V\varphi = X\varphi = \sum_0^\infty \varphi_n X^{n+1} = \sum_1^\infty \varphi_{n-1} X^n,$$

or, what is the same, the operator that shifts each sequence one place to the right,

$$(V\varphi)_n = \begin{cases} 0, & n = 0, \\ \varphi_{n-1}, & n > 0. \end{cases}$$

The adjoint operator V' in K' acts in the following way:

$$(V'f)(\varphi) = f(V\varphi) = \sum_1^\infty f_n \varphi_{n-1} = \sum_0^\infty f_{n+1} \varphi_n,$$

i.e. $(V'f)_n = f_{n+1}$. Hence for an eigenfunctional g (i.e. $V'g = \lambda g$) we must have

$$g_{n+1} = \lambda g_n \qquad (n = 1, 2, \ldots).$$

The normalization condition $g(e) = 1$ gives $g_0 = 1$. By induction

$$g_n = \lambda^n.$$

Thus

$$g_\lambda(\varphi) = \sum_0^\infty \lambda^n \varphi_n;$$

g_λ is a continuous functional on K not, in general, for all λ but for $|\lambda| \leq \lambda_0$, $0 \leq \lambda_0 \leq \infty$, where λ_0 depends on the growth of the sequences a_{np}.

A positive-definite functional T on K is such that $T(\phi \cdot \phi^*) \geq 0$, i.e.

$$T(\varphi \cdot \varphi^*) = \sum_{n=0}^\infty T_n \sum_{i=0}^\infty \varphi_i \overline{\varphi_{n-i}} = \sum_{n,\, m=0}^\infty T_{n+m} \varphi_n \overline{\varphi_m} \geqslant 0.$$

By the general scheme of §2 we have

$$T = \int_{-\lambda_0}^{\lambda_0} g_\lambda \, d\sigma(\lambda), \tag{5.4}$$

i.e.

$$T(\varphi)=\int_{-\lambda_0}^{\lambda_0} g_\lambda(\varphi)\,d\sigma(\lambda)=\int\sum_0^\infty \lambda^n\varphi_n\,d\sigma(\lambda)=\sum_0^\infty\left(\int \lambda^n\,d\sigma(\lambda)\right)\varphi_n \tag{5.4a}$$

and

$$T_n=\int \lambda^n\,d\sigma(\lambda).$$

We see that the coordinates of a positive-definite functional T on K are the power moments of some measure.

We will now give an elementary proof of a rather strong criterion, belonging to Hamburger [21], for the uniqueness of the measure σ, i.e. for the determinantion of the moment problem.

To this end we consider the countably-normed space K of sequences with the system of norms

$$|\varphi|_p=\sum_0^\infty|\varphi_n|p^n n^n;$$

and we denote the completion of K in the pth norm by K_p. We will show that for sufficiently small t the operator

$$e^{tV}=\sum_0^\infty\frac{t^kV^k}{k!}.$$

from K_q $(q>p)$ to K_p is defined and continuous.

Indeed, if $q>p$ then

$$(V^k\varphi)=\begin{cases}0, & 0\leqslant n\leqslant k-1,\\ \varphi_{n-k}, & k\leqslant n;\end{cases}$$

$$|V^k\varphi|_p=\sum_{n=0}^\infty|(V^k\varphi)_n|p^n n^n=\sum_{n=k}^\infty|\varphi_{n-k}|p^n n^n$$

$$=\sum_{n=k}^\infty|\varphi_{n-k}|q^{n-k}(n-k)^{n-k}\left(\frac{p}{q}\right)^n q^k\left(\frac{n}{n-k}\right)^{n-k}n^k\leqslant|\varphi|_q(qe)^k\sup_n(\mu^n n^k),$$

where $\mu=p/q<1$. We replace n by a continuous parameter and find the maximum of the function $y(t)=\mu^t t^k$. Since the logarithm is monotonic the maximum points of the functions $y(t)$ and $u(t)=t\ln\mu+k\ln t$ coincide,

$$u'(t) = \ln \mu + \frac{k}{t}, \qquad t_{\max} = -\frac{k}{\ln \mu}.$$

Therefore

$$y_{\max} = \left(-\frac{1}{e \ln \mu} \right)^k k^k,$$

and we obtain the estimate

$$|V^k \varphi|_p \leqslant |\varphi|_q q^k \left(-\frac{1}{\ln \mu} \right)^k k^k \qquad (k = 0, 1, \ldots);$$

It is obvious from the last inequalities that the series

$$\sum_0^\infty \frac{t^k V^k \varphi}{k!} \tag{5.5}$$

converges in the pth norm for $-eqt/\ln \mu < 1$.

The series (5.5) is a solution to the Cauchy problem

$$\frac{dw}{dt} = Vw, \quad w(0) = \varphi \in K. \tag{5.6}$$

Thus we have shown that the solution to problem (5.6) exists in any K_p for $\phi \in K$ on a sufficiently small time interval $(0, t_p)$. By Holmgren's principle ([7], Chapter 2, §2] we have the following lemma.

Lemma 5.4. *The solution to the Cauchy problem*

$$\frac{df}{dt} = V'f, \quad f(0) = \varphi \in K$$

in K'_p *is unique.*

Arguing further as in the proof of Lemma 4.2, we obtain the result that the closure of the operator V on the Hilbert space generated on K by any strongly positive-definite functional is selfadjoint. In view of Theorem 1.4 we arrive at the following result.

Theorem 5.1. *If a positive-definite functional* T *satisfies the inequalities*

$$|T_n| \leqslant C^n n^n \qquad (n = 1, 2, \ldots), \tag{5.7}$$

for some C, *then the measure in the representation* (5.4a) *is unique.*

Remark 5.1. The condition (5.7) can be rewritten in an equivalent form:

$$\overline{\lim}\ \frac{1}{n}\ T_{2n}^{\frac{1}{2n}} < \infty.$$

This is the Hamburger test.

The power moment problem on the whole line is as follows. Let T_n be a numerical sequence. Find necessary and sufficient conditions for the representation

$$T_n = \int_{-\infty}^{\infty} \lambda^n \, d\sigma(\lambda) \tag{5.8}$$

with some measure $\sigma(\lambda)$, and investigate the uniqueness problem for such a measure.

It is well known [22] that the representation (5.8) exists if and only if

$$\sum_0^N T_{n+m} \xi_n \bar{\xi}_m \geqslant 0 \tag{5.9}$$

for any ξ_n $(n = 0, 1, \cdots, N)$, or

$$\sum_0^N T_n C_n \geqslant 0 \tag{5.10}$$

for all C_n such that the polynomial $\Sigma_0^N C_n t^n$ is nonnegative for all real t. Conditions (5.9) and (5.10) are equivalent.

In our scheme, to obtain (5.8) when (5.9) is satisfied it suffices to assume as a basis the nuclear algebra of polynomials with the direct sum topology of a countable number of complex lines (this corresponds to the scheme of §5 of [23]).

The uniqueness problem for the measure is more difficult, but it also has been studied in detail. The most precise known sufficient condition for the uniqueness of the measure σ was given by Carleman [24]:

$$\sum_0^{\infty} \frac{1}{\beta_n} = \infty, \quad \text{where} \quad \beta_n = \inf_{m \geqslant n} T_{2m}^{\frac{1}{2m}}, \tag{5.11}$$

This result is obtained with the use of the powerful apparatus of analytic function theory. The Hamburger test proved above is a particular case of the

Carleman test.

We note a simple proposition.

Theorem 5.2. *If* $|T_n| \leq b^n$, $b > 0$, *then the measure* σ *is concentrated on the interval* $[-b, b]$.

Proof. $T(\phi) = \Sigma_0^\infty T_n \phi_n$ is a continuous functional on $K(a_{np})$, $a_{np} = b^n(1+n)^p$ and g_λ is continuous on $K(a_{np})$ only for $|\lambda| \leq b$. Therefore the measure $\sigma(\lambda)$ that gives the representation of a positive-definite functional T on K is concentrated on the interval $[-b, b]$, which is what we were required to prove.

The statement of the multi-dimensional moment problem is analogous to the one-dimensional problem. Let T_n, $n = (n_1, \cdots, n_k)$, be a k-fold sequence of numbers. Find necessary and sufficient conditions for the representation

$$T_{n_1, \ldots, n_k} = \int_{R^k} \lambda_1^{n_1} \ldots \lambda_k^{n_k} \, d\sigma(\lambda_1, \ldots, \lambda_k) \tag{5.12}$$

with some measure σ, and investigate the uniqueness problem for the measure.

Haviland gave [25] the following condition:

$$\sum_0^N T_{n_1, \ldots, n_k} C_{n_1, \ldots, n_k} \geqslant 0 \tag{5.13}$$

for all $C_{n_1, \cdots, n_k}$ such that the polynomial

$$P(t) = \sum_0^N C_{n_1, \ldots, n_k} t_1^{n_1} \ldots t_k^{n_k}$$

is nonnegative for all real $t = (t_1, \cdots, t_k)$.

It is not difficult to see that (5.13) implies the condition

$$\sum_0^N T_{n_1+m_1, \ldots, n_k+m_k} \xi_{n_1, \ldots, n_k} \bar{\xi}_{m_1, \ldots, m_k} \geqslant 0 \tag{5.14}$$

for any choice of numbers $\xi_{n_1, \cdots, n_k}$ $(n_i = 0, \cdots, N;\ i = 1, \cdots, k)$.

For $k \geq 2$ the converse, in general, is not true. This is shown in [9] by the construction of a sequence satisfying (5.14) for which the moment problem has no solution.

Below we give a condition on a sequence T_n, $n = (n_1, \cdots, n_k)$, with property (5.14) that guarantees the solvability of the k-dimensional moment problem for this sequence. (For the case $k = 2$ this result was obtained by Devinatz [26].)

Theorem 5.3. *Let* T_n $(n = (n_1, \cdots, n_k))$ *be a* k*-fold sequence such that*

a) *condition* (5.14) *is satisfied,*

b) *for any fixed indices* $n_1^0, \cdots, n_{j-1}^0, n_{j+1}^0, \cdots, n_k^0$, $j = 1, \cdots, k$ *the sequence*

$$\sum_{l_i=0,\, 2n_i^0} T_{l_1, \ldots, l_{j-1},\, n_j,\, l_{j+1}, \ldots,\, l_k}$$

satisfies the Carleman condition with respect to the index n_j.

Then the representation

$$T_n = \int_{R^k} \lambda_1^{n_1} \ldots \lambda_k^{n_k} \, d\sigma(\lambda), \tag{5.15}$$

is possible with a measure σ, *which is uniquely determined.*

The following lemma will be necessary for the proof of this theorem.

Lemma 5.5. *If* T_n $(n = (n_1, \cdots, n_k))$ *is a* k*-fold positive-definite sequence, i.e.* (5.14) *is satisfied, then for any* $m_1^0, \cdots, m_k^0$ *and for* $l = (l_1, \cdots, l_k)$ *such that* $0 \le l_j \le 2m_j^0$, $j = 1, \cdots, k$, *there exist* γ_ξ, $0 \le \gamma_\xi \le 1$, $\Sigma_\xi \gamma_\xi = 1$, *such that*

$$|T_l| \leqslant \prod_\xi T_\xi^{\gamma_\xi}, \tag{5.16}$$

where ξ *runs over all values of the index* n *corresponding to the vertices of the parallelepiped* $\{(0, 2m_j^0),\ j = 1, \cdots, k\}$.

Proof. The proposition is true for $k = 1$. Indeed, positive-definiteness in the one-dimensional case is sufficient for the solvability of the moment problem, and so $T_n = \int \lambda^n d\sigma(\lambda)$ for all n, where σ is some measure. Then, by Hölder's inequality,

$$|T_l| = \left| \int \lambda^l \, d\sigma(\lambda) \right| \leqslant \int 1 \cdot |\lambda^{2m}|^{\frac{l}{2m}} d\sigma(\lambda)$$

$$\leqslant \left(\int 1 \cdot d\sigma(\lambda) \right)^{1 - \frac{l}{2m}} \left(\int \lambda^{2m} \, d\sigma(\lambda) \right)^{\frac{l}{2m}} = T_0^{1 - \frac{l}{2m}} T_{2m}^{\frac{l}{2m}}.$$

for $0 \leq l \leq 2m$. Thus everything is proved for $k = 1$.

We now continue by induction on k, i.e. we assume that the lemma is valid for all $k \leq r$ and show that it is also true for $k = r + 1$.

If $T_{n_1, n_2, \cdots, n_p}$ is a p-fold positive-definite sequence, then by fixing any one index at any even value we obviously obtain a $(p-1)$-fold positive-definite sequence. Therefore if $l = (l_1, \cdots, l_{r+1})$ contains at least one even coordinate, say l_i, then by the induction hypothesis

$$|T_{l_1, \ldots, l_i, \ldots, l_{r+1}}| \leqslant \prod_{\eta} T_{\eta, l_i}^{\beta_\eta},$$

where $\Sigma_\eta \beta_\eta = 1$, and η means that the jth index, $j \neq i$, takes only the two values 0 and $2m_j^0$.

For each fixed η, T_{η, l_i} is a single-fold positive-definite sequence of index l_i; thus by (5.14)

$$|T_{\eta, l_i}| \leqslant T_{\eta, 0}^{1 - \frac{l_i}{2m_i^0}} T_{\eta, 2m_i^0}^{\frac{l_i}{2m_i^0}}.$$

Estimating the right-hand side of (5.15) with this inequality we arrive at (5.16), since $\xi = (\eta, 0)$ or $(\eta, 2m_i^0)$ and

$$\gamma_\xi = \beta_\eta \left(1 - \frac{l_i}{2m_i^0}\right) \quad \text{or} \quad \beta_\eta \frac{l_i}{2m_i^0},$$

so that $\Sigma_\xi \gamma_\xi = 1$.

If each $l_j = 2s_j + 1$ is odd (then necessarily $0 < l_j < 2m_j^0$) we first note that

$$|T_l| \leqslant T_{2s_1, \ldots, 2s_{r+1}}^{\frac{1}{2}} T_{2(s_1+1), \ldots, 2(s_{r+1}+1)}^{\frac{1}{2}}.$$

This implies that (5.16) holds when l has even coordinates and is valid for all indices, i.e. the induction assumption is justified. The lemma is completely proved.

Proof of Theorem 5.3. We consider the algebra of polynomials

$$\Phi_i = \left\{ \sum_0^N \varphi_{n_i} X_i^{n_i} \right\}, \quad i = 1, 2, \ldots, k,$$

with the direct sum topology of a countable number of lines. As the inductive limits of finite-dimensional spaces [27], the spaces Φ_i are nuclear. The usual multiplication and involution introduced in Φ_i are continuous in its topology. The tensor product

$$\Phi = \Phi_1 \hat{\otimes} \Phi_2 \hat{\otimes} \ldots \hat{\otimes} \Phi_k = \Phi_1 \otimes \Phi_2 \otimes \ldots \otimes \Phi_k$$

is the algebra of polynomials $\Sigma_0^N \xi_n X_1^{n_1} \cdots X_k^{n_k}$, or, what is the same thing, of k-fold sequences

$$\xi = \{\xi_n, \quad n = (n_1, \ldots, n_k)\}$$

with a finite number of nonzero coordinates.

Using T_n, we give a scalar product on $\Phi = \Phi_1 \otimes \cdots \otimes \Phi_k$ by

$$(\xi, \eta) = T(\xi\eta^*) = \sum T_{n+m} \xi_n \bar{\eta}_m, \tag{5.17}$$

where $m = (m_1, \cdots, m_k)$ and $n + m = (n_1 + m_1, \cdots, n_k + m_k)$.

With respect to this scalar product we complete Φ to a Hilbert space H. We will show that the closures B_i, $i = 1, 2, \cdots, k$, from Φ of the operators of multiplication by X_i are selfadjoint commuting operators on H. For this it suffices to verify the hypotheses of Theorem 1.5.

For fixed $\psi \in \Phi_2 \otimes \cdots \otimes \Phi_k$ and any $\phi, \omega \in \Phi_1$ we have

$$(\varphi \otimes \psi, \omega \otimes \psi) = \sum T_{n_1+m_1, n_1'+m_1'} \varphi_{n_1} \psi_{n_1'} \bar{\omega}_{m_1} \bar{\psi}_{m_1'},$$

where $n_1' = (n_2, \cdots, n_k)$ and $m_1' = (m_2, \cdots, m_k)$, and

$$(\varphi \otimes \psi, \omega \otimes \psi) = \sum_{n_1, m_1} Z_{n_1+m_1} \varphi_{n_1} \bar{\omega}_{m_1},$$

where

$$Z_l = \sum_{n_1', m_1'=0}^{N} T_{l, n_1'+m_1'} \psi_{n_1'} \bar{\psi}_{m_1'}, \tag{5.18}$$

which, by hypothesis b) of the theorem, satisfies the Carleman condition. Indeed, for any l, $T_{2l, n_1'}$ is a positive-definite $(k-1)$-fold sequence of index n_1'. By Lemma 5.5

$$|T_{2l_1, n_1'}| \leqslant \prod_{\zeta} T_{2l, \zeta}^{\delta_\zeta}, \tag{5.19}$$

where ζ are all possible vertices of the $(k-1)$-dimensional cube with $2N$ sides, $0 \leq \delta_\zeta \leq 1$, $\Sigma_\zeta \delta_\zeta = 1$, and δ_ζ does not depend on l. Since (5.18) has a finite number of terms, for

$$C_1(\psi)=\max_{0\leqslant n_1',\, m_1'\leqslant N}|\psi_{n_1'}\bar{\psi}_{m_1'}|$$

we have

$$|Z_{2l}|\leqslant C_1(\psi)\sum_{n_1',\, m_1'=0}^{N}|T_{2l,\, n_1'+m_1'}|$$
$$\leqslant C_1(\psi)N^{2(k-1)}\max_{0\leqslant n_1'\leqslant 2N}|T_{2l,\, n_1'}|=C_2(\psi,N)|T_{2l,\, n_1'^0}| \tag{5.20}$$

for some $n_1'^0$.

Since for any $a_i>0$, $0\leq\delta_i\leq 1$, $\Sigma_1^s\delta_i=1$ the inequality $\Pi_{i=1}^s a_i^{\delta_i}\leq\Sigma_{i=1}^s\delta_i a_i$ is satisfied ([28], Russian p. 29), (5.19) and (5.20) lead to the estimate

$$|Z_{2l}|\leqslant C_2(\psi,N)\prod_{\zeta}T_{2l,\,\zeta}^{\delta_\zeta}\leqslant C_2(\psi,N)\sum_{\zeta}\delta_\zeta T_{2l,\,\zeta}\leqslant C_2(\psi,N)\sum_{\zeta}T_{2l,\,\zeta}. \tag{5.21}$$

By hypothesis b) of the theorem the sequence $t_l=\Sigma_\zeta T_{l,\zeta}$ satisfies the Carleman condition, i.e. $\Sigma_1^\infty 1/\tau_n=\infty$, where

$$\tau_n=\inf_{l\geqslant n}t_{2l}^{\frac{1}{2l}}.$$

But by (5.21)

$$\beta_n=\inf_{l\geqslant n}Z_{2l}^{\frac{1}{2l}}\leqslant\tau_n\cdot C_3,$$

so that the series $\Sigma_1^\infty 1/\beta_n$ also converges.

Since the sequence Z_l satisfies the Carleman condition, the measure in the representation

$$Z_1(\varphi)=\int_{-\infty}^{\infty}g_{\lambda_1}(\varphi)\,d\sigma(\lambda_1)$$

of the functional Z_1 on Φ_1 is unique; thus by Theorem 1.5 the operator of multiplication by X_1 is selfadjoint on any $H_{1\psi}$.

In precisely the same way we establish that the operators of multiplication by X_i are selfadjoint on any $H_{\phi i}$, $i = 2, \cdots, k$.

The general eigenfunctional of the operator X_i', $i = 1, \cdots, k$, with eigenvalue $\lambda = (\lambda_1, \cdots, \lambda_k)$ will be g_λ, where

$$g_\lambda(\xi) = \sum \xi_n \lambda_1^{n_1} \ldots \lambda_k^{n_k}.$$

Then by the general scheme of §2 we have the representation

$$T(\xi\eta^*) = \int_{R_k} g_\lambda(\xi) \overline{g_\lambda(\eta)} \, d\sigma(\lambda),$$

and the measure σ is unique by Theorem 1.5.

Setting $\xi = X_1^{n_1} \cdots X_k^{n_k}$, $\eta = 1$, we obtain

$$T_n = \int_{R_k} \lambda_1^{n_1} \ldots \lambda_k^{n_k} \, d\sigma(\lambda),$$

where $\sigma(\lambda)$ is a uniquely determined positive measure on the k-dimensional space R^k. The theorem is proved.

We note that we have virtually obtained the solution to the following infinite-dimensional moment problem.

Theorem 5.4. *Let $T(n)$ be a function on the integral lattice $\mathfrak{A}$ in the direct sum Σ_Ω of lines, where Ω is a set of any cardinality, and be positive-definite in the sense that*

$$\sum T(n+m)\, \xi(n) \overline{\xi(m)} \geqslant 0$$

for any vector ξ in $\Sigma_{\mathfrak{A}}$, where $\mathfrak{A}$ is the integral lattice in Σ_Ω. For any p let the p-fold sequence

$$T(n_{\omega_1}, n_{\omega_2}, \ldots, n_{\omega_p}, 0, 0, \ldots)$$

satisfy hypothesis b) *of Theorem* 5.3.

Then in the direct product R_Ω of lines there exists a uniquely determined positive measure σ such that for any $n = (n_\omega) \in \mathfrak{A}$,

$$T_n = \int_{R_\Omega} \lambda_{\omega_1}^{n_{\omega_1}} \ldots \lambda_{\omega_N}^{n_{\omega_N}} \, d\sigma(\lambda).$$

Theorem 5.4 is a corollary of Theorem 5.3 and of Kolmogorov's theorem

([29], Russian p. 39] on the extension of a measure from the cylindrical sets to all the Borel sets.

Indeed, for a fixed finite set $K = (\omega_1, \omega_2, \cdots, \omega_p) \subset \Omega$ the sequence $S(n_1, \cdots, n_p) = T(n)$, where

$$n_\omega = \begin{cases} 0, & \omega \notin K, \\ n_i, & \omega = \omega_i, \quad i = 1, \ldots, p, \end{cases}$$

satisfies all the hypotheses of Theorem 5.3, and thus

$$S(n_1, \ldots, n_p) = \int \lambda_{\omega_1}^{n_1} \ldots \lambda_{\omega_p}^{n_p} \, d\sigma^k (\lambda_{\omega_1}, \ldots, \lambda_{\omega_p}). \tag{5.22}$$

Thus we obtain a system of distribution functions

$$\sigma_{\omega_1, \ldots, \omega_p}(\lambda_1, \ldots, \lambda_p),$$

for which the hypotheses of Kolmogorov's theorem hold:

1) $\sigma_{\omega_{i_1}, \ldots, \omega_{i_p}}(\lambda_{i_1}, \cdots, \lambda_{i_p}) = \sigma_{\omega_1, \ldots, \omega_p}(\lambda_1, \cdots, \lambda_p)$, where $\begin{pmatrix} 1 \cdots p \\ i_1 \cdots i_p \end{pmatrix}$ is an arbitrary permutation,

2) $\sigma_{\omega_1, \ldots, \omega_k}(\lambda_1, \cdots, \lambda_k) = \sigma_{\omega_1, \ldots, \omega_n}(\lambda_1, \cdots, \lambda_k, \infty, \cdots, \infty)$, $k < n$, and the uniqueness of σ^k in the representation (5.22) is guaranteed for any finite set K in Ω.

By Kolmogorov's theorem, the system $\{\sigma^k, K \subset \Omega\}$ uniquely determines a measure σ on the Borel sets in R_Ω.

The examples considered in §§3–5 are of interest not only for themselves but also as a concrete realization of the general scheme of §2, since each of them clearly illuminates some aspect of this scheme and indicates its capacity.

BIBLIOGRAPHY

[1] L. Schwartz, *Théorie des distributions*. Tome I, Actualités Sci. Indust., no. 1091 = Publ. Inst. Math. Univ. Strasbourg, no. 9, Hermann, Paris, 1950. MR 12, 31.

[2] I. M. Gel'fand and Shah Tao-shing, *On positive definite distributions*, Uspehi Mat. Nauk 15 (1960), no. 1 (91), 185–190. (Russian) MR 22 #2849.

[3] N. Ja. Vilenkin, *On the theory of positive definite generalized kernels*, Uspehi Mat. Nauk 15 (1960), no. 3 (93), 139–146; English transl., Amer. Math. Soc. Transl. (2) 28 (1963), 197–210. MR 22 #9849.

[4] M. G. Kreĭn, *On a general method of decomposing Hermite-positive nuclei into elementary products*, C. R. (Dokl.) Acad. Sci. URSS 53 (1946), 3–6. MR 8, 277.

[5] I. M. Gel′fand and A. G. Kostjučenko, *Expansion in eigenfunctions of differential and other operators*, Dokl. Akad. Nauk SSSR **103** (1955), 349–352. (Russian) MR 17, 388.

[6] Ju. M. Berezanskiĭ, *Generalization of Bochner's theorem to expansions according to eigenfunctions of partial differential equations*, Dokl. Akad. Nauk SSSR **110** (1956), 893–896. (Russian) MR 19, 1061.

[7] I. M. Gel′fand and G. E. Šilov, *Generalized functions*. Vol. 3: *Theory of differential equations*, Fizmatgiz, Moscow, 1958; English transl., Academic Press, New York, 1967. MR 21 #5142b; 36 #506.

[8] C. Foiaş, *Sur la décomposition spectral en opérateurs propres des opérateurs linéaires dans les espaces nucléaires*, C. R. Acad. Sci. Paris **248** (1959), 1105–1108. MR 27 #1825.

[9] R. B. Zarhina, *On the two-dimensional problem of moments*, Dissertation, Moskov. Gos. Univ., 1953; Dokl. Akad. Nauk SSSR 124 (1959), 743–746. (Russian) MR 21 #2161.

[10] M. S. Livšic, *On an application of the theory of Hermitian operators to the generalized problem of moments*, C. R. (Dokl.) Acad. Sci. URSS 44 (1944), 3–7. MR 6, 131.

[11] I. M. Gel′fand and G. E. Šilov, *Generalized functions*. Vol. 2: *Spaces of fundamental functions*, Fizmatgiz, Moscow, 1958; English transl., Academic Press and Gordon and Breach, New York, 1968. MR 21 #5142a; 37 #5693.

[12] B. S. Mitjagin, *Nuclearity and other properties of spaces of type S*, Trudy Moskov. Mat. Obšč. 9 (1960), 317–328; English transl., Amer. Math. Soc. Transl. (2) 93 (1970), (Russian) MR 23 #A2041.

[13] I. J. Schoenberg, *Metric spaces and completely monotone functions*, Ann. of Math. (2) 39 (1938), 811–841.

[14] A. Devinatz, *Integral representations of positive definite functions*, Trans. Amer. Math. Soc. 74 (1953), 56–77. MR 14, 659.

[15] S. Bochner, *Harmonic analysis and the theory of probability*, Univ. of California Press, Los Angeles, Calif., 1955. MR 17, 273.

[16] W. Gilbert, *Completely monotonic functions of several variables*, Doctoral Thesis, Princeton University, Princeton, N. J., 1951.

[17] B. M. Levitan and N. N. Meĭman, *On a uniqueness theorem*, Dokl. Akad. Nauk SSSR 81 (1951), 729–731. (Russian) MR 13, 551.

[18] E. B. Vul, *Uniqueness theorems for a certain class of functions represented by integrals*, Dokl. Akad. Nauk SSSR 129 (1959), 722–725. (Russian) MR 22 #8093.

[19] M. A. Naĭmark, *Normed rings*, GITTL, Moscow, 1956; English transl., Noordhoff, Groningen, 1959. MR 19, 870; 22 #1824.

[20] G. E. Šilov, *On regular normed rings*, Trudy Mat. Inst. Steklov. 21 (1947). (Russian) MR 9, 596.

[21] H. Hamburger, *Beiträge zur Konvergenztheorie der Stieltjesschen Kettenbrüche*, Math. Z. 4 (1919), 186–222.

[22] J. A. Shohat and J. D. Tamarkin, *The problem of moments*, Math. Surveys, vol. 1, Amer. Math. Soc., Providence, R. I., 1943. MR 5, 5; 13, 1138.

[23] M. G. Kreĭn and M. A. Krasnosel'skiĭ, *Fundamental theorems on the extension of Hermitian operators and certain of their applications to the theory of orthogonal polynomials and the problem of moments*, Uspehi Mat. Nauk 2 (1947), no. 3 (19), 60–106. (Russian) MR 10, 198.

[24] T. Carleman, *Sur le problème des moments*, C. R. Acad. Sci. Paris 174 (1922), 1680–1682.

[25] E. K. Haviland, *On the momentum problem for distribution functions in more than one dimension*, Amer. J. Math. 57 (1935), 562–568; 58 (1936), 164–168.

[26] A. Devinatz, *Two parameter moment problems*, Duke Math. J. 24 (1957), 481–498. MR 19, 1047.

[27] A. Grothendieck, *Produits tensoriels topologiques et espaces nucléaires*, Mem. Amer. Math. Soc. No. 16 (1955). MR 17, 763.

[28] G. H. Hardy, J. E. Littlewood and G. Pólya, *Inequalities*, Cambridge Univ. Press, New York, 1934; 2nd ed., 1952; Russian transl., IL, Moscow, 1948. MR 13, 727; 18, 722.

[29] A. N. Kolmogorov, *Grundbegriffe der Wahrscheinlichkeitsrechnung*, Springer, Berlin, 1933; Russian transl., ONTI, Moscow, 1936; English transl., Chelsea, New York, 1948; 2nd ed., 1956. MR 11, 374; 18, 155.

[30] F. Riesz and B. Sz.-Nagy, *Leçons d'analyse fonctionnelle*, 2nd ed., Akad. Kiadó, Budapest, 1953; Russian transl., IL, Moscow, 1954; English transl., Ungar, New York, 1955. MR 15, 132; 17, 175.

[31] A. G. Kostjučenko and B. S. Mitjagin, *Positive definite functionals on nuclear spaces*, Dokl. Akad. Nauk SSSR **131** (1960), 13–16 = Soviet Math. Dokl. **1** (1960), 177–180. MR 22 #12391.

[32] ———, *The multi-dimensional problem of moments*, Dokl. Akad. Nauk SSSR **131** (1960), 1249–1252 = Soviet Math. Dokl. **1** (1960), 415–419. MR 22 #12392.

Translated by:
A. H. Shuchat

Amer. Math. Soc. Transl.
(2) Vol. 93, 1970

NUCLEARITY AND OTHER PROPERTIES OF SPACES OF TYPE S* 1)

B. S. MITJAGIN

The spaces S_α^β play an important role in many problems of analysis (see [1, 2]). In the case of one variable S_α^β consists of the infinitely differentiable functions $\phi(t)$ on a straight line which satisfy the following inequalities for certain constants A, B and C (depending on ϕ):

$$\sup_t |t^k \varphi^{(q)}(t)| \leqslant CA^k B^q k^{k\alpha} q^{q\beta}.$$

Many questions in the theory of the spaces S_α^β as well as the more general spaces of type S, which are given by arbitrary sequences m_{kq} in place of the special sequences $k^{k\alpha} q^{q\beta}$, are considered in detail in [2], Chapter 4. But the question of the nuclearity of these spaces, which is important for spectral theory (see [2], Chapter 4), is not considered there. In the present note it is solved in the affirmative for a very wide class of spaces of type S.

1. Suppose[2)] given a countable (in $p = 0, 1, 2, \cdots$) collection $\langle m \rangle$ of sequences $m_{kq}^{(p)}$ satisfying the following conditions:

1°. $m_{kq}^{(p)}$ increases monotonically in all indices.

2°. For any p, s and l a p' and a $C(p, s, l)$ be found such that

$$k^s m_{l+k,\, s+q}^{(p)} \leqslant C(p, s, l)\, m_{kq}^{(p')} \quad \text{for all } k \text{ and } q.$$

---

* Translation of Trudy Moskov. Mat. Obšč. 9 (1960), 317–328.

1) The main results of this article were presented at a meeting of the Moscow Mathematical Society on October 27, 1959.

2) The following conventional notation will be used below. A letter denotes a collection of n (n_1 or n_2) numbers, be it the variables t or the integral indices k, q etc.,

$$D^q\varphi = \frac{\partial^{q_1+\cdots+q_n}}{\partial t_1^{q_1} \cdots \partial t_n^{q_n}} \varphi(t_1, \ldots, t_n), \quad k^a = k_1^{a_1} \ldots k_n^{a_n},$$

$$k\alpha = (k_1\alpha_1, \ldots, k_n\alpha_n), \quad t^2 = \sum_{j=1}^n t_j^2 .$$

h and p will denote natural numbers, and the inequality "$k \leq h$" will mean that $k_i \leq h$ ($i = 1, 2, \cdots, n$).

Let $S_{<m>}$ denote the space of infinitely differentiable functions $\phi(t)$ in the n-dimensional space R^n such that for certain C and p (depending on ϕ)

$$\sup_t |t^k D^q \varphi| \leqslant C m_{kq}^{(p)}.$$

We introduce on $S_{<m>}$ the topology of the inductive limit[1)] of the Banach spaces $S^p_{<m>}$ with norms

$$|\varphi|^p = \sup_{t,k,q} \frac{|t^k D^q \varphi|}{m_{kq}^{(p)}}.$$

Condition 2° on the sequence $<m>$ guarantees the continuity of the operators of multiplication by t_j and of differentiation D_j in $S_{<m>}$ (an analogous condition is given in [1], Russian page 289).

The spaces S_α^β ([1], Russian page 282) are obtained as a special case of the spaces $S_{<m>}$ if one puts, for example,

$$m_{kq}^{(p)} = p^{k+q} k^{k\alpha} q^{q\beta}.$$

Condition 2° is easily verified in this case.

We present two examples of linear topological spaces which are important in the sequel.

Example 1. M is the space of numerical sequences $x = \{x_{kq}\}$ which do not grow faster than some $m_{kq}^{(p)}$. It is the union of Banach spaces M^p with norms

$$|x|^p = \sup_{kq} \frac{|x_{kq}|}{m_{kq}^{(p)}}.$$

We introduce on M the topology of the inductive limit of the spaces M^p:

$$M = \lim_{\substack{\longrightarrow \\ p}} M^p,$$

i.e. the convergence of a sequence x_ν to zero in M is equivalent to the fact that for some p all $x_\nu \in M^p$ and $|x_\nu|^p \longrightarrow 0$ $(\nu \longrightarrow \infty)$.

Example 2 (see [4], page 89, and [1], Russian page 103). S is the space of infinitely differentiable functions $\phi(t)$ on R^n such that

---

1) For the definition of *inductive* and *projective limits* see [3].

$$|\varphi|_h = \sup_{t;\, k,\, q \leqslant h} |t^k D^q \varphi| < \infty.$$

It is the intersection of Banach spaces S_h with norms $|\phi|_h$. We introduce on S the topology of the projective limit of the spaces S_h:

$$S = \lim_{\overleftarrow{h}} S_h,$$

i.e. the convergence of a sequence ϕ_ν to zero in S is equivalent to the fact that $|\phi|_h \to 0$ $(\nu \to \infty)$ for all h.

A consideration of the spaces $S(E)$ will be very useful in the sequel. The space $S(E)$ is defined (see [5], §3, page 97) analogously to the space S of numerical functions as the space of infinitely differentiable functions $\phi(t)$ on R^n with values in a linear topological space E such that the function $(1+r^2)^k D^q \phi$ is bounded on R^n for all k and q, with the topology of uniform convergence on R^n for each of the functions $(1+r^2) D^q \phi$. If $\{\mu\}$ is a fundamental system of seminorms in E then one can take as a fundamental system of neighborhoods in $S(E)$ sets of the form

$$U_{h,p,\mu} = \left\{\varphi\colon \mu\{(1+r^2)^k D^q \varphi\} < \frac{1}{p};\quad k,\ q \leqslant h\right\},$$

$$h,\ p = 1, 2, \ldots, \mu \in \{\mu\}.$$

2. We now prove a basic lemma.

Lemma 1. *If the sequence* $\langle m \rangle$ *satisfies conditions* 1° *and* 2°, *then the space* $S_{\langle m \rangle}$ *is isomorphic to a closed subspace* L *of the space* $S(M)$.

Proof. We restrict ourselves for the sake of simplicity to the case of one variable. By definition $S(M)$ is the space of infinite matrix functions $[x_{kq}(t)]_0^\infty$ such that $x_{kq}(t) \in S$ for fixed k and q and $[t^l D^p x_{kq}(t)]_0^\infty \in M$ for any l and p and any fixed t. The subspace L is defined as the totality of matrices of the form

$$x = [t^k D^q x_{00}(t)]_0^\infty. \tag{1}$$

We will show that if $\phi(t) \in S_{\langle m \rangle}$ and $x_{00}(t) = \phi(t)$, then $x \in S(M)$.

By assumption, for some p

$$\sup_t |t^k D^q \varphi| \leqslant C(\varphi)\, m_{kq}^{(p)}.$$

Differentiating the product according to Leibniz' rule, we have

$$t^l D^s x_{kq}(t) = t^l D^s (t^k D^q \varphi) = t^l \sum_{j=0}^{\inf(k,\, s)} \frac{k!}{(k-j)!} \frac{s!}{j!\,(s-j)!} t^{k-j} D^{q+s-j} \varphi$$

$$= \sum_{j=0}^{\inf(k,\, s)} \frac{k!}{(k-j)!\,j!} \frac{s!}{(s-j)!} t^{k+l-j} D^{q+s-j} \varphi.$$

It is clear that the expression $k!/(k-j)!\,j!$ $(j = 0, 1, \cdots, \inf(k, s))$ does not exceed 2^{2s} for $2s \geq k$ and takes its greatest value for $k > 2s$ when $j = s$. This value is equal to $k!/(k-s)!\,s!$, and since $k!/k^s(k-s)! \to 1$ there exists an A_s such that

$$\frac{k!}{(k-s)!\,s!} \leqslant A_s k^s \text{ for all } k.$$

Therefore

$$\sup_t |t^l D^s x_{kq}(t)| \leqslant \begin{cases} 2s \cdot 2^{2s}\, s!\, m^{(p)}_{k+l,\, q+s} C(\varphi) & \text{for} \quad k \leqslant 2s, \\ s \cdot A_s \cdot s!\, k^s m^{(p)}_{k+l,\, q+s} C(\varphi) & \text{for} \quad k > 2s. \end{cases}$$

By virtue of condition 2° there exists a p' such that

$$\sup_t |t^l D^s x_{kq}(t)| \leqslant C_1(p, s, l)\, C(\varphi)\, m^{(p')}_{k,\, q}.$$

Thus $x \in S(M)$ and $x \in L$. The latter inequality simultaneously shows that the mapping $S_{<m>} \to L$ described above is continuous.

Conversely, if $x \in L$ then $x_{00}(t) \in S_{<m>}$, since

$$|t^k D^q x_{00}(t)| \leqslant C(x)\, m^{(p)}_{kq}.$$

This mapping of L into $S_{<m>}$ is continuous. In fact, if $x_\nu \in L$ and $x_\nu \to 0$ in $S(M)$ for $\nu \to \infty$, then, by the definition of the topology in $S(M)$, for some p

$$|t^k D^q x_{\nu,\, 00}(t)| \leqslant C(x_\nu)\, m^{(p)}_{kq}$$

and $C(x_\nu) \to 0$ $(\nu \to \infty)$. But this means that the $x_{\nu,00}(t)$ in $S_{<m>}$ tend to zero. Thus the correspondence $L \leftrightarrow S_{<m>}$ described above is one-to-one (which is obvious) and bicontinuous.

We will now show that L is a closed subspace in $S(M)$. As in the numerical space S, the operators of differentiation D and of multiplication by t in $S(M)$ are continuous. We introduce in $S(M)$ operators V, W, P and Q by means of the formulas

$$(Vx)_{kq}=\begin{cases}0, & q=0,\\ x_{k,\,q-1} & q>0,\end{cases}\qquad (Px)_{kq}=\begin{cases}x_{kq}, & k=0,\\ 0, & k>0,\end{cases}$$
$$(Wx)_{kq}=\begin{cases}0, & k=0,\\ x_{k-1,\,q}, & k>0,\end{cases}\qquad (Qx)_{kq}=\begin{cases}x_{kq}, & q=0,\\ 0, & q>0.\end{cases}$$

It is easily seen that these operators are continuous. The operator $V(W)$ is the operator which shifts the columns (rows) of a matrix $[x_{kq}(t)]_0^\infty$ to the right (downwards) by one, while the operator P (Q) projects the whole matrix onto its zero row (column). We put

$$T_1=(VD-1+Q)P,$$
$$T_2=Wt-1+P.$$

If $x \in L$, then $T_1x=0$. In fact, P projects a matrix onto its zero row $x_{0q}(t)=D^q x_{00}(t)$. The operator VD, which consists of applying the differentiation operator and then shifting to the right, does not change the terms of this sequence for $q>0$. For $q=0$ the coordinate of the image is equal to zero, so that $VD+Q$ acts on the zero row of the elements $x \in L$ as the identity operator. But this means that $T_1x=0$.

It is analogously verified that also $T_2x=0$ if $x \in L$.

Conversely, if $T_1x=0$ then

$$x_{0,\,q+1}(t)=Dx_{0,\,q}(t)=D^{q+1}x_{00}(t);$$

if also $T_2x=0$ then for each $q=0,1,\cdots$

$$x_{k+1,\,q}(t)=tx_{kq}(t)=t^{k+1}x_{0,\,q}(t)=t^{k+1}D^qx_{00}(t),$$

i.e. $x \in L$. Thus $x \in L$ if and only if

$$T_1x=0 \text{ and } T_2x=0,$$

i.e.

$$L=\{x\colon T_1x=0,\ T_2x=0\}.$$

Since the operators T_1 and T_2 are continuous in $S(M)$, the subspace L is closed in $S(M)$. By the same token Lemma 1 is completely proved.

3. We recall some concepts from [16].

Let E and F be linear topological spaces. The totality of elements

$$u=\sum_1^n e_i\otimes f_i,\quad e_i\in E,\ f_i\in F\qquad (i=1,\ldots,n),$$

n being any finite number, with the equivalence relation

$$\left(\sum_1^m a_i e_i\right) \otimes \left(\sum_1^l b_j f_j\right) = \sum_{i=1}^m \sum_{j=1}^l a_i b_j (e_i \otimes f_j)$$

and the usual rules for addition and multiplication by numbers, forms a linear manifold. Its factor manifold by the equivalence relation is called the *algebraic tensor product* of E and F, and is denoted by $E \otimes F$.

Let $\{\mu\}$ and $\{\nu\}$ be fundamental systems of seminorms on E and F respectively. μ and ν determine a seminorm $\lambda = \mu \otimes \nu$ on $E \otimes F$ by means of the formula

$$\lambda(u) = \inf_{u = \Sigma e_i \otimes f_i} \Sigma \mu(e_i) \nu(f_i).$$

The completion of $E \otimes F$ with respect to the topology defined by all seminorms of the form $\lambda = \mu \otimes \nu$ is called the *projective tensor product* of E and F, and is denoted by $E \hat{\otimes} F$.

Example. Suppose $E = F$ is the Banach space l_1 of summable sequences $\xi = \{\xi_n\}$ with norm $|\xi| = \Sigma_1^\infty |\xi_n|$. It is easily verified that $l_1 \hat{\otimes} l_2$ is the space of summable double sequences $\psi = \{\psi_{nm}\}$ with norm $|\psi| = \Sigma_{n,m} |\psi_{nm}|$.

The completion of $E \otimes F$ with respect to the topology of uniform convergence (as bilinear functionals on $E' \times F'$) on the sets $K \times H$, where K and H are equicontinuous sets in E' and F' respectively, is denoted by $E \hat{\hat{\otimes}} F$.

Lemma 2 ([15], page 111). $S(E) = S \hat{\hat{\otimes}} E$.

Lemmas 1 and 2 obviously imply

Theorem 1. *Let $\mathfrak{S}$ be a class of linear topological spaces having a property ϵ which is invariant with respect to going over to tensor products of the type $\hat{\hat{\otimes}}$ or to closed subspaces.*

If $S \in \mathfrak{S}$ and $M \in \mathfrak{S}$, then $S_{<m>} \in \mathfrak{S}$.

4. We proceed now to the question of the nuclearity of S.

Definition.[1]) A space Φ is called *nuclear* if every weakly absolutely convergent series of continuous founctionals on Φ

1) For other definitions of the notion of nuclearity see [7, 8]. These definitions are equivalent for the classes of spaces considered in this note.

$$\sum_{j=1}^{\infty} |f_j(\varphi)| < \infty$$

is such that for some neighborhood U of zero in Φ

$$\sum_{j=1}^{\infty} |f_j|_U < \infty,$$

where

$$|f|_U = \sup_{\varphi \in U} |f\varphi|.$$

We note some properties of nuclear spaces [6].

Lemma 3. *Every (closed) subspace and factor space of a nuclear space Φ is nuclear in the topology induced by the topology of Φ.*

Lemma 4. *A nuclear space is reflexive.*

The topology of $E \hat{\hat{\otimes}} F$ is generally weaker than the topology of $E \hat{\otimes} F$. However, we have [6]

Lemma 5. *If E and F are nuclear spaces, then $E \hat{\hat{\otimes}} F = E \hat{\otimes} F$ is also a nuclear space.*

Suppose now the sequence $\langle m \rangle$ satisfies the following property.

3°. For all p

$$\sum_{kq} \frac{m_{kq}^{(p)}}{m_{kq}^{(p+1)}} < \infty.$$

Then the space M is nuclear (see [6]), as is also the space S (see [6] and [2], Russian page 206). It will always be assumed below that condition 3° is satisfied.

Lemmas 2 and 5 therefore imply

Lemma 6. *$S(M) = S \hat{\otimes} M$ and is a nuclear space.*

In view of Lemmas 3 and 5 (6), a special case of Theorem 1 is

Theorem 2. *If the sequence $\langle m \rangle$ satisfies conditions $1^\circ - 3^\circ$ then the space $S_{\langle m \rangle}$ is nuclear.*

Corollary. *The spaces $S_\alpha^\beta = S_{\alpha_1 \alpha_2 \cdots \alpha_n}^{\beta_1 \beta_2 \cdots \beta_m}$ are nuclear.*

Remark. We again emphasize that we deal throughout (expect in the proof of Lemma 1) with spaces of functions of several variables. In this case the space $S_{\langle m \rangle}$ is isomorphic to a closed subspace of $S(M)$, viz. the manifold of the zeros of $2n$ continuous operators in $S(M)$, which are defined analogously to T_1 and T_2.

5. We now consider the spaces $S_{\{a\}}$ and $S^{\{b\}}$, which consist of infinitely differentiable functions on R^n and are defined with respect to collections of sequences $a_k^{(p)}$ and $b_q^{(p)}$ as follows: $\phi \in S_{\{a\}}$ if for some p

$$\sup_t |t^k D^q \varphi| \leqslant C_q a_k^{(p)} \quad \text{for all } q \text{ and } k;$$

$\phi \in S^{\{b\}}$ if for some p

$$\sup_t |t^k D^q \varphi| \leqslant C_k\, b_q^{(p)} \quad \text{for all } q \text{ and } k.$$

A topology is introduced in a corresponding manner (see [1], Chapter 4). (S_α and S^β are obtained by putting $a_k^{(p)} = p^k k^{k\alpha}$ and $b_q^{(p)} = p^q q^{q\beta}$.)

In these cases we assume that the following conditions are satisfied.

1°. $a_k^{(p)}$ and $b_q^{(p)}$ increase monotonically in all indices.

2°. For any p

$$\sum_k \frac{a_k^{(p)}}{a_k^{(p+1)}} < \infty, \quad \sum_q \frac{b_q^{(p)}}{b_q^{(p+1)}} < \infty.$$

3°. For any p, l and s it is possible to find a p' and some $C_j(p, s, l)$ $(j = 1, 2)$ such that the sequence $\{a\}$ satisfies $k^s a_{k+l}^{(p)} \le C_1 a_k^{(p')}$ for all k and the sequence $\{b\}$ satisfies $b_{q+s}^{(p)} \le C_2 b_q^{(p')}$ for all q. The nuclearity of $S_{\{a\}}$ and $S^{\{b\}}$ does not follow from Theorem 2. In order to prove the nuclearity of these spaces under conditions $1^\circ - 3^\circ$ it is necessary to repeat the plan carried out above for $S_{<m>}$, replacing the space M by a space A of numerical sequences $x = \{x_{kq}\}$, which is the direct product of the inductive limits A_q of the Banach spaces A_q^p with norms

$$|x|_q^p = \sup_k \frac{|x_{kq}|}{a_k^{(p)}},$$

or by a space B (defined analogously). The nuclearity of the spaces A and B is ensured by condition 2°.

The countably normed spaces $S_{\alpha a}^{\beta b}$, $S_{\alpha a}$, $S^{\beta b}$ etc. (see [2], Chapter 4, §§3 and 9) can be considered in exactly the same way. We therefore formulate without further ado the following theorem.

Theorem 3. *The spaces* S_α, S^β, $S_{\alpha a}$, $S^{\beta b}$, $S_\alpha^{\beta b}$, $S_{\alpha a}^{\beta}$, $S_{\alpha a}^{\beta b}$ *are nuclear.*

Remark. The space $S(M)$ never degenerates into zero, while the triviality of the space $S_{<m>}$ is equivalent to the fact that the system

$T_1 x = 0$, $T_2 x = 0$ (or the corresponding system in the multidimensional case) has only the zero solution in $S(M)$.

6. *The general form of a continuous linear functional on* $S_{<m>}$. Let $f \in S_{<m>}$. The space $S_{<m>}$ is a closed subspace of $S(M)$, and hence f can be extended to a continuous functional F on $S(M)$. We will study the conjugate space $S(M)'$. We introduce operators P_{kq} in $S(M)$:

$$(P_{kq}x)_{ij} = \begin{cases} x_{kq}(t), & i = k, \ j = q, \\ 0, & i \neq k \ \text{ or } \ j \neq q. \end{cases}$$

Every $x \in S(M)$ can be expanded in the series

$$x = \sum_{kq} P_{kq} x. \tag{2}$$

In fact, for some p

$$|x_{kq}(t)| \leqslant C(x)\, m_{kq}^{(p)} \leqslant C(x) \frac{m_{kq}^{(p)}}{m_{kq}^{(p+1)}} m_{kq}^{(p+1)}$$

and the series (2) converges in M^{p+1}, since

$$\frac{|P_{kq}(x)(t)|}{m_{kq}^{(p+1)}} \leqslant C(x) \frac{m_{kq}^{(p)}}{m_{kq}^{(p+1)}}$$

and

$$\sum \frac{m_{kq}^{(p)}}{m_{kq}^{(p+1)}} < \infty$$

by condition 3°. It follows from the expansion (2) that the representation

$$F = \sum_{kq} P_{kq}^* F, \tag{3}$$

where $P_{kq}^* F$ is a continuous functional on S_h (h generally depends on k and q), holds for $F \in S(M)'$. The nuclearity of $S(M)$ implies that for some h

$$\sum_{kq} |P_{kq}^* F|_h \, m_{kq}^{(p)} < \infty \quad \text{for all } p.$$

Each $P_{kq}^* F$ has the following representation (see [2], Russian page 144):

$$(\varphi, P_{kq}^* F) = \sum_{l \leqslant h} \int (1 + r^2)^h f_{kq,\,l}(t)\, D^l \varphi_{kq}(t)\, dt, \tag{4}$$

with

$$|P_{kq}^* F|_h = \sum_{l \leq h} \sup_t |f_{kq,l}(t)|.$$

Formulas (3), (4) and (1), Leibniz' formula and conditions $1^\circ - 3^\circ$ on the sequence $m_{kq}^{(p)}$ now permit one to represent $f \in S'_{<m>}$ in the form

$$f(\varphi) = \sum_{kq} \int t^k f_{kq}(t) D^q \varphi(t)\, dt \tag{5}$$

(the f_{kq} are obtained as finite combinations of the $f_{kq,l}$), with

$$\sum_{kq} \sup_t |f_{kq}(t)| m_{kq}^{(p)} < \infty \quad \text{for all } p. \tag{5a}$$

The spaces $S_{\{a\}}$ and $S^{\{b\}}$ are considered analogously. We have thus proved

Theorem 4. *A continuous linear functional on* $S_{<m>}$ *admits the repre-santation* (5) *with the conditions* (5a); *on* $S_{\{a\}}$ *it admits the representation* (5) *with the conditions*

$$\sum_k \sup_t |f_{kq}(t)| a_k^{(p)} < \infty \quad \textit{for all } p \textit{ and } q; \tag{5b}$$

and on $S^{\{b\}}$ *it admits the same representation* (5) *with the conditions*

$$\sum_q \sup_t |f_{kq}(t)| b_q^{(p)} < \infty \quad \textit{for all } p \textit{ and } k. \tag{5c}$$

A functional on $S_{<m>}$ is said to be *concentrated on the closed set* Γ if it is equal to zero at all fundamental functions which vanish in a neighborhood of Γ. Such a definition is correct if $S_{<m>}$ is sufficiently rich in functions with compact support. For S_α^β this means that $\beta > 1$. Otherwise there are no nonzero functions in S_α^β which vanish in a neighborhood.

Sometimes, for example, in determining uniqueness classes for the solutions of certain differential equations it is necessary to know the form of a functional on S_α^β which is concentrated at a point or on some closed set Γ. This question is solved in [9] for a point in the one-dimensional case of S_α^β. The following result can be formulated if $S_{<m>}$ is sufficiently rich in functions with compact support.

Theorem 5. *A functional* f *on* $S_{<m>}$ *which is concentrated on a bounded infinitely differentiable manifold admits the representation*

$$f(\varphi) = \sum_q \int_\Gamma D^q \varphi(t)\, d\sigma_q,$$

where σ_q *is a countably additive set function on* Γ, *with*

$$\sum_q \operatorname{var} \sigma_q \cdot m_{0q}^{(p)} < \infty \qquad \textit{for any } p.$$

In particular, a functional f_0 *which is concentrated at a point* t_0 *has the form*

$$f_0\varphi = \sum_q f_q D^q \varphi(t_0),$$

with

$$\sum_q |f_q| \, m_{0q}^{(p)} < \infty \qquad \textit{for all } p.$$

f *(or* f_0*) can be extended to a continuous functional on* $S^{\{m_{0q}\}}$.

A proof can be obtained by starting not from formula (4) but from formula (2) on Russian page 139 in [2]. It (together with a fact concerning the possibility of extending an n-times differentiable function with an n-times differentiable manifold Γ onto a neighborhood of Γ with preservation of the class of differentiability) permits one to give the following formula for the general form of a functional F on S which is concentrated on an infinitely differentiable manifold Γ:

$$F(\varphi) = \sum_0^m \int_\Gamma D^q \varphi(t) \, d\sigma_q.$$

7. We note further two propositions on tensor products [6].

Lemma 7. *Let* M_i $(i = 1, 2)$ *be a space of Example* 1 *given by a collection* $m_{i,k,q}^{(p)}$ *of* $2n_i$*-multiple sequences, and let* M *be the space of Example* 1 *given by the collection*

$$m_{k_1 k_2 q_1 q_2}^{(p)} = m_{k_1 q_1}^{(p)} \, m_{k_2 q_2}^{(p)}$$

of $2(n_1 + n_2)$*-multiple sequences, the latter satisfying condition* 3°. *Then*

$$M_1 \hat{\otimes} M_2 = M.$$

Lemma 8. *The projective tensor product of linear spaces is commutative and associative. More precisely, the spaces* $E \hat{\otimes} F$ *and* $F \hat{\otimes} E$ *are isomorphic, and* $(E \hat{\otimes} F) \hat{\otimes} K$ *is isomorphic to the space* $E \hat{\otimes} (F \hat{\otimes} K)$, *the isomorphisms being given by the mappings*

$$e \otimes f \longleftrightarrow f \hat{\otimes} e \quad \textit{and} \quad (e \otimes f) \otimes k \longleftrightarrow e \otimes (f \otimes k).$$

This proposition permits one to speak of the projective tensor product of any finite number of linear spaces.

Suppose $S_{<m_1>}$, $S_{<m_2>}$, $S_{<m>}$ are spaces of functions given on R^{n_1}, R^{n_2} and R^n respectively, and

$$n = n_1 + n_2, \quad m^{(p)}_{k_1k_2q_1q_2} = m^{(p)}_{1k_1q_1} m^{(p)}_{2k_2q_2}.$$

Theorem 6. *The set of all possible functions of the form*

$$\varphi(t) = \sum_{j=1}^{l} \varphi_{1j}(t_1)\varphi_{2j}(t_2), \quad \varphi_{1j} \in S_{<m_1>}, \quad \varphi_{2j} \in S_{<m_2>}; \quad t = (t_1, t_2)$$

is dense in $S_{<m>}$. *More precisely,*

$$S_{<m_1>} \hat{\otimes} S_{<m_2>} = S_{<m>}. \tag{6}$$

Proof. By virtue of Lemmas 7 and 8

$$S(M_1) \hat{\otimes} S(M_2) = ({}_{n_1}S \hat{\otimes} M_1) \hat{\otimes} ({}_{n_2}S \hat{\otimes} M_2)$$

$$= ({}_{n_1}S \hat{\otimes}_{n_2} S) \hat{\otimes} (M_1 \hat{\otimes} M_2) = {}_nS \hat{\otimes} M = S(M). \tag{7}$$

Suppose Φ_1 and Φ_2 are nuclear operators and A_1 is a continuous operator in Φ_1. It is easily verified that the operator $A_1 \otimes 1_2$ acting in $\Phi_1 \otimes \Phi_2$ according to the formula $(A_1 \otimes 1_2)(\phi_1 \otimes \phi_2) = A_1\phi_1 \otimes \phi_2$, extends to a continuous operator in $\Phi = \Phi_1 \hat{\otimes} \Phi_2$ (the set of zeros of an operator T is denoted below by Ker T, and its image, by Im T). By virtue of the construction of Lemma 1 and the remark following the corollary of Theorem 2 we can prove Theorem 6 by establishing the validity of the following lemma.

Lemma 9. *Under the above assumptions*

$$\operatorname{Ker}(A_1 \otimes 1_2) = \operatorname{Ker} A_1 \hat{\otimes} \Phi_2.$$

In fact, we arrive at (6) from (7) by applying this lemma $2n_1 + 2n_2$ times for the operators T_{1i} and T_{2j} $(i = 1, \cdots, 2n_1;\ j = 1, \cdots, 2n_2)$ the joint kernels of which are the spaces $S_{<m_1>}$ and $S_{<m_2>}$ in $S(M_1)$ and $S(M_2)$ respectively.

In proving Lemma 9 we make use of the following two propositions.

Lemma 10. *If* B_1 *and* B_2 *are nuclear spaces and* $B = B_1 \hat{\otimes} B_2$, *then the algebraic tensor product* $B_1' \otimes B_2'$ *is dense in* B'.

Proof.[1] We will prove a stronger assertion, namely

$$(B_1 \hat{\otimes} B_2)' = \overline{B_1' \otimes B_2'},$$

1) Due to A. S. Dynin.

where $B_1' \overline{\otimes} B_2'$ denotes the inductive tensor product.

Every nuclear space can be represented as a projective limit of Hilbert spaces with nuclear imbeddings, i.e.

$$B_1 = \lim_{\leftarrow \alpha} H_{1\alpha}, \quad B_2 = \lim_{\leftarrow \beta} H_{2\beta}.$$

In this connection

$$B_1 \hat{\otimes} B_2 = \lim_{\leftarrow \alpha \times \beta} H_{1\alpha} \hat{\otimes} H_{1\beta}$$

is also a projective limit of Hilbert spaces with nuclear imbeddings. A theorem of D. A. Raĭkov ([3], §9) establishes that the conjugate of a projective limit of Banach spaces with completely continuous imbeddings is the inductive limit of the conjugates of the Banach spaces. In our case

$$(B_1 \hat{\otimes} B_2)' = \lim_{\overrightarrow{\alpha \times \beta}} (H_{1\alpha} \hat{\otimes} H_{2\beta})'.$$

Since $H_{1\alpha}$ and $H_{2\beta}$ are Hilbert spaces it follows that $(H_{1\alpha} \hat{\otimes} H_{2\beta})' = H'_{1\alpha} \hat{\hat{\otimes}} H'_{2\beta}$, and the imbeddings are compatible and nuclear. Therefore

$$\lim_{\overrightarrow{\alpha \times \beta}} (H_{1\alpha} \hat{\otimes} H_{2\beta})' = \lim_{\overrightarrow{\alpha \times \beta}} (H'_{1\alpha} \hat{\hat{\otimes}} H'_{2\beta}) = \lim_{\overrightarrow{\alpha}} H'_{1\alpha} \overline{\otimes} \lim_{\overrightarrow{\beta}} H'_{2\beta} = B_1' \overline{\otimes} B_2'.$$

Lemma 10 is proved.

Lemma 11. *Let T be a continuous operator in a reflexive space F. Then the conditions,* a°) $\operatorname{Ker} T = \{0\}$ *and* b°) $\operatorname{Im} T'$ *is dense in F', are equivalent.*

Proof. Suppose a°) is satisfied. If $\operatorname{Im} T'$ is not dense in F' then by the Hahn-Banach Theorem there exists a functional $f \neq 0$ on F' such that $f(\operatorname{Im} T') = 0$. Since F is reflexive, it can be assumed that $f \in F$. Then $(T'b)(f) = 0$ for any $b \in F'$, $b(Tf) = 0$, and hence $f = 0$ by virtue of a°). This contradiction proves that a°) implies b°).

Suppose b°) is satisfied and $Tf = 0$. Then $b(Tf) = 0$ for all $b \in F'$, and $(T'b)(f) = 0$. Since $\operatorname{Im} T'$ is dense in F' it follows that $f = 0$. Lemma 11 is proved.

Proof of Lemma 9. It is obvious that $\operatorname{Ker} A_1 \hat{\otimes} \Phi_2 \subset \operatorname{Ker}(A_1 \otimes 1_2)$ and is a closed subspace in Φ. By virtue of Proposition 3 of [6], page 38,

$$\overline{\Phi/\operatorname{Ker} A_1 \hat{\otimes} \Phi_2} = \overline{\Phi_1/\operatorname{Ker} A_1} \hat{\otimes} \Phi_2.$$

The operators A_1 and $A_1 \otimes 1_2$ carry over in a natural way onto the corresponding factor spaces; namely, they go over into operators $\widetilde{A}_1$ and $\overline{A_1 \otimes 1_2}$, it being clear that $\overline{A_1 \otimes 1_2} = \widetilde{A}_1 \otimes 1_2$. By Lemma 3 the factor spaces are nuclear. We put

$$\overline{\Phi/\mathrm{Ker}\, A_1 \hat{\otimes} \Phi_2} = B, \quad \overline{\Phi_1/\mathrm{Ker}\, A_1} = B_1, \quad \widetilde{A}_1 = T, \quad B = B_1 \hat{\otimes} \Phi_2.$$

The operator T does not have any zeros in B_1 by virtue of the construction of this factor. It follows according to Lemmas 3 and 11 that $T'B_1'$ is dense in B_1'. Clearly, for $b_1' \otimes b_2 \in B_1' \otimes B_2'$

$$(T \otimes 1_2)' (b_1' \otimes b_2') = T' b_1' \otimes b_2'.$$

These two facts and Lemma 10 imply that $(T \otimes 1_2)' (B_1' \otimes B_2')$ is dense in B', and hence, according to Lemma 11, $T \otimes 1_2$ does not have any nontrivial zeros in B. This means that

$$\mathrm{Ker}\,(A_1 \otimes 1_2) = \mathrm{Ker}\, A_1 \hat{\otimes} \Phi_2.$$

Otherwise, under the factorization a vector $\phi \neq 0$, $\phi \in \mathrm{Ker}(A_1 \hat{\otimes} 1_2) \backslash \mathrm{Ker}\, A_1 \hat{\otimes} \Phi_2$, would go over into a nonzero vector in B and be a zero of the operator $T \times 1_2$.

Lemma 9 and by the same token Theorem 6 are completely proved.

Corollary. $S_{\alpha_1, \dots, \alpha_n}^{\beta_1, \dots, \beta_n} = S_{\alpha_1}^{\beta_1} \hat{\otimes} \cdots \hat{\otimes} S_{\alpha_n}^{\beta_n}$.

The nuclearity property of the spaces S_α^β permits one to apply the Gel'fand-Kostjučenko spectral theory (see [2], Chapter 4) to them. In this way A. G. Kostjučenko and the author obtained [10] a number of results on the representation of positive definite functionals on the spaces S_α^β (for the details see the preceding article of this volume).

In conclusion the author thanks G. E. Šilov and A. S. Dynin for essential help in writing this paper.

BIBLIOGRAPHY

[1] I. M. Gel'fand and G. E. Šilov, *Generalized functions*. Vol. 2: *Spaces of fundamental functions*, Fizmatgiz, Moscow, 1958; English transl., Academic Press and Gordon and Breach, New York, 1968.
MR 21 #5142a; 37 #5693.

[2] ———, *Generalized functions.* Vol. 3: *Some questions in the theory of differential equations,* Fizmatgiz, Moscow, 1958; English transl., Academic Press, New York, 1967. MR 21 #5142b; 36 #506.

[3] D. A. Raĭkov, *Completely continuous spectra of locally convex spaces,* Trudy Moskov. Mat. Obšč. 7 (1958), 413–438. (Russian) MR 20 #6642.

[4] L. Schwartz, *Théorie des distributions.* Tom II, Actualités Sci. Indust., no. 1122 = Publ. Inst. Math. Univ. Strasbourg, no. 10, Hermann, Paris, 1951. MR 12, 833.

[5] ———, *Espaces de fonctions différentiables à valeurs vectorielles,* J. Analyse Math. 4 (1954/55), 88–148. MR 18, 220.

[6] A. Grothendieck, *Produits tensoriels topologiques et espaces nucléaires,* Mem. Amer. Math. Soc. No. 16 (1955). MR 17, 763.

[7] D. A. Raĭkov, *On a property of nuclear spaces,* Uspehi Mat. Nauk 12 (1957), no. 5 (77), 231–236. (Russian) MR 19, 967.

[8] A. S. Dynin, *On spaces nuclear in different senses,* Dokl. Akad. Nauk SSSR 121 (1958), 790–792. (Russian) MR 20 #4168.

[9] A. G. Kostjučenko, Report at the Conference on Functional Analysis, Odessa, 1958.

[10] A. G. Kostjučenko and B. S. Mitjagin, *Positive definite functionals on nuclear spaces,* Dokl. Akad. Nauk SSSR 131 (1960), 13–16 = Soviet Math. Dokl. 1 (1960), 177–180. MR 22 #12391.

Translated by:
S. Smith

Amer. Math. Soc. Transl.
(2) Vol. 93, 1970

ON REGULAR BASES IN NUCLEAR SPACES*

UDC 513.881

M. M. DRAGILEV

The paper investigates some properties of bases for countably normed nuclear spaces. We consider mainly spaces E possessing bases $\{x_n\}$ with the following property: the quotient $|x_n|_p/|x_n|_q$ is a monotonic function of n for a certain system of norms $|\cdot|_p$ defining a topology in E and for any fixed $p, q = 0, 1, \cdots$. A basis with this property is called a regular basis. It is shown in this paper that every basis for a space E becomes regular after a suitable permutation of elements. An elementary example of such a space E is the space A_R of functions which are analytical in the disk $|z| < R \leq \infty$. On the other hand, there exist countably normed nuclear spaces without any regular basis, e. g. the topological product $A_R \times A_\infty$ $(R < \infty)$.

It will be shown that, under some additional restrictions on E, normalized regular bases are equivalent, i.e. one can transform one into another by an invertible linear operator. Therefore the theorem on quasi-equivalence of bases, proved in some special cases in papers [1] and [2], can be extended to corresponding spaces. An analogous statement is true for conjugate spaces.

The additional restriction ensuring the equivalence of normalized regular bases for E can be reduced to the following requirement: Let Σ be an arbitrary system of neighborhoods of zero U_p in E $(U_p \supseteq U_{p+1}, p = 0, 1, \cdots)$ associated with the system of norms $|\cdot|_p$ by the relations

$$U_p = \{x \in E: |x|_p \leqslant 1\}, \quad |x|_p = |x|_{U_p} = \{\inf t: \frac{x}{t} \in U_p\} \quad (p = 0,1,\ldots).$$

Furthermore, let $d_n(V, U)$ be the n-dimensional diameter of an arbitrary set $V \subset E$ with respect to the neighborhood of zero U of the form λU_p $(\lambda > 0)$:

$$d_n(V, U) = \inf_{L_n} \sup_{x \in V} \inf_{y \in L_n} |x - y|_U \quad (n = 0,1,\ldots),$$

where the infimum is taken over all the n-dimensional spaces $L_n \subset E$ (see [3], [4]).

Denote the sets of sequences of numbers $\{\gamma_n\}$ satisfying one of the

* Translation of Mat. Sb. 68 (110) (1965), 153–173.

following conditions:

a) $\gamma_n d_n(U_q, U_p) \to 0$ $(n \to \infty)$ for any p, and for a q dependent on p;

b) $\gamma_n d_n(U_q, U_p) \to 0$ $(n \to \infty)$ for a certain p and for any q (see [8]) by $\Gamma_i (i = 1, 2)$.

Finally, let $\gamma_n = 2/d_n(U_q, U_p)$. A countably normed nuclear space E will be said to be of class (d_1) if the condition $\{\gamma_n\} \in \Gamma_1(E)$ holds for a certain p and for an arbitrary q. If for every p there exists a q such that $\{\gamma_n\} \in \Gamma_2(E)$, then the corresponding space will be said to be of class (d_2). In particular, $A_\infty \in (d_1)$ and $A_R \in (d_2)$ for $R < \infty$. Centers (cocenters) of nuclear Riesz scales, introduced in note [5], are spaces $E \in (d)$ (respectively, their conjugate spaces). A larger class of such spaces is formed by the spaces L_f introduced in [6]. The article shows some simple invariants of the spaces L_f defining a topology and gives a natural characterization of these spaces (Theorem 13). Not all nuclear spaces with regular bases belong to a class (d). Notice that spaces with regular bases cannot be described in terms of approximate dimensions.

§1. Theorem on regular bases

Let $\{x_n\}$ be a basis for a countably normed nuclear space E, $\{x_n'\}$ a biorthogonal sequence of functions and $x \in E$ an arbitrary element

$$x = \sum_{n=0}^{\infty} x_n'(x)\, x_n. \tag{1}$$

It is known that all bases for a nuclear space are absolute [7]. According to the definition, this means that

$$|x|_p^* = \sum_n |x_n'(x)|\, |x_n|_p < \infty \tag{2}$$

for every x and $p = 0, 1, \cdots$. Moreover, the following statement is valid: for any p one can find a q such that

$$\sum_n \frac{|x_n|_p}{|x_n|_q} < \infty. \tag{3}$$

The last inequality can be considered as a condition for a countably normed space possessing a basis to be a nuclear space. Note that the system of norms $|\cdot|_p^*$ defines the initial topology for E (see [2]).

Theorem 1. *Let $\{x_n\}$ be a regular basis for the space E and let $\{y_n\}$ be an arbitrary basis for E. Then there exists a permutation of the sequence of natural numbers (j_n) such that the basis $\{y_{j_n}\}$ is a regular basis.*

Proof. Let us introduce the following notation: $x_n'(y_k) = a_{nk}$, $y_n'(x_k) = b_{nk}$ $(n, k = 0, 1, \cdots)$. Since $\Sigma_k a_{kn} b_{nk} = y_n'(y_n) = 1$ $(n = 0, 1, \cdots)$, for any

n one can find a k_n such that

$$|a_{k_n n} b_{n k_n}| > \frac{1}{(k_n+1)^2 \sigma} \quad (n = 0, 1, \ldots),$$

where $\sigma = \sum_k (k+1)^{-2} = \pi^2/6$. If the norms $|\cdot|_p^*$ are defined by (2), then

$$|x|_p^{**} = \sum_n |y_n'(x)| |y_n|_p^* < \infty \quad (p = 0, 1, \ldots)$$

for every $x \in E$. The systems of norms $|\cdot|_p$, $|\cdot|_p^*$ and $|\cdot|_p^{**}$ are equivalent. Consequently for any p one can find q and C_1 dependent on p such that

$$|x|_p^{**} < C_1 |x|_q \quad (x \in E)$$

and, in particular,

$$|x_k|_p^{**} = \sum_n |b_{nk}| |y_n|_p^* < C_1 |x_k|_q \quad (k = 0, 1, \ldots).$$

In turn, for every q (in view of the condition (3) for the nuclearity of a space) one can find an r and next an s dependent on r such that

$$\sum_k \frac{|x_k|_q}{|x_k|_r} < \infty \text{ and } \sum_k \frac{|x_k|_r}{|x_k|_s} < \infty.$$

Since the terms of these series decrease monotonically, the following inequalities hold:

$$|x_k|_q < \frac{C_2}{k+1} |x_k|_r, \quad |x_k| < \frac{C_3}{k+1} |x_k|_s \quad (k = 0, 1, \ldots),$$

where C_2 and C_3 do not depend on k. Hence it follows that

$$|b_{nk_n}| |y_n|_p^* < \frac{C_1 C_2 C_3}{(k_n+1)^2} |x_{k_n}|_s \quad (n = 0, 1, \ldots).$$

On the other hand,

$$|y_n|_p^* = \sum_k |a_{kn}| |x_k|_p \geqslant |a_{k_n n}| |x_{k_n}|_p \quad (n = 0, 1, \ldots).$$

Setting $|a_{k_n n}| = \alpha_n$, $\sigma C_1 C_2 C_3 = C_p$, $s = s_p$ and making use of the definition of k_n, we find

$$\alpha_n |x_{k_n}|_p \leqslant |y_n|_p^* < C_p \alpha_n |x_{k_n}|_{s_p} \quad (p, n = 0, 1, \ldots). \tag{4}$$

Hence it follows that every value in the sequence $\{k_n\}$ is admitted at most a finite number of times. In fact,

$$\alpha_n |x_{k_n}|_q \leqslant |y_n|_q^{**} < C_q \alpha_n |x_{k_n}|_{s_q} \quad (n = 0, 1, \ldots)$$

for any q. The contrary assumption would bring us to the inequality

$$\sum_n \frac{|y_n|_p^*}{|y_n|_q^*} \geqslant \frac{1}{C_q} \sum_n \frac{|x_{k_n}|_p}{|x_{k_n}|_{s_q}} = \infty,$$

which contradicts condition (3). Therefore there exists a permutation of the sequence of natural numbers (j_n) such that $k_{j_n} \uparrow \infty$ for $n \to \infty$. The permutation (j_n) is a permutation which is desired. In fact, let us numerate the elements of the basis $\{y_{j_n}\}$ according to the increase of n and let us introduce one more system of norms for E, namely

$$|x|_p^0 = \sum_n \alpha_n |y_n'(x)| |x_{k_n}|_p \quad (p = 0, 1, \ldots; x \in E).$$

Inequality (4) yields

$$|x|_p^0 \leqslant \sum_n |y_n'(x)| |y_n|_p^* = |x|_p^{**} < C_p \sum_n \alpha_n |y_n'(x)| |x_{k_n}|_{s_p} = C_p |x|_{s_p}^0 \quad (x \in E).$$

Thus the systems of norms $|\cdot|_p^0$ and $|\cdot|_p^{**}$, and consequently also the initial system of norms $|\cdot|_p$, are equivalent. Simultaneously, the quotient $|y_n|_p^0/|y_n|_q^0 = |x_{k_n}|_p/|x_{k_n}|_q$ is a monotone function of n for every fixed p, $q = 0, 1, \cdots$. Thus the theorem is proved.

We shall consider, simultaneously with E, an isomorphic space E_1 with the topology given by a system of norms $|\cdot|_p^1$. In the sequel we shall apply the following lemma.

Lemma 1. *Let $\{x_n\}$ and $\{y_n\}$ be arbitrary bases for the spaces E and E_1, respectively. The bases $\{x_n\}$ and $\{y_n\}$ are equivalent if and only if for any p there exists a q such that*

$$\sup_n \frac{|y_n|_p^1}{|x_n|_q} < \infty, \tag{5'}$$

and, conversely, such that for any r there exists an s such that

$$\sup_n \frac{|x_n|_r}{|y_n|_s^1} < \infty. \tag{5''}$$

In fact, in view of the remark made at the beginning of this section, we may assume, without loss of generality, that the norms $|\cdot|_p^1$ satisfy the condition

$$|y|_p^1 = \sum_n |y_n'(y)| |y_n|_p^1 \quad (y \in E_1; p = 0, 1, \ldots).$$

If the bases $\{x_n\}$ and $\{y_n\}$ are equivalent, there exists a linear operator T from E into E_1 defined by the equality $Tx = \sum_n x_n'(x) y_n$ $(x \in E)$. Obviously, for any p there exists a q such that $|Tx|_p^1 \leq C_p |x|_q$ $(x \in E)$, where C_p does not depend on x. Setting $x = x_n$ $(n = 0, 1, \cdots)$ in this inequality, we obtain $|y|_p^1 \leq C_p |x_n|_q$, and hence (5′) follows. By symmetry (5″) should hold also. To prove this it suffices to reproduce the appropriate considerations in the reverse order.

The definition of a regular basis yields immediately the following theorem.

Theorem 2. *An isomorphic mapping of the space E onto E_1 transforms regular bases onto regular ones.*

In particular, the following corollary holds.

Corollary. *If $\{x_n\}$ is a regular basis for the space E, then all the bases for E equivalent to $\{x_n\}$ are regular.*

The converse statement does not hold (i.e. not all the regular bases for a given space are equivalent), because the equivalence of bases is essentially connected with the normalization of its elements (Lemma 1), which plays no role in the definition of a regular basis. However, the following question is of a considerable interest: Which conditions on a space admit a normalization of all regular bases for this space such that all the (normalized) bases are equivalent? This question is considered in §2. An example of a countably normed nuclear space without any regular basis is given there.

§2. Equivalence of bases

Let E be a countably normed nuclear space. If Γ_1 and Γ_2 are two classes of sequences of numbers $\{\gamma_n\}$ defined above, then the following theorem holds.

Theorem 3. *For isomorphic spaces E the corresponding classes Γ_1 and Γ_2 coincide.*

The proof makes use of the following properties of n-diameters implied immediately by their definition. For any neighborhoods of zero U, U', V, $V' \subset E$ and for any constants $\lambda, \mu > 0$

a) $d_n(V, U) \leq d_n(V', U')$ if $V \subseteq V'$ or if $U \subseteq U'$,

b) $d_n(\lambda V, \mu U) = (\lambda/\mu) d_n(V, U)$ (see [2], Russian p. 81).

Hence it follows also that neither of the classes Γ_i $(i = 1, 2)$ is empty. For instance, the class Γ_1 contains the sequence $(\gamma_n = 1, n = 0, 1, \cdots)$, and the class Γ_2 contains a sequence of the form $(\gamma_n = d_n^{-1}(U_{q_n}, U_p);\ n = 0, 1, \cdots)$, where p is an arbitrary natural number and $q_n \to \infty$ when $n \to \infty$.

Assume that the space E has a basis $\{x_n\}$. We shall consider also two other classes M_i $(i = 1, 2)$ of sequences $\{\gamma_n\}$ satisfying the condition

1. For any p there exists a q such that

$$\gamma_n \frac{|x_n|_p}{|x_n|_q} \to 0 \quad (n \to \infty). \tag{6'}$$

and, respectively, the condition

2. There exists a p such that for all q

$$\gamma_n \frac{|x_n|_p}{|x_n|_q} \to \infty \quad (n \to \infty). \tag{6''}$$

The classes M_i are determined uniquely by the basis, and (similarly to the classes Γ_i) they do not depend on the choice of norms $|\cdot|_p$. The intersection $M_1 \cap M_2$ is not empty.

Lemma 2. *If $\{x_n\}$ is a regular basis for the space E, there exists a system of norms $|\cdot|_p$ defining the initial topology for E such that*

$$d_n(U_q, U_p) = \frac{|x_n|_p}{|x_n|_q} \quad (q \geqslant p;\ p = 0, 1, \ldots;\ n = 0, 1, \ldots). \tag{7}$$

In fact, starting from any system of norms satisfying the condition $|x_n|_p/|x_n|_q \downarrow 0$ when $n \to \infty$ and $q > p$, and using the construction (2), we can find an equivalent system with the same property and such that

$$|x|_p = \sum_{n=0}^{\infty} |x'_n(x)|\,|x_n|_p$$

for any p. The completion of the space E with respect to the norm $|\cdot|_p$ can be considered as the Banach space l_{1p} of all sequences $\xi = \{\xi_n\}$ such that $||\xi|| = \Sigma_n |\xi_n|\,|x_n|_p < \infty$. Then the neighborhoods of zero U_p and $U_q\ (q \geq p)$ can be interpreted as dense subsets of the unit sphere $S \subset l_{1p}$ and of the "ellipsoid" $\mathfrak{E} \subset l_{1p}$ with the semiaxes $a_n = |x_n|_p/|x_n|_q$ $(n = 0, 1, \cdots)$, respectively. Equality (7) follows from the following estimate of the n-diameter of the ellipsoid $\mathfrak{E}$ with respect to the sphere S: $d_n(\mathfrak{E}, S) = a_n$ $(n = 0, 1, \cdots)$ (see [2], Russian p. 81).

Corollary 1. *Under the conditions of the lemma $\Gamma_i = M_i$ $(i = 1, 2)$.*

Corollary 2. *The corresponding classes M_i $(i = 1, 2)$ determined by different regular bases coincide.*

The class M_i $(i = 1, 2)$ determined by any regular basis for E will be denoted by $M_i(E)$. The following statement holds.

Corollary 3. *If E and E_1 are isomorphic, then $M_i(E) = M_i(E_1)$ $(i = 1, 2)$.*

For any system of norms $|\cdot|_p$ defining a topology for E we set $\gamma_n = |x_n|_q/|x_n|_p$ $(n = 0, 1, \cdots)$.

Actually we have proved the following theorem.

Theorem 4. *A necessary and sufficent condition for a countably normed nuclear space E with a regular basis $\{x_n\}$ to be a d-space is that $\{\gamma_n\} \in M_1(E)$ for a certain p and for any q, or that $\{\gamma_n\} \in M_2(E)$ for any p and for*

q dependent on p. If the corresponding condition holds, $E \in (d_i)$ $(i = 1, 2)$.

We shall introduce the following notation:

$$\gamma_n^0 = \frac{|x_n|_q^0}{|x_n|_p^0}, \quad \mu_n^0 = \frac{|x_n|_{q_n}^0}{|x_n|_{p_n}^0} \quad (n = 0, 1, \dots), \tag{8}$$

where $p_n, q_n \uparrow \infty$ when $n \to \infty$. In the previous notation we can formulate the following theorem.

Theorem 4′. $E \in (d_i)$ $(i = 1, 2)$ *if and only if there exists a system of norms* $|\cdot|_p^0$ *equivalent to the initial system of norms* $|\cdot|_p$ *in* E *such that* $\{\gamma_n^0\} \in M_1(E)$ *for any* p, q *and correspondingly* $\{\mu_n^0\} \in M_1(E)$ *for any* p_n, q_n.

Proof. 1. If $E \in (d_1)$, there exists a $p = p_0$ such that $\{\gamma_n\} \in M_1(E)$ for any q. All the more $\{\gamma_n\} \in M_1(E)$ for all $p, q \geq p_0$. The system of norms $|\cdot|_p^0$, defined by the equality $|x|_p^0 = |x|_{p_0+p}$ $(x \in E;\ p = 0, 1, \cdots)$, is equivalent to the initial one and satisfies the required condition. The converse statement is obvious.

2. Let $E \in (d_2)$. In view of the previous theorem, for every $p = 0, 1, \cdots$ there exist q_p and r_p such that $(|x_n|_{q_p} / |x_n|_p)(|x_n|_{r_p} / |x_n|_{s_p}) \to \infty$ when $n \to \infty$ for any $s_p = s$. Obviously, for fixed p and q_p the set of all numbers r_p satisfying the given relation is unbounded from above. Making use of this we can determine r_p and s_p such that the inequalities $r_p < s_p < r_{p+1}$ $(p = 0, 1, \cdots)$ are fulfilled. Since $(|x_n|_{q_0} / |x_n|_0)(|x_n|_{r_0} / |x_n|_{s_0}) \to \infty$ when $n \to \infty$, there exists an $n_0 > 0$ such that for all $n \geq n_0$ we have $|x_n|_{q_0} / |x_n|_0 > |x_n|_{s_0} / |x_n|_{r_0}$ $(n \geq n_0)$.

Analogously, starting from the relations

$$\frac{|x_n|_{q_0}}{|x_n|_0} \frac{|x_n|_{r_0}}{|x_n|_{s_1}} \to \infty, \quad \frac{|x_n|_{q_1}}{|x_n|_1} \frac{|x_n|_{r_1}}{|x_n|_{s_1}} \to \infty \quad (n \to \infty),$$

we can find $n_1 > n_0$ such that the inequalities

$$\frac{|x_n|_{q_0}}{|x_n|_0} > \frac{|x_n|_{s_1}}{|x_n|_{r_0}}, \quad \frac{|x_n|_{q_1}}{|x_n|_1} > \frac{|x_n|_{s_1}}{|x_n|_{r_1}} \quad (n \geqslant n_1).$$

are satisfied simultaneously for all $n > n_1$.

Generally, if the numbers $n_0, n_1, \cdots, n_{k-1}$ are already chosen, we determine $n_k > n_{k-1}$ such that

$$\frac{|x_n|_{q_i}}{|x_n|_i} > \frac{|x_n|_{s_k}}{|x_n|_{r_i}} \quad (n \geqslant n_k;\ i = 0, 1, \dots, k;\ k = 0, 1, \dots). \tag{9}$$

Furthermore, let

$$|x_n|_p' = |x_n|_p, \quad 0 \leqslant n < n_0, \quad p = 0, 1, \dots,$$

$$|x_n|'_p = \begin{cases} |x_n|_p, & p \leqslant s_k, \\ |x_n|_{s_k}, & p > s_k, \end{cases} \quad n_k \leqslant n < n_{k+1}, \quad k = 0, 1, \ldots, \tag{10}$$

and $|x|_p^0 = \sum_{n=0}^{\infty} |x_n'(x)|\,|x_n|'_p$ $(p = 0, 1, \cdots)$.

The systems of norms $|\cdot|_p^0$ and $|\cdot|_p$ are equivalent. In fact, for any p there exists a k such that $s_k \geq p$, and then for $n \geq n_k$ we have $|x_n|_p^0 = |x_n|'_p = |x_n|_p$. Consequently the inequalities

$$\sup_n \frac{|x_n|_p^0}{|x_n|_p} < \infty, \quad \sup_n \frac{|x_n|_p}{|x_n|_p^0} < \infty,$$

(analogous to (5′) and (5″)) hold. Now we shall show that, for any sequence of natural numbers $\{t_n\}$, $t_n \to \infty$ $(n \to \infty)$ and for any p the inequality $|x_n|_{t_n}^0 / |x_n|_{r_p}^0 < |x_n|_{q_p} / |x_n|_p$ $(n \geq n_p)$ holds.

In fact, for every $n \geq n_p$ there exists a $k \geq p$ such that $n_k \leq n < n_{k+1}$. If $t_n > s_k$, then, on the basis of (9) and (10),

$$\frac{|x_n|_{t_n}^0}{|x_n|_{r_p}^0} = \frac{|x_n|'_{t_n}}{|x_n|'_{r_p}} = \frac{|x_n|_{t_n}}{|x_n|_{r_p}} < \frac{|x_n|_{s_k}}{|x_n|_{r_p}} < \frac{|x_n|_{q_p}}{|x_n|_p}.$$

Let $\{t_n'\}$ and $\{t_n''\}$ be two arbitrary sequences, $t_n = \min(t_n', t_n'')$ and $t_n \to \infty$ when $n \to \infty$. Obviously, for any p there exists an $N > n_p$ such that $t_n > r_p$ for $n > N$. Then the inequality $|x_n|_{t_n'}^0 / |x_n|_{t_n''}^0 < |x_n|_{t_n}^0 / |x_n|_{r_p}^0 < |x_n|_{q_p} / |x_n|_p$ holds for all $n > N$. In other words, if we set $\mu_n^0 = |x_n|_{t_n'}^0 / |x_n|_{t_n''}^0$ $(n = 0, 1, \cdots)$, then for any p there exists a $q = q_p$ such that $\sup_n \mu_n^0 |x_n|_p / |x_n|_{q_p} < \infty$.

Consequently there also exists a $q > q_p$ such that $\mu_n^0 |x_n|_p / |x_n|_q \to 0$ $(n \to \infty)$.

Thus $\{\mu_n^0\} \in M_1(E)$, which proves the necessity.

The sufficiency will be proved by contradiction. Suppose that $\{\mu_n^0\} \in M_1(E)$ for any sequence $\{\mu_n^0\}$ while $\{\gamma_n^0\} \notin M_2(E)$ for a certain $p = p_0$ and for any q. The last statement implies that for any q and r there exists an s such that $\underline{\lim}_{n\to\infty} (|x_n|_q^0 / |x_n|_{p_0}^0)(|x_n|_r^0 / |x_n|_s^0) < 1$.

Given two arbitrary sequences $\{q_k\}$ and $\{r_k\}$, $q_k, r_k \to \infty$ $(k \to \infty)$, we choose s_k and n_k such that

$$\frac{|x_{n_k}|_{q_k}^0}{|x_{n_k}|_{p_0}^0} \frac{|x_{n_k}|_{r_k}^0}{|x_{n_k}|_{s_k}^0} < 1 \quad (k = 0, 1, \ldots).$$

Then the inequality $(|x_{n_k}|^0_{s_k}/|x_{n_k}|^0_{r_k})(|x_{n_k}|^0_{p_0}/|x_{n_k}|^0_q) > 1$ $(k > k_q)$ holds for any q and $k > k_q$, where $k_q = \{\inf k\colon q < q_k\}$.

Let $\{\mu_n^0\}$ be an arbitrary sequence of the form (8) such that $q_n = s_k$ and $p_n = r_k$ for $n = n_k$. According to what we have proved, $\{\mu_n^0\} \notin M_1(E)$, which contradicts the initial supposition. Thus the theorem is proved.

In view of Lemma 1, Theorem 4′ implies

Theorem 4″. *$E \in (d_i)$, $i = 1, 2$, if and only if for a certain system of norms $|\cdot|_p$, defining a topology for E, all the bases $\{x_n/|x_n|_p\}$ $(p = 0, 1, \cdots)$ and, respectively, all the bases of the form $\{x_n/|x_n|_{p_n}\}$, where $p_n \to \infty$ when $n \to \infty$, are equivalent.*

To prove this it suffices to notice that a basis of the form $\{\lambda_n x_n\}$ $(\lambda_n > 0,$ $n = 0, 1, \cdots)$ is equivalent to the basis $\{x_n\}$ if and only if the sequences $\{\lambda_n\}$ and $\{\lambda_n^{-1}\}$ belong to the corresponding class M_1.

Definition 1. A regular basis $\{x_n\}$ for a space $E \in (d_i)$, $i = 1, 2$, will be called a *normalized* basis if for a system of norms $|\cdot|_p$, defined by Theorems 4′ and 4″, $|x_n|_p = 1$ $(n = 0, 1, \cdots)$ for a certain p and, correspondingly, $|x_n|_{p_n} = 1$ $(n = 0, 1, \cdots)$ for a sequence $\{p_n\}$, $p_n \to \infty$ $(n \to \infty)$.

We pass to the basic theorem of this section.

Theorem 5. *Normalized regular bases for a space of class (d) are equivalent.*

Proof. Let $E \in (d_1)$, and let $\{x_n\}$ and $\{y_n\}$ be two normalized regular bases for E; then

$$|x_n|_p = |y_n|^1_{p_1} = 1 \quad (n = 0, 1, \ldots) \tag{11}$$

according to the definition. If the bases $\{x_n\}$ and $\{y_n\}$ are not equivalent, then at least one of the conditions (5′), (5″) of Lemma 1, e.g. (5″), does not hold. Then $\sup_n |x_n|_r/|y_n|^1_s = \infty$ for a certain r and for any s.

Making use of this, we isolate two sequences $\{s_k\}$ and $\{n_k\}$, $s_k, n_k \to \infty$ $(k \to \infty)$, such that

$$\frac{|x_{n_k}|_r}{|y_{n_k}|^1_{s_k}} > 1$$

for all $k = 0, 1, \cdots$. Hence and from (11) it follows that

$$\frac{|y_{n_k}|^1_{s_k}}{|y_{n_k}|^1_{p_1}} < \frac{|x_{n_k}|_r}{|x_{n_k}|_p} \qquad (k = 0, 1, \ldots).$$

Let us introduce the notation

$$\gamma_n = \frac{|x_n|_r}{|x_n|_p} \quad (n = 0, 1, \ldots), \quad \gamma_n^1 = \frac{|y_{n_k}|_{s_k}^1}{|y_{n_k}|_{p_1}^1} \quad (k = 0, 1, \ldots), \quad \gamma_n^1 = 0 \quad (n \neq n_k).$$

By Theorem 4′, $\{\gamma_n\} \in M_1(E)$. Since $\gamma_n' < \gamma_n$ $(n = 0, 1, \cdots)$, we have $\{\gamma_n'\} \in M_1(E)$. Consequently one can find a q_1 such that $\gamma_n^1 |y_n|_{p_1}^1 / |y_n|_{q_1}^1 \to 0$ $(n \to \infty)$.

Simultaneously, for all k such that $s_k > q$ we have

$$\gamma_{n_k}^1 \frac{|y_{n_k}|_{p_1}^1}{|y_{n_k}|_{q_1}^1} = \frac{|y_{n_k}|_{s_k}^1}{|y_{n_k}|_q^1} \geqslant 1.$$

This contradiction proves the theorem in the case under consideration.

2. Let $E \in (d_2)$ and, according to the definition of a normalized basis, $|x_n|_{p_n} = |y_n|_{q_n}^1 = 1$ $(n = 0, 1, \cdots)$, where $p_n, q_n \to \infty$ $(k \to \infty)$. Repeating the previous considerations, we shall select an r and two sequences $\{s_k\}$ and $\{n_k\}$, $s_k, n_k \to \infty$ $(k \to \infty)$, such that the inequality (12)

$$\frac{|x_{n_k}|_r}{|y_{n_k}|_{s_k}^1} > 1 \quad (k = 0, 1, \ldots)$$

holds. Hence it follows that

$$\frac{|x_{n_k}|_{p_{n_k}}}{|x_{n_k}|_r} < \frac{|y_{n_k}|_{q_{n_k}}^1}{|y_{n_k}|_{s_k}^1} \quad (k = 0, 1, \ldots).$$

Let $\{t_n\}$ be an arbitrary sequence such that $t_n \to \infty$ when $n \to \infty$ and $t_{n_k} = s_k$ for $k = 0, 1, \cdots$. We shall introduce the notation

$$\gamma_n^1 = \frac{|y_n|_{q_n}^1}{|y_n|_{t_n}^1} \quad (n = 0, 1, \ldots), \quad \gamma_{n_k} = \frac{|x_{n_k}|_{p_{n_k}}}{|x_{n_k}|_r}, \quad \gamma_n = 0 \quad (n \neq n_k), \quad k = 0, 1, \ldots.$$

According to Theorem 4′, $\{\gamma_n^1\} \in M_1(E)$. Now we perceive that $\{\gamma_n\} \notin M_1(E)$ while $\gamma_n < \gamma_n'$ for all n; this is a contradiction. Thus the proof is finished.

It follows from Theorems 1 and 5 that

Theorem 6. *All bases for a space of class* (d) *with a regular basis are quasi-equivalent.*

We shall make a remark concerning the problem of the isomorphism of the spaces considered. It follows from the definition that spaces which belong to distinct classes (d_i), $i = 1, 2$, are not isomorphic. Hence and from Theorem 5 we obtain

Theorem 7. *Two d-spaces with regular bases are isomorphic if and only if they belong to the same class* d_i $(i = 1, 2)$ *and have the same approximate dimension.*

As an example let us consider the Cartesian product $G = E_1 \times E_2$ of two nuclear spaces $E_i \in (d_i)$, $i = 1, 2$, with regular bases.

Corollary. *The space* G *is not isomorphic to either of the spaces* E_i $(i = 1, 2)$ *(see* [5] *and* [9]).

In fact, suppose for instance that G and E_1 are isomorphic. If $\{x_n\}$ and $\{y_m\}$ are bases for E_1 and E_2 respectively, and if

$$z_k = \begin{cases} (x_n, 0), & k = k_n,\ n = 0, 1, \ldots, \\ (0, y_m), & k = k_m,\ m = 0, 2, \ldots, \end{cases} \qquad k = 0, 1, \ldots, \tag{13}$$

where $\{k_n\}$ and $\{k_m\}$ are two supplementary sequences of natural numbers, then $\{z_k\}$ is a basis for G. Theorem 6 and the isomorphism of the spaces E_1 and G imply that the basis $\{x_n\}$ is equivalent to a certain $\{z_k\}$. But then the subspace $E \subset E_1$ of all x for which $x'_{k_n}(x) = 0$ $(n = 0, 1, \cdots)$ is isomorphic to E_2. However, the latter contradicts Theorem 7, since the spaces E and E_2 belong to distinct classes (d_i) $(i = 1, 2)$ and each of them has a regular basis.

It is interesting to compare the statement proved above and the fact that the space G and one of the spaces E_i $(i = 1, 2)$ can have, in general, the same approximate dimension. For instance $\Gamma(A_R \times A_\infty) = \Gamma(A_R)$ for $R < \infty$.

An example of a countably normed nuclear space without any regular basis is given by Theorem 8 (see below). Before stating it, we recall that the basic topology of uniform convergence in the space A_R $(0 < R \leq \infty)$ can be given by the system of norms $|\cdot|_p$ defined as follows:

$$|x|_p = \max_{|z| \leq R_p} |x(z)| \qquad (x \in A_R;\ p = 0, 1, \ldots;\ R_p \uparrow R\ (p \to \infty)). \tag{14}$$

Lemma 3. *For arbitrary systems of norms* $|\cdot|_p^1$ *and* $|\cdot|_p^2$ *defining the initial topology for a space* A_R $(R < \infty)$ *(respectively* A_∞*) one can find two pairs of natural numbers* p_i, q_i $(i = 1, 2)$ *satisfying the following conditions:*

a) *There exist constants* α *and* a, $1 < \alpha < a < \infty$, *such that for every* $q > q_1$ *the (asymptotic) inequality*

$$\alpha^n < \frac{|z^n|_q^1}{|z^n|_{p_1}^1} < a^n \tag{15'}$$

holds.

b) *For any* $\beta > 1$ *there exist constants* $q > q_2$ *and* b, *dependent, in general, on* q, *such that*

$$\beta^n < \frac{|z^n|_q^2}{|z^n|_{p_2}^2} < b^n. \tag{15''}$$

Proof. a) Since $|z^n|_p = R_p^n$, making use of Lemma 1, we can find p_1 and r such that the relations

$$\sup_n \frac{R_0^n}{|z^n|_{p_1}^1} < \infty, \quad \sup_n \frac{|z^n|_{p_1}^1}{R_r^n} < \infty$$

hold.

Let $R_s > R_r$. With help of the same lemma we shall find a q_1 such that

$$\sup_n \frac{R_s^n}{|z^n|_{q_1}^1} < \infty.$$

Now, to satisfy inequality (15′) it suffices to set

$$1 < \alpha < \frac{R_s}{R_r}, \quad \frac{R}{R_0} \leqslant a < \infty.$$

b) Let us take an arbitrary q_2 and evaluate p_2 and r from the conditions

$$\sup_n \frac{R_0^n}{|z^n|_{p_2}^2} < \infty, \qquad \sup_n \frac{|z^n|_{p_2}^2}{R_r^n} < \infty.$$

If β is a given number, we choose $R_s > \beta R_r$. Then there exists a $q > q_2$ such that $\sup_n R_s^n/|z^n|_q^2 < \infty$, and consequently the left-hand side of inequality (15″) is satisfied. On the other hand, there exists a t such that $\sup_n |z^n|_q^2/R_t^n < \infty$, and if one takes $b > R_t/R_0$ the right-hand inequality (15″) also holds.

Theorem 8. *The space* $G = A_R \times A_\infty$ $(R < \infty)$ *has no regular basis.*

Proof. In view of Theorem 1, it suffices to show that G has no regular basis of the form (13), where $E_1 = A_R$ and $E_2 = A_\infty$. Assume the contrary; let $\{z_k\}$ be such a basis. Then $\{x_n\}$ and $\{y_m\}$ are regular basis for A_R and A_∞, respectively. By Theorem 5 they can be normalized so that each of them is equivalent to the power basis for the corresponding space. Consequently there exists an isomorphism $G \to G$ of the space G onto itself such that $z_k \to z_k^0$ $(k = 0, 1, \cdots)$, where

$$z_k^0 = \begin{cases} (z^n, 0), & k = k_n, \\ (0, z^m), & k = k_m, \end{cases} \qquad m, n = 0, 1, \ldots.$$

According to Theorem 2, $\{z_k^0\}$ is a regular basis. Therefore there exists a system of norms $|\cdot|_p$ defining the initial topology in G such that for any fixed p and q the quotient $|z_k^0|_q/|z_k^0|_p$ is a monotone function of k. We shall bring the last statement to contradiction. Let us introduce the notation

$$|(x,0)|_p = |x|_p^1 \ (x \in A_R), \quad |(0,y)|_p = |y|_p^2 \ (y \in A_\infty), \quad p = 0, 1, \dots .$$

Without loss of generality we may assume that the corresponding constants p_i and q_i $(i = 1, 2)$, defined by Lemma 3, coincide, i.e. $p_1 = p_2 = p_0$ and $q_1 = q_2 = q_0$.

Since $|z_k^0|_q / |z_k^0|_p \uparrow \infty$ for $k \to \infty$, there exist two sequences $\{n_i\}$ and $\{m_i\}$ such that $k_{n_{i-1}} \le k_{n_{i-1}} < k_{m_i} < k_{n_i}$ $(i = 1, 2, \cdots)$.

The following two cases are possible:

a) $\overline{\lim}_{i\to\infty} m_i/n_i > \delta > 0$,

b) $\lim_{i\to\infty} m_i/n_i = 0$.

In the first case it follows from Lemma 3 that for a $\beta > a^{1/\delta}$ one can find a $q > q_1$ such that for an infinite subsequence of the sequence $\{i\}$ the inequality

$$\frac{|z_{k_{n_i}}^0|_q}{|z_{k_{n_i}}^0|_{p_0}} = \frac{|z^{n_i}|_q^1}{|z^{n_i}|_{p_0}^1} < a^{n_i} < \beta^{\delta n_i} < \beta^{m} < \frac{|z^{m_i}|_q^2}{|z^{m_i}|_{p_0}^2} = \frac{|z_{k_{m_i}}^0|_q}{|z_{k_{m_i}}^0|_{p_0}}$$

holds.

Analogously, in the case b) for any $q > q_0$ and for any b satisfying the lemma

$$\frac{|z_{k_{n_i-1}}^0|_q}{|z_{k_{n_i-1}}^0|_{p_0}} = \frac{|z^{n_i-1}|_q^1}{|z^{n_i-1}|_{p_0}^1} > a^{n_i-1} = \left(a^{\frac{n_i-1}{m_i}}\right)^{m_i} > b^{m_i} > \frac{|z^{m_i}|_q^2}{|z^{m_i}|_{p_0}^2} = \frac{|z_{k_{m_i}}^0|_q}{|z_{k_{m_i}}^0|_{p_0}},$$

for all i except, at most, a finite number of them.

Both the above inequalities contradict our supposition, which proves the theorem.

Finally, we shall construct an example of a countably normed nuclear space with a regular basis which does not belong to class (d). We apply the following definition (see [2]).

Definition 2. Let $b_{pn} > 0$ and $b_{pn} \le b_{p+1,n}$ $(p, n = 0, 1, \cdots)$. The space of all the sequences $\xi = \{\xi_n\}$ for which $|\xi|_p = \Sigma_n |\xi_n| b_{pn} < \infty$ for any p with the topology given by a system of norms $|\cdot|_p$ is called the *Köthe space* $L(b_{pn})$.

Obviously, the sequence $\{x_n\}$, $x_n = (\delta_{kn}, k = 0, 1, \cdots)$, $n = 0, 1, \cdots$, is a basis for $L(b_{pn})$.

Example. Let $\lambda_n > 0$ $(n = 0, 1, \cdots)$, $\lambda_n \downarrow 0$ $(n \to \infty)$, $\Sigma_n \lambda_n < \infty$ and $\beta_n^0 = e^{\lambda_0 n}$ $(n = 0, 1, \cdots)$. We define two sequences of numbers μ_k^1 and

n_k^1 $(k = 0, 1, \cdots)$ so that the relations

$$e^{\mu_0^1} > \beta_0^0, \quad e^{\mu_0^1+\lambda_1 n} < e^{2\lambda_1 n} \quad (n \geqslant n_0^1 - 1), \quad e^{\mu_k^1+\lambda_1 n_{k-1}^1} > \beta_{n_{k-1}^1}^0,$$

$$e^{\mu_k^1+\lambda_1 n} < e^{2\lambda_1 n} \quad (n \geqslant n_k^1 - 1), \quad k = 1, 2, \ldots$$

hold. Next we set

$$\beta_n^1 = e^{\mu_0^1+\lambda_1 n} \, (0 \leqslant n < n_0^1), \quad \beta_n^1 = e^{\mu_k^1+\lambda_1 n} \, (n_{k-1}^1 \leqslant n < n_k^1; \; k = 1, 2, \ldots).$$

If the sequences $\{\beta_n^s\}$ $(s < p)$ are already chosen, we construct $\{\beta_n^p\}$ as follows: we find μ_k^p and n_k^p $(k = 0, 1, \cdots)$ from the conditions

$$e^{\mu_0^p} > \beta_0^0 \beta_0^1 \ldots \beta_0^{p-1}, \quad e^{\mu_0^p+\lambda_p n} < e^{2\lambda_p n} \, (n \geqslant n_0^p - 1), \quad e^{\mu_k^p+\lambda_p n_{k-1}^p} > \beta_{n_{k-1}^p}^0 \ldots \beta_{n_{k-1}^p}^{p-1},$$

$$e^{\mu_k^p+\lambda_p n} < e^{2\lambda_p n} \quad (n \geqslant n_k^p - 1), \quad k = 1, 2, \ldots,$$

and next we set

$$\beta_n^p = e^{\mu_0^p+\lambda_p n} \, (0 \leqslant n < n_0^p), \quad \beta_n^p = e^{\mu_k^p+\lambda_p n} \, (n_{k-1}^p \leqslant n < n_k^p), \; k = 1, 2, \ldots.$$

After this we define the numbers b_{pn} $(p, n = 0, 1, \cdots)$ using the recursion relation

$$b_{0n} > 0, \qquad \frac{b_{pn}}{b_{p-1, n}} = \beta_n^p \quad (n = 0, 1, \ldots), \quad p = 1, 2, \ldots.$$

We shall show that the Köthe space $L(b_{pn})$ satisfies the required conditions. In fact, the basis $\{x_n\}$ for $L(b_{pn})$ is a regular basis since we have

$$\frac{|x_n|_q}{|x_n|_p} = \frac{b_{qn}}{b_{pn}} = \beta_n^q \beta_n^{q-1} \ldots \beta_n^{p+1}$$

for $p > q$. Here, on the right-hand side, a product of nondecreasing functions appears. Since for any p

$$\sum_n \frac{|x_n|_p}{|x_n|_{p+1}} = \sum_n \beta_n^{-p-1} = \sum_n e^{-\mu_k^{p+1} - \lambda_{p+1} n} < \sum_n e^{-\lambda_{p+1} n} < \infty,$$

it follows that $L(b_{pn})$ is a nuclear space. This space does not belong to class (d_2), although for any q and for all $n = n_{k-1}^{q+1}$

$$\frac{|x_n|_{q+1}}{|x_n|_q} = \beta_n^{q+1} > \beta_n^0 \beta_n^1 \cdots \beta_n^q = \frac{b_{qn}}{b_{0n}} = \frac{|x_n|_q}{|x_n|_0},$$

i.e. $d_n(U_q, U_0) > d_n(U_{q+1}, U_q)$ $(n = n_{k-1}^{q+1};\; k = 1, 2, \cdots)$, which contradicts the corresponding definition. It remains to verify that $L(b_{pn}) \notin (d_1)$. Let $p \geq 0$ be an arbitrary natural number and let $q > p$ be such that

$$\sum_{p+1}^{q} \lambda_s > 2 \sum_{q+1}^{\infty} \lambda_s.$$

Then for any $r > q$ and for all $n = n_k^r - 1$ we have

$$\frac{|x_n|_r}{|x_n|_q} = \beta_n^r \beta_n^{r-1} \cdots \beta_n^{q+1} < e^{2n \sum\limits_{q+1}^{r} \lambda_s} < e^{n \sum\limits_{p+1}^{q} \lambda_s}.$$

On the other hand, for the same n

$$\frac{|x_n|_q}{|x_n|_p} = \beta_n^q \beta_n^{q-1} \cdots \beta_n^{p+1} > e^{n \sum\limits_{p+1}^{q} \lambda_s}.$$

Thus $d_n(U_q, U_p) < d_n(U_r, U_q)$ ($n = n_k^r - 1$; $k = 0, 1, \cdots$), and consequently $L(b_{pn}) \notin (d_1)$.

§3. Conjugate spaces

In this section we shall prove the theorem on quasi-equivalence of bases for the conjugate space E' of a nuclear d-space E with a regular basis. The strong topology induced by that of the space E (see [10]) is taken as the initial topology for E'.

Let $\{x_n'\}$ be a basis for E', $\{x_n\}$ a biorthogonal sequence of functionals, $x' \in E'$ an arbitrary element,

$$x' = \sum_{n=0}^{\infty} x_n(x') x_n'. \tag{16}$$

As is known (see [2]), the functionals x_n are continuous and $x_n \in E$ ($n = 0, 1, \cdots$) in view of the reflexivity of the space E ([10]). It follows from (16) that for an arbitrary $x \in E$

$$x'(x) = \sum_{n=0}^{\infty} x'(x_n) x_n'(x) \quad (x \in E,\ x' \in E').$$

Thus, for any $x \in E$, series (1) converges weakly in the space E, and consequently it converges in the initial topology [10]. In addition, the expansion (1) is unique, i.e. the sequence $\{x_n\}$ is a basis for E. Since the initial topology for E and the strong topology, induced by that for the space E', coincide, the argument can be inverted. By the same token the following lemma (admitting the dual formulation) is proved.

Lemma 4. *A sequence $\{x_n\}$ of elements of a space E is a basis for E if and only if there exists a biorthogonal sequence $\{x_n'\}$ which is a basis for the conjugate space E'.*

For an arbitrary basis $\{x_n'\}$ for E' we shall denote the set of all the sequences $\eta = \{\eta_n\}$, $\eta_n = x_n(x')$ ($n = 0, 1, \cdots$; $x' \in E'$), by G'.

Lemma 5. *Two bases $\{x_n'\}$ and $\{y_n'\}$ for the space E' are equivalent if and only if the corresponding sets G' coincide.*

In fact, the necessity of the condition is obvious. To prove the sufficiency let us consider two sequences of operators $\{T_n\}$ and $\{R_n\}$ defined as follows:

$$T_n x' = \sum_{k=0}^{n} x_k(x')\, y'_k, \quad R_n x' = \sum_{k=0}^{n} y_k(x')\, x'_k \quad (x' \in E';\ n = 0, 1, \ldots).$$

All the operators are continuous by virtue of the continuity of the functionals $x_n, y_n \in E$. It follows from the condition of the lemma that $T_n x' \to Tx'$ and $R_n x' \to Rx'$ for any fixed $x' \in E'$ and $n \to \infty$. By the Banach-Steinhaus Theorem the limit operators T and R are linear, and since $Tx'_n = y'_n$ and $Ry'_n = x'_n$ $(n = 0, 1, \cdots)$ we have $R = T^{-1}$, which proves the lemma.

Let G be the set of all the sequences $\xi = \{\xi_n\}$, $\xi_n = x'_n(x)$ $(x \in E$; $n = 0, 1, \cdots)$, determined by an arbitrary basis $\{x_n\}$ for E. Obviously the following statement, dual with respect to Lemma 5 and equivalent to Lemma 1, holds.

Lemma 5′. *Two bases $\{x_n\}$ and $\{y_n\}$ for the space E are equivalent if and only if the corresponding sets G coincide.*

Finally, it is easy to verify the following remark: if $\{x'_n\}$ is a basis for the space E and $\{x'_n\}$ is a biorthogonal sequence of functionals of E', the set G' and the set of all the η for which $\Sigma_n |\xi_n \eta_n| < \infty$ for any $\xi \in G$ coincide.

Consider now two arbitrary bases for E'. By permutation and normalization of their elements one can attain that the conjugate biorthogonal sequences are normalized regular bases for the space E. Since the latter are equivalent, the corresponding sets G coincide, and, in view of the last remark, the corresponding sets G' coincide also. Then Lemma 5 yields the equivalence of the bases considered. Thus the following theorem holds.

Theorem 6′. *All the bases for the strong conjugate space E' of a d-space with a regular bases are quasi-equivalent.*

Let E be an arbitrary countably normed nuclear space possessing a basis. Actually a more general statement is proved.

Theorem 9. *All the bases for a space E are quasi-equivalent if and only if the same condition holds for the conjugate space E'.*

§4. Spaces $L_f(b, r)$

Definition 3 (see [5]). The Köthe space $L(b_{pn})$, $b_{pn} = \exp(r_p b_n)$, where $b_n > 0$ $(b_n \in b;\ n = 0, 1, \cdots)$, $b_n \uparrow \infty$ $(n \to \infty)$, $r_p \uparrow r$ $(p \to \infty)$ and $-\infty < r < \infty$ (respectively $r = \infty$) is called the *finite (infinite) center of the Riesz scale* $L(b, r)$.

It is easy to see that the nuclearity of the space $L(b, r)$ is equivalent to the condition $\sum_n e^{-\delta b_n} < \infty$. In the case of a finite center the series converges for any $\delta > 0$ and in the case of an infinite center it converges at least for one δ. Clearly $\{x_n\}$ (see §2) is a regular basis for $L(b, r)$. Finally $L(b, r)$ belongs to the class (d_1) for $r = \infty$ and to (d_2) for $-\infty < r < \infty$. In fact, here the corresponding conditions can be reduced to the following requirements:

1) For any q there exists an s such that (for a fixed p) $r_p + r_s > 2r_q$. Clearly this requirement is satisfied if $r = \infty$.

2) For every p there exists a q such that $r_p + r_s < 2r_q$ for an arbitrary s. This latter holds if $r < \infty$.

Thus the following theorem is valid.

Theorem 10 (see [2]). *All bases for a nuclear center of Riesz scale are quasi-equivalent.*

The following statement (which is implied also by Theorem 7) is cited in [5].

Theorem 11. *Two nuclear spaces $L(b', r')$ and $L(b'', r'')$ are isomorphic if and only if they are both centers of either finite or infinite scales and the following condition holds*:

$$0 < \varliminf_{n\to\infty} \frac{b'_n}{b''_n} \leqslant \varlimsup_{n\to\infty} \frac{b'_n}{b''_n} < \infty.$$

In this way, conditions for two arbitrary spaces to be isomorphic are formulated very simply in terms of the sequences $\{b_n\}$ and of the constant r. The same holds for a larger class of spaces L_f which were introduced in [6] as an essential generalization of centers of Riesz scales.

Definition 4. Let $f(u)$ be a nondecreasing odd function of a real variable u. Moreover, let $f(u)$ be logarithmically convex for $u \geq 0$ and not identically equal to a constant. The Köthe space $L(b_{pn})$, where $b_{pn} = \exp f(r_p b_n)$ $(p, n = 0, 1, \cdots)$ is called a *space* $L_f(b, r)$.

It follows from this definition that $L(b, r) = L_f(b, r)$ for $f(u) \equiv u$. Let us introduce the notation

$$\psi(w) = f(e^w)\ (-\infty < w < \infty), \rho_p = \ln|r_p|\ (p = 0, 1, \ldots), \beta_n = \ln b_n\ (n = 0, 1, \ldots).$$

It is easy to see that $\{x_n\}$ is a regular basis for the space $L_f(b, r)$. In fact,

$$\frac{|x_n|_p}{|x_n|_q} = \begin{cases} \exp\,[\psi(\rho_p + \beta_n) - \psi(\rho_q + \beta_n)], & 0 < r \leqslant \infty, \\ \exp\,[\psi(\rho_q + \beta_n) - \psi(\rho_p + \beta_n)], & -\infty < r \leqslant 0, \end{cases} \tag{17}$$

and the right-hand side is monotonic for fixed p and q because the function $\psi(w)$ is a convex function. Since, moreover, the function $\ln \psi(w)$ is convex, the limit

$$\tau(\alpha) = \tau_f(\alpha) = \lim_{w\to\infty} [\ln \psi(w+\alpha) - \ln \psi(w)] \qquad (-\infty < \alpha < \infty)$$

exists for any real α.

Relying on known properties of convex functions, we can easily show that one and only one of the following possibilities occurs:

$$\text{a) } \tau(\alpha) < \infty \quad (0 < \alpha < \infty), \qquad \text{b) } \tau(\alpha) = \infty \quad (0 < \alpha < \infty),$$

In the first case $\tau(\alpha)$ is a continuous odd strongly increasing function. We shall write $\tau < \infty$ if the case a) occurs and $\tau = \infty$ in the contrary case. Sometimes we shall say also that the function $f(u)$ has slow or, respectively, rapid growth when $u \to \infty$.

We shall prove that the space L_f belongs to class (d). In fact, if $\tau < \infty$, then for every p there exists a q such that $|e^{\tau(\rho_p)} - e^{\tau(\rho_q)}| > |e^{\tau(\rho_q)} - e^{\tau(\rho_s)}|$ for any s. Hence for $0 < r < \infty$ (respectively for $-\infty < r \leq 0$) the following asymptotic inequalities can be obtained:

$$\psi(\rho_p + \beta_n) - \psi(\rho_q + \beta_n) < \psi(\rho_q + \beta_n) - \psi(\rho_s + \beta_n),$$
$$\psi(\rho_q + \beta_n) - \psi(\rho_q + \beta_n) < \psi(\rho_s + \beta_n) - \psi(\rho_q + \beta_n),$$

It is easy to see that the last inequality is satisfied also when $\tau = \infty$ and $r \leq 0$. In all three cases considered it follows from (17) and (7) that $L_f \in (d_2)$. In the two other cases (i.e. if $\tau < \infty$ and $r = \infty$ or $\tau = \infty$ and $r > 0$) $L_f \in (d_1)$, which can be verified analogously. Thus the following theorem holds.

Theorem 12. *All bases for a nuclear space $L_f(b, r)$ are quasi-equivalent.*

We shall find the approximate dimension (class Γ_1) of an arbitrary space $L_f(b, r)$ under the following different assumptions on τ and r:

a) $\tau < \infty$, $r < \infty$, b) $\tau < \infty$, $r = \infty$, c) $\tau = \infty$, $r > 0$, d) $\tau = \infty$, $r \leqslant 0$.

Lemma 6. *In all the cases* a) – d) *the class Γ of the space $L_f(b, r)$ coincides with the set of these and only these sequences $\{\gamma_n\}$ for which*

$$\text{a) } \overline{\lim_{n\to\infty}} \frac{\ln \gamma_n}{f(b_n)} \leqslant 0, \quad \text{b) } \overline{\lim_{n\to\infty}} \frac{\ln \gamma_n}{f(b_n)} < \infty, \quad \text{c, d) } \overline{\lim_{n\to\infty}} \frac{\ln \gamma_n}{f(|r_p| b_n)} \leqslant 0, \quad (18)$$

respectively. In the case d) *the last relation holds for any p, and in the case* c), *for at least one p.*

In the proof we make use only of the properties of the function $f(u)$ mentioned above and we apply relations (7) and (17). In particular, the lemma implies the following criterion for the nuclearity of the space L_f.

Corollary 1. *A space $L_f(b, r)$ is a nuclear space if and only if in each of the cases* a) – d) *the corresponding relation* (18) *is satisfied for $\gamma_n = n$* $(n = 1, 2, \cdots)$.

We shall note two more corollaries which follow from the lemma and from Theorem 7. In what follows all spaces are assumed to be nuclear spaces.

Corollary 2. *A space* L_f *is isomorphic to some center of a Riesz scale if and only if* $\tau < \infty$.

Corollary 3. *Two spaces* $L_f(b', r')$ *and* $L_f(b'', r'')$ *are isomorphic if and only if one of the following conditions holds*:

$$\text{a) } r' = r'' = 0, \quad \text{or} \quad r' = r'' = \infty \text{ and } 0 < \underline{\lim}_{n \to \infty} \frac{b'_n}{b''_n} \leqslant \overline{\lim}_{n \to \infty} \frac{b'_n}{b''_n} < \infty,$$

$$\text{b) } 0 < r', \ r'' < \infty, \quad \text{or} \quad -\infty < r', \ r'' < 0 \text{ and } \lim_{n \to \infty} \frac{b'_n r'}{b''_n r''} = 1.$$

(The proof is actually reproduced below).

Furthermore, let $\phi(u)$ be a function defined for $u \geq 0$ and satisfying the basic assumptions on $f(u)$.

Corollary 4. *If* $f_1(u)$ *and* $f_2(u)$ *are functions for which* $\tau = \infty$ *and* $f_2(u) = f_1[\phi(u)]$, *where* $\phi(u)$ *increases slowly, then the corresponding spaces* $L_{f_1}(b', r')$ *and* $L_{f_2}(b'', r'')$ *are isomorphic at any rate in the following cases*:

a) $0 < r', \ r'' < \infty$, or $-\infty < r', r'' < 0$ and $|r'| b'_n = \varphi(|r''| b''_n)$ $(n = 0, 1, \ldots)$,

b) $r' = r'' = 0$, or $r' = r'' = \infty$ and $b'_n = \varphi(b''_n)$ $(n = 0, 1, \ldots)$.

In fact, let for instance $r' > 0$ and $r'' > 0$. Since for any $a > 0$ the limit $\lim_{u \to \infty} \phi(au)/\phi(u) = \exp \tau_\phi(\ln a)$ exists, for every $r'_p < r'$ one can find an $r''_q < r''$ such that the (asymptotic) inequality

$$r'_p \varphi(r'' b''_n) < r' \varphi(r''_q b''_n)$$

holds. Consequently

$$r'_p b'_n = r' b'_n \frac{r'_p}{r'} = \varphi(r'' b''_n) \frac{r'_p}{r'} < \varphi(r''_q b''_n),$$

whence $f_1(r'_p b'_n) < f_2(r''_q b''_n)$.

Conversely, for any $r''_q < r''$ one can find an $r'_s < r'$ such that $r' \phi(r''_q b''_n) < \phi(r'' b''_n) r'_s$, and consequently $f_2(r''_q b''_n) = f_1[\phi(r''_q b''_n)] < f_1[\phi(r'' b''_n) r'_s / r'] = f_1(r'_s b'_n)$.

Hence and from (18, c) the coincidence of the corresponding classes Γ_1 follows, and Theorem 7 yields isomorphism of the corresponding spaces. In the other cases the proof is analogous.

Remark. Actually, the spaces considered, as well as the classes Γ_1, coincide in the set-theoretic sense.

Corollary 5. *If, under the same assumptions,* $\phi(u)$ *increases quickly, no two spaces* $L_{f_1}(b', r')$ *and* $L_{f_2}(b'', r'')$ *are isomorphic.*

To prove this we shall introduce the function $\omega(v)$ which is the inverse of $v = f(u)$. It is easy to see that the class Γ_1 of an arbitrary space $L_f(b, r)$ for $r = \infty$ can be described as the set of all sequences $\{\gamma_n\}$ for which in each of the cases a) $0 < r < \infty$, b) $-\infty < r < 0$, c) $r = \infty$, d) $r = 0$ the following relations hold, respectively:

$$\text{a)}\ \overline{\lim_{n\to\infty}} \frac{\omega(\ln \gamma_n)}{b_n} < r, \qquad \text{b)}\ \overline{\lim_{n\to\infty}} \frac{\omega(\ln \gamma_n)}{b_n} \leqslant |r|,$$

$$\text{c)}\ \overline{\lim_{n\to\infty}} \frac{\omega(\ln \gamma_n)}{b_n} < \infty, \qquad \text{d)}\ \overline{\lim_{n\to\infty}} \frac{\omega(\ln \gamma_n)}{b_n} \leqslant 0. \tag{19}$$

Hence and from the fact that $L_f(b, r) \in (d_i)$ $(i = 1, 2)$ respectively for $r > 0$ and $r \le 0$ it follows that the spaces $L_{f_1}(b', r')$ and $L_{f_2}(b'', r'')$ can be isomorphic only in the following four cases:

$$\text{a)}\ 0 < r', r'' < \infty, \qquad \text{b)}\ -\infty < r', r'' < 0,$$

$$\text{c)}\ r' = r'' = 0, \qquad \text{d)}\ r' = r'' = \infty. \tag{20}$$

In fact, let for instance $0 < r' < \infty$, $r'' = \infty$ and $L_{f_1}(b', r') \sim L_{f_2}(b'', \infty)$. (Thus we suppose that the spaces considered are isomorphic.) Then (19, a) and (19, c) imply that $L_{f_1}(\lambda b', r') \sim L_{f_2}(\lambda b'', \infty) \sim L_{f_2}(b'', \infty)$ for an arbitrary $\lambda > 0$. Consequently $L_{f_1}(b', r') \sim L_{f_1}(\lambda b', r')$, which is impossible in view of (19, a).

Consider one of the cases (20), e. g. the first one, in detail. If we assume that $L_{f_1}(b', r') \sim L_{f_2}(b'', r'')$, then for any $r'_p < r'$ there exists an $r''_q < r''$ such that $f_1(r'_p b'_n) < f_2(r''_q b''_n)$, and consequently $r'_p b'_n < \phi(r''_q b''_n)$ (both inequalities, as well as the next ones, are asymptotic). Since in this case $\lim_{u\to\infty} \phi(au)/\phi(u) = \infty$ for any $a > 1$, one can find an r''_s, $r''_q < r''_s < r''$, such that $r' b'_n < \phi(r''_s b''_n)$, and consequently $f_1(r' b'_n) < f_2(r''_s b''_n)$.

Then the sequence $\{\gamma_n\}$, where $\gamma_n = \exp f_2(r''_s b''_n)$ $(n = 0, 1, \cdots)$, belongs to $\Gamma(L_{f_2})$, while $\{\gamma_n\} \in \Gamma(L_{f_1})$, which contradicts the assumption.

This proves the theorem.

Let F be an arbitrary linearly-ordered [12] set of functions $f(u)$ with the following order relation between any two elements:

$$f_1 \leqslant f_2, \quad \text{if} \quad f_2(u) = f_1[\varphi(u)] \quad \text{for} \quad 0 \leqslant u < \infty \quad (f_1, f_2 \in F).$$

We shall denote by (f) the class of all the nuclear spaces $L_f(b, r)$ defined by a given function $f(u) \in F$.

Theorem 13. *For any functions* $f_1, f_2 \in F$, $f_1 \leq f_2$, *the corresponding classes* (f_1) *and* (f_2) *of spaces* L_f *either coincide (set-theoretically) or are essentially different in the sense that no two spaces belonging to different classes are isomorphic. The first case occurs if and only if* $\phi(u)$ *is a slowly increasing function.*

Remark. According to Corollaries 1 and 2 every class (f) joins either two (for $\tau < \infty$) or four (for $\tau = \infty$) subclasses which are essentially different in the above sense.

We shall adduce an example illustrating Theorem 13.

Example. For $w > w_\psi$ let

$$\psi_{1,\alpha,\lambda}(w) = \exp(\lambda w^\alpha), \quad 1 \leqslant \alpha < \infty, \quad 0 < \lambda < \infty,$$

$$\psi_{n,\alpha,\lambda}(w) = \exp^n(\lambda w^\alpha), \quad 0 < \alpha, \ \lambda < \infty, \quad n = 2, 3, \ldots,$$

where $\exp^n$ is the corresponding iteration of the exponential function and w_ψ is chosen such that the function $\ln \psi_{n,\alpha,\lambda}(w)$ can be extended to the whole x-axis so that convexity and monotonicity are preserved. Let $f_{n,\alpha,\lambda}(u) = \psi_{n,\alpha,\lambda}(\ln u)$ $(u > 0)$ and let us extend the function $f_{n,\alpha,\lambda}(u)$ as an odd function. The set of functions $f_{n,\alpha,\lambda}(u)$, defined in this way, will be taken for F. According to the definition $f_{n_1,\alpha_1,\lambda_1} \leq f_{n_2,\alpha_2,\lambda_2}$ if and only if one of the following relations holds:

$$\text{a) } n_1 < n_2; \quad \text{b) } n_1 = n_2, \ \alpha_1 < \alpha_2;$$

$$\text{c) } n_1 = n_2, \ \alpha_1 = \alpha_2, \ \lambda_1 < \lambda_2; \quad \text{d) } n_1 = n_2, \ \alpha_1 = \alpha_2, \ \lambda_1 = \lambda_2.$$

According to Theorem 13, the classes $(f_{n_1,\alpha_1,\lambda_1})$ and $(f_{n_2,\alpha_2,\lambda_2})$ coincide for $n_1 = n_2$ and $\alpha_1 = \alpha_2$, and they are essentially different in the other cases. The class $(f_{1,1,1}) = (f_{1,1,\lambda})$ contains all the nuclear centers of Riesz scales.

BIBLIOGRAPHY

[1] M. M. Dragilev, *Standard form of basis for the space of analytic functions*, Uspehi Mat. Nauk 15 (1960), no. 2 (92), 181–188. (Russian) MR 22 #6992.

[2] B. S. Mitjagin, *Approximate dimension and bases in nuclear spaces*, Uspehi Mat. Nauk 16 (1961), no. 4 (100), 63–132 = Russian Math. Surveys 16 (1961), no. 4, 59–127. MR 27 #2927.

[3] A. Kolmogoroff, *Über die beste Annäherung von Funktionen einer*

gegebenen Funktionenklasse, Ann. of Math. (2) 37 (1936), 107–111.

[4] V. M. Tihomirov, *On n-dimensional diameters of certain functional classes*, Dokl. Akad. Nauk SSSR 130 (1960), 734–737 = Soviet Math. Dokl. 1 (1960), 94–97. MR 22 #8267.

[5] B. S. Mitjagin, *Nuclear Riesz scales*, Dokl. Akad. Nauk SSSR 137 (1961), 519–522 = Soviet Math. Dokl. 2 (1961), 309–312. MR 24 #A3489.

[6] M. M. Dragilev, *On bases in nuclear spaces*, Theses of Reports at the Seventh All-Union Conference Theory of Functions of a Complex Variable, Rostov-on-Don, 1964.

[7] A. Dynin and B. Mitjagin, *Criterion for nuclearity in terms of approximate dimension*, Bull. Acad. Polon. Sci. Sér. Sci. Math. Astronom. Phys. 8 (1960), 535–540. MR 24 #A2221.

[8] Cz. Bessaga, A. Pełczyński and S. Rolewicz, *Approximate dimension of linear topological spaces and some of its applications*, Studia Math. (Ser. Specjalna) Zeszyt 1 (1963), 27–29. MR 26 #4157.

[9] S. Rolewicz, *On spaces of holomorphic functions*, Studia Math. 21 (1961/62), 135–160. MR 27 #4097.

[10] I. M. Gel'fand and G. E. Šilov, *Generalized functions*. Vol. 2: *Spaces of fundamental functions*, Fizmatgiz, Moscow, 1958; English transl., Academic Press and Gordon and Breach, New York, 1968. MR 21 #5142a; 37 #5693.

[11] L. V. Kantorovič and G. P. Akilov, *Functional analysis in normed spaces*, Fizmatgiz, Moscow, 1959; English transl., Internat. Series of Monographs in Pure and Appl. Math., vol. 46, Macmillan, New York, 1964. MR 22 #9837; 35 #4699.

[12] N. Dunford and J. T. Schwartz, *Linear operators*. I: *General theory*, Pure and Appl. Math., vol. 7, Interscience, New York, 1958; Russian transl., IL, Moscow, 1962. MR 22 #8302.

Translated by:
H. Swiatak

Amer. Math. Soc. Transl.
(2) Vol. 93, 1970

BESSEL AND HILBERT SYSTEMS IN BANACH SPACES AND QUESTIONS OF STABILITY

B. E. VEĬC

In §1 we study Bessel and Hilbert systems in arbitrary Banach spaces, and generalize results of N. K. Bari [1] for these systems. The interest in such systems comes from the fact that in any Banach space there exist Bessel systems, in any reflexive space there exist Hilbert systems, and in every L^p-space there are Hilbert and Bessel bases. In §2 we introduce the concept of weak quadratic stability and establish the weak quadratic stability of Bessel systems. We also show that quadratic stability is a characteristic property of Bessel bases. We also consider other stability properties of these bases. In §3 we introduce the concept of q-stability and weak q-stability, generalizing results presented by the author at the Fourth All-Union Congress of Mathematics.

1. Let E be a separable Banach space over the field of complex numbers, $(x_k) \subset E$ a minimal complete system of elements, and (f_k) a system of biorthogonal functionals: $f_i(x_k) = \delta_{ik}$, $i, k = 1, 2, \cdots$. Following N. K. Bari [1], we call the sequence (x_k) a *Bessel system* if for any $x \in E$ the series $\sum_{k=1}^{\infty} |f_k(x)|^2$ converges. The sequence is a Hilbert system if for any square-summable sequence of complex numbers $(\alpha_k) \in l^2$ there exists a unique element $x \in E$ such that $f_k(x) = \alpha_k$, $k = 1, 2, \cdots$. Since a universal space C always exists, we may assume that E is dense in some Hilbert space H: $E \subset H$. For simplicity, we assume that $\|x\|_H \leq \|x\|_C$ for $x \in E$. Since $|(x, y)| \leq \|y\|_H \|x\|_E$ for $x \in E$ and $y \in H$, each element $y \in H$ generates a functional $f_y \in E^*$ of the form $f_y(x) = (x, y)$. Moreover, $\|f_y\|_{E^*} \leq \|y\|_H$. Therefore in the sequel we assume that $E \subset H \subset E^*$, where the inclusion is topological.

Theorem 1. *A system $(x_k) \subset E$ is a Bessel system in a Banach space E if and only if there exists a bounded linear operator T: $E \to H$ such that the system of elements $e_k = Tx_k \in H$, $k = 1, 2, \cdots$, forms an orthogonal basis in the space H.*

Proof. Necessity. Let (e_k) be a complete orthonormal basis in the

* Translation of Izv. Vysš. Učebn. Zaved. Matematika 1965, no. 2 (45), 7–23.

space H, and (x_k) a Bessel system in E. Then for any $x \in E$ the series $\sum_{k=1}^{\infty} |f_k(x)|^2$ converges, so that the series $\sum_{k=1}^{\infty}{}_H f_k(x) e_k = y$, where convergence is the norm of the space H. These equalities define a linear operator $T(E \to H)$: $y = Tx$ that is bounded:

$$\| Tx \|_H = \Big\| \sum_{k=1}^{\infty} f_k(x) e_k \Big\|_H \leqslant \sqrt{\sum_{k=1}^{\infty} |f_k(x)|^2} \leqslant a \| x \|_E \quad (a > 0).$$

The last inequality follows by a lemma of Gel'fand [2] (also see [3], Russian page 432). Finally we note that $Tx_k = e_k$, $k = 1, 2, \cdots$.

Sufficiency. Let $\|Tx\|_H \leq a\|x\|_E$ and $Tx_k = e_k$, $k = 1, 2, \cdots$, where (e_k) is an orthonormal basis in H. For any $y \in H$ we have $|(Tx, y)| \leq \|Tx\|_H \|y\|_H \leq a\|y\|_H\|x\|_E$. Consequently each element $y \in H$ generates a linear functional $\phi_y \in E^*$: $\phi_y(x) = (Tx, y) = (x, T_* y)$. The operator T_*: $H \to E^*$ carries each element $y \in H$ into a functional $\phi_y \in E^*$. From the inequalities $|\phi_y(x)| = |(Tx, y)| \leq \|T\| \, \|y\|_H \|x\|_E$ and $\|\phi_y\| = \|T_* y\| \leq \|T\| \, \|y\|_H$ we find that T_* is bounded. From the equation $\delta_{ik} = (e_i, e_k) = (Tx_i, e_k) = (x_i, T_* e_k)$, $i, k = 1, 2, \cdots$, we obtain $T_* e_k = f_k$ and $\sum_{k=1}^{\infty} |f_k(x)|^2 = \sum_{k=1}^{\infty} |(x, T_* e_k)|^2 = \sum_{k=1}^{\infty} |(Tx, e_k)|^2 < +\infty$. This also shows that (x_k) is a Bessel system in the space E.

Theorem 2. *If (x_k) is a Bessel system in E, then the system (f_k) of biorthogonal functionals forms a Hilbert system in its linear closure $[f_k]_{E^*}$.*

Proof. By Theorem 1, there exists a continuous linear operator T: $E \to H$ such that $Tx_k = e_k$, $k = 1, 2, \cdots$, where (e_k) is any orthonormal basis in H. If $(\alpha_k) \in l^2$, then there exists a unique element $y \in H$ such that $(e_k, y) = (Tx_k, y) = \alpha_k$, $k = 1, 2, \cdots$. Then $T_* y = \phi_y \in E^*$ and $\phi_y(x_k) = (x_k, T_* y) = (Tx_k, y) = \alpha_k$. If there exists at least one functional $\phi \in E^*$ such that $\phi(x_k) = \alpha_k$, $k = 1, 2, \cdots$, then $(\phi_y - \phi)(x_k) = 0$ for all $k = 1, 2, \cdots$, and in view of the completeness of the system (x_k), $\phi_y = \phi$. It is clear that $\phi_y \in [f_k]_{E^*}$, since

$$y = \sum_{k=1}^{\infty}{}_H \alpha_k e_k \text{ and } T_* y = \sum_{k=1}^{\infty}{}_{E^*} \alpha_k T_* e_k = \sum_{k=1}^{\infty}{}_{E^*} \alpha_k f_k.$$

Theorem 3. *If (x_k) is a Hilbert system in the space E, then (f_k) is a Bessel system in $[f_k]_{E^*}$.*

Proof. By the definition of a Hilbert system, every sequence $\alpha = (\alpha_k) \in l^2$ corresponds to a unique element $x \in E$ that satisfies the conditions

$f_k(x) = \alpha_k$, $k = 1, 2, \cdots$, and so defines a linear operator $A(l^2 \to E)$: $Ax = x$. We prove that it is closed. Let $\alpha^{(n)} = (\alpha_1^{(n)}, \alpha_2^{(n)}, \cdots) \to \alpha = (\alpha_1, \alpha_2, \cdots)$ in the norm of l^2 and $A\alpha^{(n)} = x^{(n)} \to x$ in the norm of E. Then $f_k(x^{(n)}) = \alpha_k^{(n)}$, $k = 1, 2, \cdots$, and $f_k(x) = \lim_{n\to\infty} f_k(x^{(n)}) = \lim \alpha_k^{(n)} = \alpha_k$, $k = 1, 2, \cdots$. Consequently $A\alpha = x$ and the operator A is closed. Since it is defined on all of l^2, it is bounded, so that there exists a number $m > 0$ such that for all $\alpha \in l^2$

$$\|A\alpha\|_E \leqslant m\|\alpha\|_{l^2} = m\sqrt{\sum_{k=1}^{\infty}|\alpha_k|^2}.$$

In particular, if $\alpha = (\alpha_k) \in l^2$ and $\alpha^{(n)} = (\alpha_1, \alpha_2, \cdots, \alpha_n, 0, 0, \cdots)$, then $\alpha^{(n)} \xrightarrow{l^2} \alpha$ and

$$A\alpha^{(n)} = \sum_{k=1}^{n}\alpha_k x_k \xrightarrow{E} x = \sum_{k=1}^{\infty}{}_E\, \alpha_k x_k \in E. \tag{1}$$

Moreover,

$$\|x\|_E = \|A\alpha\|_E \leqslant m\sqrt{\sum_{k=1}^{\infty}|\alpha_k|^2}. \tag{2}$$

From (1) it follows that for any $f \in E^*$ the series $\Sigma_{k=1}^{\infty}\alpha_k f(x_k) = f(x)$ converges, and according to a theorem of Landau [3] this means that for any $f \in E^*$ the series $\Sigma_{k=1}^{\infty}|f(x_k)|^2$ converges. This proves that (f_k) is a Hilbert system in the space $[f_k]_{E^*}$.

Theorem 4. *In order that a complete minimal system $(x_k) \subset E$ be a Hilbert system it is necessary and sufficient that there exist a continuous linear operator R: $H \to E$ such that $x_k = Re_k$, where (e_k) is some orthonormal basis in H.*

Proof. Necessity. Let (x_k) be a Hilbert system in the space E. If (e_k) is an orthonormal basis in H, then $\Sigma_{k=1}^{\infty}|(y, e_k)|^2 < \infty$, $y \in H$. By definition of the Hilbert system we can find, in the space E, a unique element x such that $f_k(x) = (y, e_k)$, $k = 1, 2, \cdots$. From the proof of Theorem 3, we see that in this case the series $\sum_{k=1}^{\infty}{}_E\, f_k(x)\,x_k = \sum_{k=1}^{\infty}{}_E\,(y, e_k)x_k$ converges in the norm of E. Therefore there is a linear operator $R(H \to E)$: $Ry = \sum_{k=1}^{\infty}{}_E\,(y, e_k)x_k$.

Moreover (see the proof of Theorem 3), $\|Ry\|_E \le m\sqrt{\Sigma_{k=1}^{\infty}|(y, e_k)|^2} =$

$m\|y\|_H$, and since $Re_k = x_k$, the necessity is proved.

Sufficiency. Let $x_k = Re_k$, where (e_k) is some orthonormal basis in H, and $R(H \to E)$ is a bounded linear operator from H to E. For any sequence $(\alpha_k) \in l^2$ there exists a unique element $y \in H$ such that $(y, e_k) = \alpha_k$. Moreover, $y = \sum_{k=1}^{\infty} {}_H \alpha_k e_k$, where convergence is in the norm of H. But then $x = Ry = \sum_{k=1}^{\infty} {}_E \alpha_k Re_k = \sum_{k=1}^{\infty} {}_E \alpha_k x_k$. Therefore $\alpha_k = f_k(x)$, which is what we wished to prove.

Theorem 5. *If (x_k) is simultaneously a Hilbert and a Bessel system in a Banach space E, then this space is linearly isomorphic to a Hilbert space.*

Proof. Since (x_k) is a Banach system in E, the series $\sum_{k=1}^{\infty} |f_k(x)|^2$ converges and $\sum_{k=1}^{\infty} |f_k(x)|^2 \le M^2 \|x\|_E^2$. Hence the series $\sum_{k=1}^{\infty} f_k(x)\overline{f_k(y)}$ converges for every $x, y \in E$. In the space E we introduce a scalar product by means of the equation

$$(x, y) = \sum_{k=1}^{\infty} f_k(x)\overline{f_k(y)},$$

and the induced norm $|x| = \sqrt{\sum_{k=1}^{\infty} |f_k(x)|^2}$. It is clear that

$$|x| \leqslant M\|x\|_E. \tag{3}$$

The completion of the space E with respect to the norm we designate as H, $E \subset H$. Since (x_k) is a Hilbert system, by Theorem 4 there exists a continuous linear operator $R\colon H \to E$ such that $Re_k = x_k$ where (e_k) is some orthonormal basis in H. Let $x \in E$. Then

$$\Big\|\sum_{k=p}^{q} f_k(x) x_k\Big\|_E = \Big\|\sum_{k=p}^{q} f_k(x) Re_k\Big\|_E$$

$$= \Big\|R\Big(\sum_{k=p}^{q} f_k(x) e_k\Big)\Big\|_E \leqslant \|R\| \Big\|\sum_{k=p}^{q} f_k(x) e_k\Big\|_H.$$

In view of the convergence of the series $\sum_{k=1}^{\infty} f_k(x)e_k$ in H, the last inequality shows that for each $x \in H$, the series $\sum_{k=1}^{\infty} {}_E f_k(x)x_k$ converges in the norm of the space E. But then

$$\|x\|_E = \Big\| \sum_{k=1}^{\infty}{}_E f_k(x)\, x_k \Big\|_E = \Big\| R\Big(\sum_{k=1}^{\infty}{}_H f_k(x)\, e_k \Big) \Big\|_E$$

$$\leqslant \|R\| \Big\| \sum_{k=1}^{\infty} f_k(x)\, e_k \Big\|_H = \|R\| \sqrt{\sum_{k=1}^{\infty} \|f_k(x)\|^2} = \|R\|\, |x| . \tag{4}$$

Inequalities (3) and (4) show that E and H are set-theoretically and topologically equivalent.

Corollary 1. *If the conditions of Theorem 5 are fulfilled, then the system (x_k) forms a Riesz basis [1] (an unconditional basis) in the space $E = H$.*

Theorem 6. *Any unconditional basis that is bounded above ($\|x_k\| \le b$, $k = 1, 2, \cdots$) in L^p for $p > 2$ is a Hilbert basis. An unconditional basis that is bounded below ($\|x_k\| \ge a$; $k = 1, 2, \cdots$) in L^p, $1 < p \le 2$, is a Bessel basis for $1 < p \le 2$.*

Proof. If (x_k) is an unconditional basis in L^p for $p > 2$ and $\|x_k\| \le b$, then the biorthogonal functionals (f_k) form an unconditional basis in L^q $(1/p + 1/q = 1)$ and since $1 = f_k(x_k) \le \|f_k\|\ \|x_k\|$, it follows that $\|f_k\| \ge 1/\|x_k\| \ge 1/b$. By Orlicz's theorem [4] the series $\sum_{k=1}^{\infty} |f(x_k)|^2\, \|f_k\|$ converges for any $f \in L^q$ $(q < 2)$, and consequently the series $\sum_{k=1}^{\infty} |f(x_k)|^2$ also converges. Thus (f_k) is a Bessel system in L^q, and by the reflexive character of L^p the system (x_k), according to Theorem 4, forms a Hilbert system in $L^q \cong (L^q)^*$. Passing to the conjugate space, we obtain the second part of the theorem.

Corollary 2. *If $p < 2$, there exists no unconditional Bessel basis for L^p that is bounded above (in particular, no orthonormal one), and if $p > 2$ there exists no unconditional Hilbert basis that is bounded from below (in particular, no orthonormal one).*

Theorem 7. *In the spaces $C([a, b])$ and $L^p([a, b])$ for $p \ge 2$ there exist Bessel bases, and for $1 \le p \le 2$ Hilbert bases consisting of orthonormal elements (in the sense of the scalar product in L^2).*

Proof. As is known [5], there exists a basis for the space of functions on $[a, b]$ that is also an orthonormal basis for L^2. We assume for simplicity that $[a, b] = [0, 1]$ and show that the system of functions $\{e_k(t)\}$ forms a basis in L^p for any $p \ge 1$. We consider the Lebesgue functions for the system $\{e_k(t)\}$ ([3], Russian page 180):

$$L_n(t) = \int_0^1 \Big| \sum_{k=1}^{n} e_k(t)\, e_k(s) \Big|\, ds = \int_0^1 |K_n(t, s)|\, ds.$$

Since $e_k(t)$ is continuous on $[0, 1]$ the norms of the operators that correspond to the partial sums of the Fourier series $S_n(x; t) = \int_0^1 K_n(t, s)x(s)\,ds$ by [6] (Russian page 105) are equal to

$$\| S_n \|_{C\to C} = \max_{(t)} \int_0^1 | K_n(t, s) | \, ds = \max_{(s)} \int_0^1 |K_n(t; s)| \, dt = \| S_n \|_{L^1\to L^1}.$$

Consequently $\|S_n\|_{L^1\to L^1} = \|S_n\|_{C\to C} = \max_{(t)} L_n(t)$, and therefore the system $\{e_k(t)\}$ forms a basis in L^1.

By the Riesz convexity theorem [7], for any $p > 2$,

$$\| S_n \|_{L^p\to L^p} \leqslant \| S_n \|^{\alpha}_{L^\infty\to L^\infty} \| S_n \|^{1-\alpha}_{L^2\to L^2} \leqslant M^\alpha \cdot 1 \quad \left(\alpha = \frac{p-2}{2}\right),$$

where M is an upper bound for the norms of the partial sums in the space C. This means that the operators S_n mapping L^p to L^p are uniformly bounded, and since the system (e_k) is minimal and complete in any L^p, it follows that (e_k) forms a basis in L^p for $p > 2$. By the reflexive character of L^p, (e_k) is also a basis in the conjugate space of L^q, where $q < 2$. It remains to note that since the functionals $f_k = f_{e_k}$ are biorthogonal to the system (e_k), it follows that for $p > 2$

$$\sum_{k=1}^{\infty} | f_k(x) |^2 = \sum_{k=1}^{\infty} | f_{e_k}(x) |^2 = \sum_{k=1}^{\infty} | (x, e_k) |^2 < +\infty,$$

and, from this, that (e_k) forms a Bessel basis in L^p, $p \geq 2$, and C. Since $L^q \cong (L^p)^*$ $(1/p + 1/q = 1)$, this system also forms a Hilbert basis for $q \leq 2$ in view of Theorem 2.

Remark. It is easy to show that in a separable Banach space there exists a Bessel system, and that if it is reflexive, the system is also Hilbert.

In fact, it is sufficient to recall the construction of A. I. Markuševič [8]: from the elements of the given E, we may construct an orthonormal basis in H: $E \subset H$. This system is complete and minimal in E with biorthogonal functionals $f_k = d_{e_k}$. It is clear that the system just constructed is a Bessel system, since for any $x \in E$

$$f_k(x) = f_{e_k}(x) = (x, e_k); \quad k = 1, 2, \ldots$$

2. In this section we consider questions of stability and Bessel bases. The system of elements (x_k) of the Banach space E is said to be ω-linearly

independent if the equation $\Sigma_{k=1}^{\infty} c_k x_k = \theta$ implies $c_k = 0$, $k = 1, 2, \cdots$. The system is said to be l^2-linearly independent if the conclusion $c_k = 0$, $k = 1, 2, \cdots$, holds under the additional condition that $(c_k) \in l^2$. A similar definition is given to l^p-linear independence. A complete minimal system $(x_k) \in E$ is said to be quadratically stable if any l^2-linearly independent system of elements $(u_k) \subset E$ close to the system (x_k) in mean square $(\Sigma_{k=1}^{\infty} \|x_k - u_k\|^2 < +\infty)$ is also a complete minimal system in the space E. We introduce the following definition:

Definition 1. A complete minimal system (x_k) is said to be *weakly quadratically stable* if any l^2-linearly independent system $(u_k) \subset E$ such that the series

$$\sum_{k=1}^{\infty} |f(u_k - x_k)|^2 \tag{5}$$

converges uniformly for $f \in E^*$, $\|f\| \leq 1$, is also minimal and complete in the space E.

Definition 2. If, under the conditions of Definition 1, there exists a bounded linear inverse operator A, where $Ax_k = u_k$, $k = 1, 2, \cdots$, then the system (x_k) is called *equivalently stable*.

Remark. The uniform convergence condition for series (5) is weaker than mean-square approximation. In fact, let (v_k) be a normalized unconditional basis in L^p for $p > 2$. If the series $\Sigma_{k=1}^{\infty} \alpha_k v_k$ converges, then by a theorem of Orlicz [4], the series $\Sigma_{k=1}^{\infty} |\alpha_k|^q = \Sigma_{k=1}^{\infty} \|\alpha_k v_k\|^q$ also converges. Furthermore, by Corollary 2, we may assume that $\Sigma_{k=1}^{\infty} |\alpha_k|^2 = +\infty$. If we write $y_k = \alpha_k v_k$, $k = 1, 2, \cdots$, then $\Sigma_{k=1}^{\infty} \|y_k\|^2 = +\infty$, but because of the unconditional convergence of the series $\Sigma_{k=1}^{\infty} y_k$ the series $\Sigma_{k=1}^{\infty} |f(y_k)|$ also converges, and $\Sigma_{k=1}^{\infty} |f(y_k)|^2$ is uniformly convergent with respect to $f \in E^*$, $\|f\| \leq 1$. We now prove two lemmas (these were proved in [9] for two bases in a Hilbert space).

Lemma 1. *Let (x_k) and (u_k) be two minimal, complete systems with biorthogonal systems (f_k) and (ϕ_k) that are total. The linear operator A defined on the linear manifold of (x_k) by the equations $Ax_k = u_k$, $k = 1, 2, \cdots$, is closed.*

Proof. Let $z_n = \Sigma_{k=1}^{p_n} \alpha_k^{(n)} x_k$ and $\lim_{n\to\infty} z_n = \theta$. If $\lim_{n\to\infty} Az_n = \lim_{n\to\infty} \Sigma_{k=1}^{p_n} \alpha_k^{(n)} u_k = y$, then

$$0=\lim_{n\to\infty} f_i(z_n)=\lim_{n\to\infty}\sum_{k=1}^{p_n}\alpha_k^{(n)} f_i(x_k)=\lim \alpha_i^{(n)},\quad i=1,\ 2,\ldots,$$

and consequently $\phi_i(y)=\lim_{n\to\infty}\sum_{k=1}^{p_n}\alpha_k^{(n)}\phi_i(u_k)=\lim_{n\to\infty}\alpha_i^{(n)}=0$, $i=1, 2, \cdots$. Therefore, from the totality of the system (ϕ_k), we see that $y=\theta$. The lemma is proved.

In what follows, the operator A is assumed to be closed.

Lemma 2. *Under the conditions of the preceding lemma, the operator A has an inverse (which is unbounded, in general).*

Proof. Let $x\in D_A$ be the solution of the equation $Ax=\theta$. If $y_n\in D_A$ and $y_n=\sum_{k=1}^{p_n}\alpha_k^{(n)}x_k \xrightarrow[n\to\infty]{} x$; $Ay_n\to Ax$, then

$$\theta=Ax=\lim_{n\to\infty} Ay_n=\lim_{n\to\infty}\sum_{k=1}^{p_n}\alpha_k^{(n)}u_k.$$

Therefore

$$0=\varphi_i(Ax)=\lim_{n\to\infty}\sum_{k=1}^{p_n}\alpha_k^{(n)}\varphi_i(u_k)=\lim_{n\to\infty}\alpha_i^{(n)}$$

and

$$f_i(x)=\lim_{n\to\infty} f_i(y_n)=\lim_{n\to\infty}\sum_{k=1}^{p_n}\alpha_k^{(n)} f_i(x_k)=\lim_{n\to\infty}\alpha_i^{(n)}=0;\qquad i=1,\ 2,\ldots .$$

In view of the totality of the system (f_k) it follows that $x=\theta$. The lemma is proved.

Theorem 8. *If (x_k) is a Bessel system with a biorthogonal system of functionals (f_k), then it is equivalently quadratically stable.*

This theorem is a consequence of the following.

Theorem 9. *If (x_k) is a Bessel system with a total system of biorthogonal functions (f_k), then it is weakly equivalently quadratically stable.*

Proof. Let $(u_k)\subset E$ be an l^2-linearly independent system, and suppose that $\sum_{k=1}^{\infty}|f(u_k-x_k)|^2$ converges uniformly relative to $f\in E^*$ and $\|f\|\le 1$. For any two integers m and n, there exists a functional $f_{mn}\in E^*$ ($\|f_{mn}\|=1$) such that

$$\Big\|\sum_{k=m}^{n} f_k(x)(u_k-x_k)\Big\|=\sum_{k=m}^{n} f_k(x) f_{mn}(u_k-x_k).$$

Consequently

$$\left\|\sum_{k=m}^{n} f_k(x)(u_k - x_k)\right\| \leqslant \left(\sum_{k=m}^{n} |f_k(x)|^2\right)^{1/2} \left(\sum_{k=m}^{n} |f_{mn}(u_k - x_k)|^2\right)^{1/2}$$

$$\leqslant \left(\sum_{k=1}^{\infty} |f_k(x)|^2\right)^{1/2} \varepsilon \leqslant M\|x\|\varepsilon \quad \text{for } m, n > N_\varepsilon. \tag{6}$$

Hence for any $x \in E$ the series $\Sigma_{k=1}^{\infty} f_k(x)(u_k - x_k) = Bx$ defines a completely continuous linear operator B. The complete continuity of this operator follows from the estimate (6), since, as we show, the operator B is the uniform limit (with respect to the operator norm) of the finite-dimensional operators $B_n x = \Sigma_{k=1}^{n} f_k(x)(u_k - x_k)$. The operator $A = I + B$ has a continuous inverse since $(I + B)x = \theta$ implies $\Sigma_{k=1}^{\infty} f_k(x) u_k = \theta$, so that by the l^2-linear independence of the system (u_k) we have $f_k(x) = 0$ for all k. Then, because of the completeness of the system (f_k), $x = \theta$. Thus $Ax_k = u_k$, $k = 1, 2, \cdots$, where the operator A is linear, bounded, and continuously invertible. Consequently the system (u_k) is complete, minimal, and equivalent to (u_k).

Remark. Theorem 9 is also valid under the assumption of l^2-linear independence if we assume that the series $\Sigma_{k=1}^{\infty} |f(u_k - x_k)|^2 < 1$ relative to $f \in E^*$ and $\|f\| \leq 1$.

The converse of Theorem 8 was presented by the author at the Fourth All-Union Mathematical Congress. A stronger theorem (without the assumption of equivalent stability) for Hilbert spaces has been given in [9]. Theorems 10 and 11 of the present article are generalizations of results on systems and Bessel bases in arbitrary Banach spaces.

Theorem 10. *If a minimal and complete system (x_k) with a total system of biorthogonal functionals (f_k) is quadratically stable, then it is equivalently stable.*

Proof. Let (u_k) be a system of l^2-linearly independent elements that are quadratically close to the quadratically stable system (x_k). On the linear manifold generated by the elements (x_k) define the linear operator A by the equations $Ax_k = u_k$, $k = 1, 2, \cdots$. By Lemma 1 we may assume that A is closed. We show that it is bounded. It is sufficient to show that $D_A = E$. Assume the converse; that is, there exists an element $x_0 \in E$ and $x_0 \notin D_A$. By the theorem of Men′šov and Markuševič [11], the element x_0 may be represented as the sum of two elements $y_0 + z_0$, where the series $\Sigma_{k=1}^{\infty} f_k(y_0) x_k$ and $\Sigma_{k=1}^{\infty} f_k(z_0) x_k$ have subsequences of partial sums that converge to y_0 and z_0 respectively. Since $x_0 \notin D_A$, we may assume that $y_0 \notin D_A$. But then the limit

$$\lim_{n\to\infty}\sum_{k=1}^{p_n} f_k(y_0)\,Ax_k=\lim_{n\to\infty}\sum_{k=1}^{p_n} f_k(y_0)\,u_k,$$

fails to exist, and also there is no limit for

$$\lim_{n\to\infty}\sum_{k=1}^{p_n} f_k(y_0)\,(A-I)\,x_k=\lim_{n\to\infty}\sum_{k=1}^{p_n} f_k(y_0)\,(u_k-x_k).$$

Let L be the linear manifold of the functions (f_k). Let $f_0 \in L$ and $f_0(y_0) \neq 0$. Clearly, $\Sigma_{k=1}^{\infty}|f_0(x_k)|^2 < +\infty$.

We consider the system (y_k):

$$y_k = x_k - \frac{y_0}{f_0(y_0)} f_0(x_k), \quad k = 1, 2, \ldots \tag{7}$$

Here

$$\sum_{k=1}^{\infty}\|y_k - x_k\|^2 = \sum_{k=1}^{\infty}\frac{\|y_0\|^2}{|f_0(y_0)|^2}|f_0(x_k)|^2 < +\infty.$$

Thus the system (y_k) is quadratically close to (x_k). Now suppose that $(c_k) \in l^2$ and $\Sigma_{k=1}^{\infty} c_k y_k = \theta$. Then, by taking account of (7), we obtain

$$\sum_{k=1}^{\infty} c_k x_k = \frac{y_0}{f_0(y_0)}\sum_{k=1}^{\infty} c_k f_0(x_k), \tag{8}$$

$$c_k = \frac{f_k(y_0)}{f_0(y_0)}\sum_{k=1}^{\infty} c_k f_0(x_k) = \frac{\gamma}{f_0(y_0)} f_k(y_0); \quad k = 1, 2, \ldots, \tag{9}$$

where γ indicates the sum of the series on the right-hand side of inequality (8). As is well known, $(c_k) \in l^2$ and the series $\Sigma_{k=1}^{\infty} c_k(A - I)x_k = \Sigma_{k=1}^{\infty} c_k(u_k - x_k)$; consequently the limit $\lim_{n\to\infty}\Sigma_{k=1}^{p_n} c_k(A - I)x_k$ exists, but by our assumptions, a limit for the series $\lim_{n\to\infty}\Sigma_{k=1}^{p_n} f_k(y_0)\,(A - I)x_k$ fails to exist. Therefore equation (9) is possible only if $\gamma = 0$; that is, for $c_k = 0$, $k = 1, 2, \cdots$. Under the assumptions of the theorem, the system (y_k) is minimal and complete; however, it is clearly not complete, since

$$f_0(y_k) = f_0(x_k) - \frac{f_0(y)}{f_0(y_0)} f_0(x_k) = 0; \quad k = 1, 2, \ldots$$

Thus the operator A is bounded. In view of Lemma 2 this operator has an inverse A^{-1} that carries the complete minimal system (u_k) with total system of biorthogonal functionals (ϕ_k) to the complete minimal system (x_k) with total system of biorthogonal functionals (f_k). The system (u_k) clearly is also quadratically stable. From this, by applying the above reasoning, we find that the operator A^{-1} is also bounded. The theorem is proved.

Theorem 11. *If the normed basis* $(x_k) \subset E$ *is quadratically stable (or, more generally, if it is weakly quadratically stable), then it is a Bessel basis.*

Proof. Assume the converse. Suppose that there exists an element x_0 for which $\sum_{k=1}^{\infty} |f_k(x_0)|^2 = +\infty$. If $1 < 2\gamma < 2$, then, as is known ([3], Russian page 10),

$$\sum_{k=1} \frac{|f(x_0)|^2}{\left(\sum_{i=1}^{k} |f_i(x_0)|^2\right)^{\gamma}} = +\infty, \text{ but } \sum_{k=1}^{\infty} \frac{|f_k(x_0)|^2}{\left(\sum_{i=1}^{k} |f_i(x_0)|^2\right)^{2\gamma}} < +\infty.$$

For brevity, we introduce the following notation:

$$\xi_k = f_k(x_0); \quad \alpha_k = \frac{\overline{f_k(x_0)}}{\left(\sum_{i=1}^{k} |f_i(x_0)|^2\right)^{\gamma}}; \quad k = 1, 2, \ldots \tag{10}$$

and set

$$u_1 = x_1; \quad u_k = \alpha_k \sum_{i=1}^{k-1} \xi_i x_i + x_k; \quad k = 2, 3, \ldots \tag{11}$$

Then

$$\|u_k - x_k\|^2 = |\alpha_k|^2 \left\| \sum_{i=1}^{k-1} \xi_i x_i \right\|^2 \leqslant M^2 \|x_0\|^2 |\alpha_k|^2; \quad k = 2, 3, \ldots,$$

where M is the uniform bound on the operators

$$S_n x = \sum_{k=1}^{n} f_k(x) x_k.$$

Since $\sum_{k=1}^{\infty} |\alpha_k|^2 < +\infty$, the series $\sum_{k=1}^{\infty} \|u_k - x_k\|^2$ converges, and consequently the system (u_k) is quadratically close to the given basis.

We prove that this system is l^2-linearly independent. Let $(c_k) \in l^2$ and $\sum_{k=1}^{\infty} c_k u_k = \theta$. Then the series $\sum_{k=1}^{\infty} c_k (u_k - x_k)$ converges, and consequently so does $\sum_{k=1}^{\infty} c_k x_k$. We have

$$\theta = \sum_{k=1}^{\infty} c_k u_k = c_1 x_1 + \sum_{k=2}^{\infty} c_k \alpha_k \sum_{l=1}^{k-1} \xi_i x_i + \sum_{k=2}^{\infty} c_k x_k.$$

Therefore

$$\sum_{k=2}^{\infty} c_k \alpha_k \sum_{i=1}^{k-1} \xi_i x_i = -\sum_{k=1}^{\infty} c_k x_k.$$

Consequently

$$-c_j = f_j\Big(\sum_{k=2}^{\infty} c_k \alpha_k \sum_{l=1}^{k-1} \xi_i x_l\Big) = \sum_{k=2}^{\infty} c_k \alpha_k \sum_{l=1}^{k-1} \xi_i \delta_{jl} = \xi_j \sum_{k=j+1}^{\infty} c_k \alpha_k, \tag{12}$$

since $i < j$ and $\delta_{ji} = 0$ for $k \leq j$, and $\sum_{i=1}^{k-1} \xi_i \delta_{ji} = \xi_j$ for $k > j$. From relation (12) it follows that if some $\xi_k = 0$, then the corresponding c_k also is zero. Suppose ξ_{n_k} $(k = 1, 2, \cdots)$ are all different from zero by ξ_i, that is, $\xi_m = 0$, implying that $c_m = 0$ for $m \neq n_k$. Then the following equalities are valid:

$$\frac{c_{n_1}}{\xi_{n_1}} + c_{n_2}\alpha_{n_2} + c_{n_3}\alpha_{n_3} + \ldots = 0,$$

$$\frac{c_{n_2}}{\xi_{n_2}} + c_{n_3}\alpha_{n_3} + \ldots = 0$$

and, in general,

$$\frac{c_{n_k}}{\xi_{n_k}} + c_{n_{k+1}}\alpha_{n_{k+1}} + \ldots = 0, \quad k = 1, 2, \ldots$$

From this it follows that

$$\frac{c_{n_{k-1}}}{\xi_{n_{k-1}}} + c_{n_k}\alpha_{n_k} - \frac{c_{n_k}}{\xi_{n_k}} = 0, \quad k = 1, 2, \ldots,$$

and

$$c_{n_k}\left(\frac{1}{\xi_{n_k}} - \alpha_{n_k}\right) = \frac{c_{n_{k-1}}}{\xi_{n_{k-1}}}, \quad k = 1, 2, \ldots \tag{13}$$

Since

$$\alpha_{n_k}\xi_{n_k} = \frac{|f_{n_k}(x_0)|^2}{\left(\sum_{l=1}^{n_k} |f_l(x_0)|^2\right)^{\tau}} \to 0 \text{ as } k \to \infty,$$

for $k \geq k_0$ we have

$$\frac{1}{\xi_{n_k}} - \alpha_{n_k} = \frac{1 - \alpha_{n_k} \xi_{n_k}}{\xi_{n_k}} \neq 0. \tag{14}$$

Let k_0 be the smallest of the numbers that satisfies inequality (14). Let c_{n_k}, ξ_{n_k} and α_{n_k} be written as c_k, ξ_k and α_k. From (13) we obtain

$$c_{k_0+1} = \frac{\xi_{k_0+1}}{\xi_{k_0}(1 - \alpha_{k_0+1}\xi_{k_0+1})} c_{k_0},$$

$$c_{k_0+2} = \frac{\xi_{k_0+2}}{\xi_{k_0+1}(1 - \alpha_{k_0+2}\xi_{k_0+2})} c_{k_0+1} = \frac{\xi_{k_0+2}}{\xi_{k_0}(1 - \alpha_{k_0+1}\xi_{k_0+1})(1 - \alpha_{k_0+2}\xi_{k_0+2})} c_{k_0}.$$

and, in general,

$$c_{k_0+n} = \frac{\xi_{k_0+n}}{\xi_{k_0}(1 - \alpha_{k_0+1}\xi_{k_0+1})(1 - \alpha_{k_0+2}\xi_{k_0+2})\cdots(1 - \alpha_{k_0+n}\xi_{k_0+n})} c_{k_0}. \tag{15}$$

This means that

$$|c_{k_0+n}| = \left|\frac{c_{k_0}}{\xi_{k_0}}\right| \frac{|\xi_{k_0+n}|}{(1 - \alpha_{k_0+1}\xi_{k_0+1})(1 - \alpha_{k_0+2}\xi_{k_0+2})\cdots(1 - \alpha_{k_0+n}\xi_{k_0+n})}$$

$$> \left|\frac{c_{k_0}}{\xi_{k_0}}\right| |\xi_{k_0+n}| \tag{16}$$

and, since $(c_k) \in l^2$, it follows from the inequality just obtained that either $c_{k_0} = 0$ or $(\xi_{k_0+n}) \in l^2$; that is, $(\xi_k) \in l^2$ (recall that the ξ_k are not all members of the sequence $(\xi_{n_{k_0+m}}) = (\xi_{k_0+m})$, all of which are equal to zero). But this last conclusion is impossible by our assumption: $\xi_k = f_k(x_0) \notin l^2$. Thus $c_{k_0} = 0$, and from this, by (15), it follows that $c_k = 0$ for $k \geq k_0$.

Now let m be the largest integer such that $1/\xi_m - \alpha_m = 0$. Then $m = k_0 - 1$ and from relation (13) it follows that $c_k = 0$ for $k \leq m = k_0 - 1$. Thus the system (u_k) is l^2-linearly independent and quadratically close to the basis (x_k). Consequently the series

$$\sum_{k=1}^{\infty} f_k(x_0)(u_k - x_k). \tag{17}$$

must converge. However, it is easy to see that this series diverges. In fact, under the conditions $f_0 \in E^*$ and $f_0(x_0) = \|x_0\|$ we have

$$\|x_0\| = f_0(x_0) = \lim_{n\to\infty} \sum_{k=1}^{n} f_k(x_0) f_0(x_k) = \lim_{n\to\infty} f_0(S_n x_0).$$

This means that for $\epsilon > 0$ and $\epsilon < f_0(x_0) = \|x_0\|$ there is an N_ϵ such that for $n > N_\epsilon$,

$$0 < f_0(x_0) - \varepsilon < f_0(S_n x_0),$$

where $S_n x_0 = \sum_{k=1}^{n} \xi_n x_k$. If we apply the functional f_0 to the series (17), we obtain

$$f_0\left\{\sum_{k=1}^{\infty} f_k(x_0)(u_k - x_k)\right\} = f_0\left\{\sum_{k=2}^{\infty} f_k(x_0)\,\alpha_k \sum_{i=1}^{k-1} \xi_i x_i\right\}$$

$$= f_0\left\{\sum_{k=2}^{\infty} \frac{|f_k(x_0)|^2}{\left(\sum_{i=2}^{k} |f_i(x_0)|^2\right)} S_{k-1}x_0\right\} = \sum_{k=2}^{N_\varepsilon} \frac{|f_k(x_0)|^2}{\left(\sum_{i=1}^{k} |f_i(x_0)|^2\right)^\gamma} f_0(S_{k-1}x_0)$$

$$+ \sum_{k=N_\varepsilon+1}^{\infty} (\cdots) > \sum_{k=1}^{N_\varepsilon} (\cdots) + [f_0(x_0) - \varepsilon] \sum_{k=N_\varepsilon+1}^{\infty} \frac{|f_k(x_0)|^2}{\left(\sum_{i=1}^{k} |f_i(x_0)|^2\right)^\gamma} = +\infty.$$

Thus we obtain a contradiction that proves the theorem.

Theorems 8, 10 and 11 show that quadratic stability is a characteristic property of Bessel bases. Let (x_k) be a Bessel basis in a Banach space. Then

$$\sqrt{\sum_{k=1}^{\infty} |f_k(x)|^2} \leqslant \alpha \|x\|. \tag{18}$$

If the system (u_k) satisfies the condition $\sum_{k=1}^{\infty} |f(u_k - x_k)|^2 < +\infty$, then, by the lemma of Gel'fand referred to above,

$$\sqrt{\sum_{k=1}^{\infty} |f(u_k - x_k)|^2} \leqslant \beta \|f\|. \tag{19}$$

The numbers α and β in the inequalities (18) and (19) are assumed to be best possible.

Theorem 12. *If the system $(x_k) \subset E$ is a Bessel basis, then any complete system $(u_k) \subset E$ that satisfies condition* (19) *for $\alpha\beta < 1$ is equivalent to the given one.*

Proof. On the linear manifold of the elements (x_k) we define the linear operator A by the equalities $Ax_k = u_k$, $k = 1, 2, \cdots$. We let z_n be the sum $\sum_{k=1}^{\infty} \alpha_k x_k = \sum_{k=1}^{n} f_k(z_n) x_k$, $n = 1, 2, \cdots$. If the functional $f^{(n)} \in E^*$, $\|f^{(n)}\| = 1$, is such that

$$\Big\| \sum_{k=1}^{n} \alpha_k (u_k - x_k) \Big\| = \sum_{k=1}^{n} \alpha_k f^{(n)} (u_k - x_k),$$

then

$$\Big\| \sum_{k=1}^{n} \alpha_k A x_k \Big\| = \Big\| \sum_{n=1}^{n} \alpha_k u_k \Big\| = \Big\| \sum_{k=1}^{n} \alpha (u_k \quad x_k) + \sum_{k=1}^{n} \alpha_k x_k \Big\|$$

$$\leqslant \Big\| \sum_{k=1}^{n} \alpha_k (u_k - x_k) \Big\| + \Big\| \sum_{k=1}^{n} \alpha_k x_k \Big\| = \sum_{k=1}^{n} \alpha_k f^{(n)} (u_k - x_k) + \Big\| \sum_{k=1}^{n} \alpha_k x_k \Big\|,$$

$$\leqslant \Big(\sum_{k=1}^{n} |f_k(z_n)|^2 \Big)^{1/2} \Big(\sum_{k=1}^{n} |f^{(n)} (u_k - x_k)|^2 \Big)^{1/2} + \|z_n\| \leqslant (1 + \alpha\beta) \Big\| \sum_{k=1}^{n} \alpha_k x_k \Big\|. \tag{20}$$

Thus $\|Az_n\| \leq (1 + \alpha\beta) \|z_n\|$.

The last inequality allows us to extend the operator A to the whole space E by setting

$$Ax = \lim_{n\to\infty} A\Big(\sum_{k=1}^{n} f_k(x) x_k \Big) = \lim_{n\to\infty} \sum_{k=1}^{n} f_k(x) u_k.$$

Moreover, A is bounded: $\|Ax\| \leq (1 + \alpha\beta)\|x\|$. If $f_0 \in E^*$, $\|f_0\| = 1$ and $f_0(x - Ax) = \|x - Ax\|$, then

$$\|(I - A) x\| = \Big\| \sum_{k=1}^{\infty} f_k(x) x_k - \sum_{k=1}^{\infty} f_k(x) u_k \Big\| = \Big\| \sum_{k=1}^{\infty} f_k(x) (u_k - x_k) \Big\|$$

$$= f_0 \Big(\sum_{k=1}^{\infty} f_k(x) (u_k - x_k) \Big) = \sum_{k=1}^{\infty} f_k(x) f_0 (u_k - x_k)$$

$$\leqslant \sum_{k=1}^{\infty} |f_k(x)| \, |f_0 (u_k - x_k)|$$

$$\leqslant \Big(\sum_{k=1}^{\infty} |f_k(x)|^2 \Big)^{1/2} \Big(\sum_{k=1}^{\infty} |f_0 (u_k - x_k)|^2 \Big)^{1/2} \leqslant \alpha \|x\| \beta \|f_0\| = \alpha\beta \|x\|.$$

Consequently $\|I - A\| < 1$, and therefore the operator A has a continuous inverse. The theorem is proved.

Remark. If, together with condition (19), we assume weak quadratic closeness, then the theorem is valid without the assumption of completeness for the system (u_k).

3. Definition 3. We say that the complete minimal system $(x_k) \subset E$ is *q-stable* if any l^p-linearly independent system (u_k) $(1/p + 1/q = 1)$ that satisfies the condition of q-closeness

$$\sum_{k=1}^{\infty} \| u_k - x_k \|^q < +\infty$$

is also a complete minimal system.

Definition 4. The complete minimal system $(x_k) \subset E$ is *weakly q-stable* if any l^p-linearly independent system (u_k) that satisfies the condition of weak q-closeness (the series $\Sigma_{k=1}^{\infty} |f(u_k - u_k)|^q$ converges uniformly with respect to $f \in E^*$, $\|f\| \leq 1$, is also a complete minimal system.

Let $L_{(x_k)}$ be the space of coefficients of a complete minimal system (x_k). In [9], the q-stability of the basis (x_k) is proved for $L_{(x_k)} \subset l^p$. We also call such systems and bases *p-Bessel systems and bases*.

Theorem 13. *A p-Bessel system with a total system of biorthogonal functionals is weakly q-stable* $(1/p + 1/q = 1)$.

The proof follows in the same way as Theorem 9.

Theorem 14. *If a normalized basis (x_k) has the property of q-stability, then it is a p-Bessel basis; that is, $L_{(x_k)} \in l^p$.*

Proof. 1). Any q-stable complete minimal system with total system of biorthogonal functionals is equivalently q-stable. The proof follows the lines of the proof of Theorem 10, with obvious changes.

2). We assume that there exists an element $x_0 \in E$ such that $\Sigma_{k=1}^{\infty} |f_k(x_0)|^p = \infty$, Then for $1 < q\gamma < q$ and $1/p + 1/q = 1$ ([3], Russian page 10), we have

$$\sum_{k=1}^{\infty} \frac{|f_k(x_0)|^p}{\left(\sum_{l=1}^{k} |f_l(x_0)|^p\right)^{\gamma}} = +\infty, \text{ but } \sum_{k=1}^{\infty} \frac{|f_k(x_0)|^p}{\left(\sum_{l=1}^{k} |f_l(x_0)|^p\right)^{q\gamma}} < +\infty.$$

As in Theorem 11, we set

$$u_1 = x_1; \quad u_k = \alpha_k \sum_{l=1}^{k-1} \xi_l x_l + x_k, \quad k = 2, 3, \ldots,$$

where

$$\xi_k = f_k(x_0);\quad \alpha_k = \frac{\overline{f_k(x_0)}\,|f_k(x_0)|^{\frac{p-q}{q}}}{\left(\sum_{i=1}^{k} |f_i(x_0)|^p\right)^{\gamma}},\quad k = 2, 3, \ldots$$

Then

$$\sum_{k=1}^{\infty} \|u_k - x_k\|^q = \sum_{k=2}^{\infty} |\alpha_k|^q \left\|\sum_{i=1}^{k-1} \xi_i x_i\right\|^q \leqslant M^q \|x_0\|^q \sum_{k=2}^{\infty} \frac{|f_k(x_0)|^p}{\left(\sum_{i=1}^{k} |f_i(x_0)|^p\right)^{q\gamma}} < \infty$$

and consequently the system (u_k) is q-close to the basis (x_k). If we assume that $\Sigma_{k=1}^{\infty} c_k u_k = \theta$ for $(c_k) \in l^p$, then, also as in Theorem 11, we find that $c_k = 0$; $k = 1, 2, \cdots$. This means that (u_k) is a basis equivalent to (x_k), and therefore the series $\Sigma_{k=1}^{\infty} f_k(x)\,(u_k - x_k)$ converges. However, as in Theorem 11 the series may be shown to diverge. This contradiction proves the theorem.

Corollary. *Let* (x_k) *be a normed unconditional basis in* L^p $(p > 1)$. *If an* l^s*-linearly independent system* (u_k) *is weakly* r*-close to the basis* (x_k), *that is, the series* $\Sigma_{k=1}^{\infty} |f(u_k - x_k)|^r$ *converges uniformly relative to* $f \in (L^p)^* \cong L^q$, $\|f\|_q \leq 1$, *where* $1/p + 1/q = 1$;

$$S = \begin{cases} p/(p-1) & \text{for } p > 2, \\ 2 & \text{for } p \leqslant 2, \end{cases}$$

and r *is the exponent conjugate to* s *(that is,* $1/r + 1/s = 1$*), then the system* (u_k) *is also an unconditional basis in* L^p.

Proof. By the theorem of Orlicz [4], the series $\Sigma_{k=1}^{\infty} |f_k(x)|^s$ converges for $x \in L^p$. To conclude the proof, we apply Theorem 13 for the case $p > 2$ and Theorems 11 and 6 for $p \leq 2$.

Remark. The preceding corollary is also valid for the space l^p. We won't dwell on this point in detail here.

If the system (x_k) is a p-Bessel basis, that is, if $\Sigma_{k=1}^{\infty} |f_k(x)|^p < \infty$ for all $x \in E$, then $\Sigma_{k=1}^{\infty} |f_k(x)|^p \leq \alpha^p \|x\|^p$. Suppose that for the system (u_k) the series

$$\sum_{k=1}^{\infty} |f(u_k - x_k)|^q \quad \left(\frac{1}{p} + \frac{1}{q} = 1\right)$$

converges for any $f \in E^*$. Then

$$\sum_{k=1}^{\infty} |f(u_k - x_k)|^q \leqslant \beta^q \|f\|^q.$$

If we assume that the numbers α and β are best possible, then the proof follows as in Theorem 12.

Theorem 15. *If the system (x_k) is a p-Bessel basis, then any complete system $u_k \subset E$ such that the series*

$$\sum_{k=1}^{\infty} |f(u_k - x_k)|^q; \quad f \in E^* \tag{21}$$

converges for $\alpha\beta < 1$ is also a p-Bessel basis equivalent to the given one.

Remark. The theorem is valid without the assumption of completeness of the system (u_k) if we assume that the series (21) converges uniformly relative to $f \in E^*$, $\|f\| < 1$; that is, the weak q-closeness of the system (u_k) to the basis (x_k).

BIBLIOGRAPHY

[1] N. K. Bari, *Biorthogonal systems and bases in Hilbert space*, Moskov. Gos. Univ. Učen. Zap. 148, Mat. 4 (1951), 69–107. (Russian) MR 14, 289.

[2] I. Gel'fand, *Sur un lemme de la théorie des espaces linéaires*, Zap. Nauk.-Doslid. Inst. Mat. Meh. Harkiv. Derž. Univ. Harkiv. Mat. Tovar. (4) 13 (1936), 35–40.

[3] S. Kaczmarz and H. Steinhaus, *Theorie des Orthogonalreihen*, Monografie Mat., Band 6, PWN, Warsaw, 1935; reprint, Chelsea, New York, 1951; Russian transl., Fizmatgiz, Moscow, 1958. MR 20 #1148.

[4] W. Orlicz, *Über unbedingte Konvergenz in Funktionenräumen*. II, Studia Math. 4 (1933), 41–47.

[5] K. M. Šaĭdukov, *On the existence of an orthonormal basis for a class of polynomials*, Naučn. Trudy Kazan. Inst. Inž.-Stroit. Prom. 1957, no. 5, 119–151. (Russian)

[6] L. V. Kantorovič and G. P. Akilov, *Functional analysis in normed spaces*, Fizmatgiz, Moscow, 1959; English transl., Internat. Series of Monographs in Pure and Appl. Math., vol. 46, Macmillan, New York, 1964. MR 22 #9837; 35 #4699.

[7] M. A. Krasnosel′skiĭ, *On a theorem of M. Riesz*, Dokl. Akad. Nauk SSSR 131 (1960), 246–248 = Soviet Math. Dokl. 1 (1960), 229–231. MR 22 #9852.

[8] A. I. Markuševič, *Sur les bases (au sens large) dans les espaces linéaires*, C. R. (Dokl.) Acad. Sci. URSS 41 (1943), 227–229. MR 6, 69.

[9] I. C. Gohberg and A. S. Markus, *Stability of bases of Banach and Hilbert spaces*, Bull. Akad. Štiince RSS Moldoven. 1962, no. 5, 17–35. (Russian) MR 37 #1955.

[10] F. W. Schäfke, *Über einige unendliche lineare Gleichungssysteme*, Math. Nachr. 3 (1949), 40–58. MR 11, 601.

[11] A. I. Markuševič, *Generalization of a theorem of D. E. Menchov*, Mat. Sb. 15 (57) (1944), 433–436. (Russian) MR 6, 274.

Translated by:
R. F. Gundy

Amer. Math. Soc. Transl.
(2) Vol. 93, 1970

INTRODUCTION TO THE GEOMETRY OF INDEFINITE J-SPACES AND TO THE THEORY OF OPERATORS IN THOSE SPACES* [1)]

M. G. KREĬN

Introduction

An indefinite metric is said to be given in a Hilbert space $\mathfrak{H}$ if, in addition to the usual scalar product $(\cdot,\cdot)$, a so-called indefinite scalar product $[\cdot,\cdot]$ is given in $\mathfrak{H}$ which possesses the following properties:

1) $[x, y] = \overline{[y, x]}$ $(x, y \in \mathfrak{H})$;

2) $[\cdot,\cdot]$ is linear with respect to its first argument, i.e.

$$[\alpha_1 x_1 + \alpha_2 x_2, y] = \alpha_1 [x_1, y] + \alpha_2 [x_2, y],$$

where $x_1, x_2, y \in \mathfrak{H}$ and α_1, α_2 are arbitrary complex numbers;

(It follows from 1) and 2) that $[\cdot,\cdot]$ is conjugate linear with respect to its second argument.)

3) The functional $[x, x]$ $(x \in \mathfrak{H})$ takes values of opposite sign,

4) $[x, y]$ $(x, y \in \mathfrak{H})$ is a continuous function of two variables x and y in the topology defined by the metric of the Hilbert space.

These properties directly imply the existence of a bounded indefinite selfadjoint operator G such that

$$[x, y] = (Gx, y). \qquad (1)$$

Our theory takes its simplest form when $G = J$, where J denotes an operator with the special properties $J = J^*$, $J^2 = I$ (an involution).

We note that the scalar product $(\cdot,\cdot)$ is only given in order to introduce the topology in $\mathfrak{H}$, whereas the direction of our investigation is determined by the indefinite scalar product, which is used to define the basic concepts of this theory as well as to classify the operators in a manner similar to that in Hilbert space theory. For example, a linear operator A is said to be

---

* Translation of Second Math. Summer School, part I Naukova Dumka, Kiev, 1965, pp. 15–92.

1) Notes of individual lectures were made available to the author by students of the school M. L. Gorbačuk, V. I. Gorbačuk, Ju. L. Daleckiĭ and A. S. Markus. Valuable remarks and suggestions concerning the lectures were subsequently made by Ju. P. Ginzburg and I. S. Iohvidov. To all these persons the author expresses his sincere gratitude.

J-symmetric if $[Ax, y] = [x, Ay]$ for all x and y from the domain $\mathfrak{D}(A)$ of A. A bounded operator U is said to be *J-unitary* if $U\mathfrak{H} = \mathfrak{H}$ and $[Ux, Uy] = [x, y]$ for all $x, y \in \mathfrak{H}$. [1)]

There arises the question of whether the given scalar product $(\cdot, \cdot)$ can be replaced by one which defines the same topology in $\mathfrak{H}$ but with respect to which the indefinite scalar product will be written in the form (1) with $G = J$.

It is possible to prove the following proposition.

Theorem 1 (see, for example, [27a, 8]). *In order for it to be possible to replace the scalar product $(\cdot, \cdot)$ by a new scalar product that is topologically equivalent to it and is such that*

$$[x, y] = (Jx, y)_1 \quad (x, y \in \mathfrak{H}),$$

it is necessary and sufficient that the operator G have a bounded inverse.

The case we will be considering is precisely that in which G^{-1} exists and is bounded. It can therefore be assumed at once that $\mathfrak{H}$ is a Hilbert space with a J-metric, i.e.

$$[x, y] = (Jx, y), \; J = J^*, \; J^2 = I.$$

One can easily expose the structure of the operator J. Put $P_+ = (I+J)/2$. Then $P_+^* = P_+$ and $P_+^2 = P_+$. This means that P_+ is an orthogonal projection. Set $P_- = I - P_+ = (I - J)/2$. It follows that $J = P_+ - P_-$, with $P_+ + P_- = I$. But this could have been deduced at once from the spectral properties of J. The equality $J^2 = I$ shows that the spectrum of J consists of not more than the two points ± 1. We denote by $\mathfrak{H}_+$ and $\mathfrak{H}_-$ the eigenspaces of the indefinite operator J corresponding to the eigenvalues 1 and -1 respectively. Then $J = P_+ - P_-$, where P_+ and P_- are the orthogonal projections onto $\mathfrak{H}_+$ and $\mathfrak{H}_-$.

Each vector $x \in \mathfrak{H}$ can now be written in the form

$$x = P_+x + P_-x = x_+ + x_-, \; P_+x = x_+, \; P_-x = x_-,$$

with

$$Jx = P_+x - P_-x = x_+ - x_-$$

and

$$(x, x) = (x_+, x_+) + (x_-, x_-), \; [x, x] = (x_+, x_+) - (x_-, x_-).$$

---

1) The requirement in this definition that U be bounded can be dropped [13c] (see problem 2 on page 125 below).

Until recently not even the theory of finite-dimensional spaces with an indefinite metric was included in courses of linear algebra; the book *Fundamentals of linear algebra* by A. I. Mal'cev constitutes an exception. The first serious results in the theory of operators in infinite-dimensional spaces with an indefinite metric were obtained by Soviet mathematicians. S. L. Sobolev first encountered a J-metric with $\dim \mathfrak{H}_- = 1$ in connection with certain problems in ballistics. But his results were not published. In 1944 there appeared a paper of L. S. Pontrjagin [32] that dealt with J-hermitian operators for the case when $\min \dim \mathfrak{H}_\pm = \kappa (\kappa < \infty)$. In this paper a remarkable theorem, which provided the foundation for many subsequent investigations connected with an indefinite metric, was proved. Various generalizations of it were given by M. G. Kreĭn with the use of topological methods. The subsequent development of the theory of spaces with an indefinite metric and the operators acting in them was obtained in papers by M. G. Kreĭn, I. S. Iohvidov, H. Langer, Ju. P. Ginzburg, R. Phillips and other Soviet and foreign mathematicians. At the same time, in theoretical physics in connection with the construction of a nonlinear theory of elementary particles (Heisenberg, Pauli and others), there appeared papers in which questions connected with an indefinite metric were touched on.

The theory which will be set forth below has a number of applications in mathematical physics. One can point out even old problems in the study of which the ignorance of the theory of spaces with an indefinite metric led to incomplete results. The application of this theory has made it possible to establish a whole series of interesting facts, which were either not generally known or required a very complicated proof.

We cite an example.

Consider a system with n degrees of freedom performing small oscillations about a position of stable equilibrium. Its kinetic energy $T = \sum_{j,k=1}^n a_{jk} \dot{q}_j \dot{q}_k$ while the potential energy $\Pi = \sum_{j,k=1}^n c_{jk} q_j q_k$, where the $q_j (j = 1, 2, \cdots, n)$ are generalized coordinates. Both forms are positive definite. Lagrange's equations for this system have the form

$$\sum_{k=1}^n a_{jk}\ddot{q}_k + \sum_{k=1}^n c_{jk}q_k = 0 \quad (j = 1, 2, \ldots, n) \tag{2}$$

or in operator notation

$$A\ddot{q} + Cq = 0, \quad q = (q_1, \ldots, q_n), \tag{2'}$$

where A and C are positive operators defined by the matrices $\|a_{jk}\|_{j,k=1}^{n}$ and $\|c_{jk}\|_{j,k=1}^{n}$.

The first integral of the system (2) can be written as follows:

$$E=(A\dot{q},\dot{q})+(Cq,q)\quad(E=\text{const}).\tag{3}$$

Physically, this equality expresses the law of conservation of total mechanical energy.

We introduce new variables

$$x=A^{\frac{1}{2}}q,\qquad y=C^{\frac{1}{2}}q.\tag{4}$$

Then

$$\frac{dy}{dt}=C^{\frac{1}{2}}A^{-\frac{1}{2}}x,$$

$$\frac{dx}{dt}=A^{-\frac{1}{2}}C^{\frac{1}{2}}x.$$

For the direct sum $z=x\dot{+}y$ we have

$$\frac{dz}{dt}=S_0z,\quad\text{where } S_0=\begin{pmatrix}0 & -A^{-\frac{1}{2}}C^{\frac{1}{2}}\\ C^{\frac{1}{2}}A^{-\frac{1}{2}} & 0\end{pmatrix}$$

is a skew-symmetric matrix: $S_0^*=-S_0$. Hence

$$z=e^{S_0t}z_0,\qquad z_0=z(0).$$

Suppose now that a resistance is present that depends linearly on the velocity. The equation for the oscillations of such a system takes the form

$$A\ddot{q}+B\dot{q}+Cq=0\tag{5}$$

with some nonnegative operator B. By means of a simple differentiation we can see that in this case

$$\frac{dE}{dt}=-2(B\dot{q},\dot{q})\leqslant 0.$$

Returning to the transformations (4), we get

$$\frac{dy}{dt}=C^{\frac{1}{2}}A^{-\frac{1}{2}}x,$$

$$\frac{dx}{dt}=-A^{-\frac{1}{2}}BA^{-\frac{1}{2}}x-A^{-\frac{1}{2}}C^{\frac{1}{2}}y,$$

and the equation for the damped oscillations of a system with resistance reduces to

$$\frac{dz}{dt} = Sz,$$

where

$$S = \begin{pmatrix} -A^{-\frac{1}{2}}BA^{-\frac{1}{2}} & -A^{-\frac{1}{2}}C^{\frac{1}{2}} \\ C^{\frac{1}{2}}A^{-\frac{1}{2}} & 0 \end{pmatrix}.$$

One can easily verify that

$$\frac{1}{2}(S + S^*) = \begin{pmatrix} -A^{-\frac{1}{2}}BA^{-\frac{1}{2}} & 0 \\ 0 & 0 \end{pmatrix}.$$

This means that S is dissipative. It is essential that S also be J-symmetric if one takes an operator of the form $J = \begin{bmatrix} I & 0 \\ 0 & -I \end{bmatrix}$ as J; in other words, that $JS = S^*J$.

Up to 1955 there was nothing known in the theory of small damped oscillations except the fact that the spectrum of equation (5) is confined to the left half-plane, which follows in particular from the fact that S is dissipative. In 1955 the American applied mathematician R. J. Duffin [12] obtained a series of interesting results in the theory of "strongly damped" systems. In this endeavor he did not make use of the theory of spaces with an indefinite metric. The use of the J-symmetry of S permits one to carry out a significantly more complete investigation.

An examination of equation (5) naturally leads to the problem of investigating the matrix quadratic equation

$$AZ^2 + BZ + C = 0, \tag{6}$$

which is the analog of the characteristic equation for a scalar differential equation with constant coefficients. As is well known, the general solution of the latter equation can be written in the form

$$q = C_1e^{z_1t} + C_2e^{z_2t},$$

where z_1 and z_2 are the roots of the characteristic equation. In connection with this there naturally arises the question of whether it is possible to represent the general solution of equation (5) in the form

$$q = C_1e^{Z_1t} + C_2e^{Z_2t},$$

where Z_1 and Z_2 are matrix roots of the quadratic equation (6) with various characteristic properties (the number of distinct roots of (6) is equal in general to C_n^{2n*}).

It turns out that this can be done by using the theory of J-symmetric operators in spaces with an indefinite metric.

Later on we will return to an examination of equation (5) and the corresponding quadratic equation (6), but in a much more general setting, when q is a vector function with values in a Hilbert space and A, B and C are nonnegative (and in general unbounded) linear operators acting in this space. In the course of this examination we will develop some new mathematical principles for the investigation of small oscillations of systems with an infinite number of degrees of freedom.

Other striking examples of the application of the theory of operators in spaces with an indefinite metric can be found in the theory of stability of the solutions of canonical systems with a periodic hamiltonian.

Some difficult questions relating to the theory of J-indefinite operators arise in the study of acoustic and electromagnetic wave guides. These questions will not be dealt with here.

§1. The geometry of spaces with an indefinite metric

In this and the next section we present results due mainly to Ju. P. Ginzburg [7b,c,8] and R. S. Phillips [36a,b] on the geometry of spaces with a J-metric. Questions concerning the geometry of spaces (not necessarily Hilbert spaces) with an arbitrary bilinear metric are discussed in the survey [8], which also contains a history of the subject and an extensive bibliography.

Let $\mathfrak{H}$ be a Hilbert space with a definite scalar product (x, y) $(x, y \in \mathfrak{H})$ and an indefinite scalar product

$$[x, y] = (Jx, y), \quad J = P_+ - P_-,$$

where P_+ and P_- are orthogonal projections such that $P_+ + P_- = I$. We set $\mathfrak{H}_+ = P_+\mathfrak{H}$ and $\mathfrak{H}_- = P_-\mathfrak{H}$. Then an arbitrary element of $\mathfrak{H}$

$$x = P_+x + P_-x = x_+ + x_- \quad (x_+ = P_+x,\ x_- = P_-x).$$

Definition 1. An element $x \in \mathfrak{H}$ is said to be *J-positive* if $[x, x] = \|x_+\|^2 - \|x_-\|^2 > 0$, *$J$-negative* if $[x, x] < 0$, and *neutral* if $[x, x] = 0$.

We denote by $\mathfrak{P}_+$ $(\mathfrak{P}_-)$ the set of all J-nonnegative (J-nonpositive) elements of $\mathfrak{H}$:

---

* *Translator's note.* The Russian text has C_{2n}^n.

$$\mathfrak{P}_+ = \{x \in \mathfrak{H} : [x, x] \geqslant 0\}, \quad \mathfrak{P}_- = \{x \in \mathfrak{H} : [x, x] \leqslant 0\}.$$

Definition 2. a) A subspace $\mathfrak{L} \subset^* \mathfrak{H}$ is said to be *J-nonnegative* if $\mathfrak{L} \subset \mathfrak{P}_+$;

b) $\mathfrak{L}$ is said to be *J-positive* if $[x, x] > 0$ for arbitrary $x \in \mathfrak{L}$, $x \neq 0$;

c) $\mathfrak{L}$ is said to be *J-uniformly positive* if there exists a constant $\mu > 0$ such that $[x, x] \geq \mu(x, x)$ for all $x \in \mathfrak{L}$.

The concepts of J-nonpositive, J-negative and J-uniformly negative subspaces are introduced analogously.

We note that if a subspace $\mathfrak{L}$ is J-uniformly positive then the indefinite scalar product is topologically equivalent on $\mathfrak{L}$ to the definite scalar product and converts $\mathfrak{L}$ into a Hilbert space.

We will denote by $\mathfrak{L}_+$ the totality $\{P_+x\}_{x \in \mathfrak{L}}$:

$$\mathfrak{L}_+ = \{P_+x\}_{x \in \mathfrak{L}}.$$

Lemma 1 [7c]. *Suppose a subspace $\mathfrak{L} \subset \mathfrak{P}_+$. Then the mapping $\mathfrak{L} \to \mathfrak{L}_+$ is linear, one-to-one and bicontinuous.*

Proof. Since $\mathfrak{L} \subset \mathfrak{P}_+$, we have $\|x_+\|^2 \geq \|x_-\|^2$ for arbitrary $x \in \mathfrak{L}$. On the other hand, $\|x\|^2 = \|x_+\|^2 + \|x_-\|^2 \leq 2\|x_+\|^2$. Thus

$$\frac{1}{2}\|x\|^2 \leqslant \|x_+\|^2 \leqslant \|x\|^2.$$

Q. E. D.

Definition 3. A subspace $\mathfrak{L} \subset \mathfrak{P}_+$ is said to be *maximal* if it is not a proper part of any other nonnegative subspace of $\mathfrak{H}$, i.e. $\mathfrak{L} \subset \mathfrak{L}_1 \subset \mathfrak{P}_+$ implies $\mathfrak{L}_1 = \mathfrak{L}$.

Theorem 2 [7c]. *In order for a subspace $\mathfrak{L} \subset \mathfrak{P}_+$ to be maximal it is necessary and sufficient that $P_+\mathfrak{L} = \mathfrak{H}_+$.*

Proof. Sufficiency. Suppose $\mathfrak{L} \subset \mathfrak{L}_1 \subset \mathfrak{P}_+$. Then $\mathfrak{H}_+ \supset P_+\mathfrak{L}_1 \supset P_+\mathfrak{L} = \mathfrak{H}_+$. Hence $P_+\mathfrak{L}_1 = P_+\mathfrak{L}$. By virtue of Lemma 1 it follows that $\mathfrak{L}_1 = \mathfrak{L}$.

Necessity. Suppose $\mathfrak{L}_+ = P_+\mathfrak{L} \neq \mathfrak{H}_+$. The linear closure of $\mathfrak{L}$ implies by virtue of Lemma 1 the linear closure of $\mathfrak{L}_+$, i.e. $\mathfrak{L}_+$ is a subspace of $\mathfrak{H}_+$. Since $\mathfrak{L}_+ \neq \mathfrak{H}_+$ there exists a vector $y \in \mathfrak{H}_+$ that is orthogonal to $\mathfrak{L}_+$. Consider the space $\mathfrak{L}_1 = \mathfrak{L} \dotplus \{\lambda y\}$ ($\dotplus$ is the symbol for a direct sum). Clearly, $\mathfrak{L}_1 \supset \mathfrak{L}$**. It remains for us to verify that $\mathfrak{L}_1$ is J-nonnegative. But for arbitrary $z = \lambda y + x$ ($x \in \mathfrak{L}$) we have

* *Translator's note.* The Russian text has $\in$.

** *Translator's note.* The Russian text has $\mathfrak{L}_+$.

$$[\lambda y + x, \lambda y + x] = (\lambda y + x_+, \lambda y + x_+) - (x_-, x_-) = |\lambda|^2 (y, y) + (x_+, x_+) - (x_-, x_-) \geq 0,$$

Q. E. D.

From Lemma 1 and the theorem just proved there follows the law of inertia.

Law of inertia [7c,8]. *The dimension of any maximal nonnegative (nonpositive) subspace of* $\mathfrak{H}$ *coincides with the dimension of* $\mathfrak{H}_+$ $(\mathfrak{H}_-)$.

Thus the dimension of any nonnegative (nonpositive) subspace of $\mathfrak{H}$ *does not exceed the dimension of* $\mathfrak{H}_+$ $(\mathfrak{H}_-)$.

We consider all subspaces $\mathfrak{L} \subset \mathfrak{H}$ which can be one-to-one and continuously projected onto $\mathfrak{H}_+$: $P_+\mathfrak{L} = \mathfrak{H}_+$. By a theorem of Banach the restriction of P_+ to $\mathfrak{L}$ has a continuous inverse, which we denote by $F: x = Fx_+$, $x \in \mathfrak{L}$, $x_+ \in \mathfrak{H}_+$. Therefore $x_- = P_-x = P_-Fx_+ = Kx_+$, where $K = P_-F$ is a continuous operator.

The operator K is called the *angular operator* [1] of $\mathfrak{L}$ with respect to $\mathfrak{H}_+$. A natural notation for it is $\operatorname{tg}(\widehat{\mathfrak{L}, \mathfrak{H}_+})$: $K = \operatorname{tg}(\widehat{\mathfrak{L}, \mathfrak{H}_+})$. We can now consider $\mathfrak{L}$ as the totality of elements of the form $x_+ + Kx_+$, where x_+ ranges over $\mathfrak{H}_+$. The operator K acts continuously from $\mathfrak{H}_+$ into $\mathfrak{H}_-$. Thus

$$\mathfrak{L} = \{x: x = x_+ + Kx_+, \; x_+ \in \mathfrak{H}_+\}. \tag{8}$$

Conversely, if a subspace $\mathfrak{L} \subset \mathfrak{H}$ can be written in the form (8) with a continuous operator K from $\mathfrak{H}_+$ into $\mathfrak{H}_-$, then, as can easily be seen, it can be mapped one-to-one and continuously onto $\mathfrak{H}_+$.

Consequently, formula (8) gives a description of all subspaces $\mathfrak{L}$ of $\mathfrak{H}$ which can be isomorphically and orthogonally projected onto $\mathfrak{H}_+$.

Suppose $\mathfrak{L} \subset \mathfrak{P}_+$ and is maximal. Then by Theorem 2 it can be isomorphically projected onto $\mathfrak{H}_+$. Therefore $\mathfrak{L}$ can be written in the form (8) with an operator K for which $\|K\| \leq 1$ by virtue of the fact that $\mathfrak{L}$ is J-nonnegative. Conversely, formula (8) with the condition $\|K\| \leq 1$ gives a nonnegative maximal subspace. We have arrived at a theorem.

Theorem 3 [36a, 7b,c]. *In order for a set* $\mathfrak{L}$ *to be a maximal nonnegative subspace of* $\mathfrak{H}$ *it is necessary and sufficient that* $\mathfrak{L}$ *have the form*

$$\mathfrak{L} = \{x \in \mathfrak{H}: x = x_+ + Kx_+, \; x_+ \in \mathfrak{H}_+\},$$

---

1) We note that in the papers of R. Phillips and Ju. P. Ginzburg the notion of an angular operator was introduced only for definite subspaces.

where K is an operator mapping $\mathfrak{H}_+$ into $\mathfrak{H}_-$ nonexpansively.

The set of all maximal subspaces of $\mathfrak{P}_+$ will be denoted below by $\mathfrak{M}_+$.

Theorem 3 can be generalized to the case of any nonnegative subspace $\mathfrak{L}$:

$$\text{if} \quad \mathfrak{L} \subset \mathfrak{P}_+, \text{ then } \mathfrak{L} = \{y \in \mathfrak{H} : y = Kx + x,\ x \in \mathfrak{L}_+\},$$

where $\mathfrak{L}_+$ is a subspace of $\mathfrak{H}_+$ and K is an operator mapping $\mathfrak{L}_+$ into $\mathfrak{H}_-$ nonexpansively.

It follows that any subspace of $\mathfrak{P}_+$ can be extended to a maximal nonnegative one, the number of maximal subspaces of $\mathfrak{P}_+$ containing $\mathfrak{L}$ being equal to the number of different extensions of K to a nonexpansive operator on $\mathfrak{H}_+$.

By making use of the notion of an angular operator, we can assert that if a subspace $\mathfrak{L}$ is positive then $\|Kx_+\| < \|x_+\|$, $x_+ \neq 0$, $x_+ \in \mathfrak{L}_+$, and if it is uniformly positive then $\|K\| < 1$.

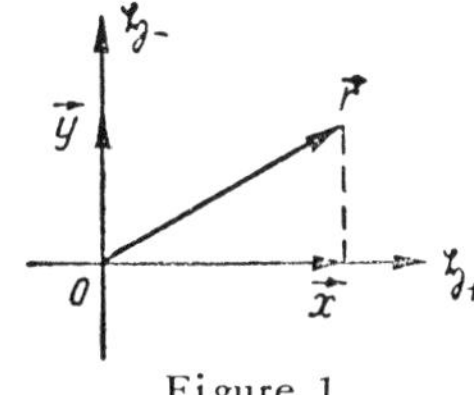

Figure 1

We consider an example which clarifies the notion of an "angular" operator.

Let $\mathfrak{H}$ be a two-dimensional euclidean space. For $\mathfrak{H}_+$ and $\mathfrak{H}_-$ we take the x- and y-axes respectively.

Then an arbitrary vector $\vec{r} = \vec{x} + \vec{y} = \vec{x} + K\vec{x}$, where K rotates the vector $\vec{x}$ through the angle $\pi/2$ and expands it by a factor $k = \|K\|$.

Addenda and problems

1. By a space Π_κ (a Pontrjagin space) we mean an abstract set satisfying the following system of axioms:

(i) Π_κ is a complex linear space;

(ii) $[\cdot, \cdot]$ is a conjugate bilinear functional on Π_κ;

(iii) the metric $[\cdot, \cdot]$ is nondegenerate, i. e. the equality $[x, x_0] = 0$ for some x_0 and all x from Π_κ implies $x_0 = 0$;

(iv) Π_κ contains at least one κ-dimensional subspace Π_+ on which $[x, x] > 0$ for $x \neq 0$ (κ is a natural number);

(v) for any $\kappa + 1$ vectors $x_1, x_2, \cdots, x_{\kappa+1}$ from Π_κ the form

$$\sum_{j,k=1}^{\kappa+1} [x_j, x_k] \xi_j \bar{\xi}_k$$

reduces to a canonical form of not more than κ positive squares;

(vi) in at least one of the decompositions $\Pi_\kappa = \Pi_+ \dot{+} \Pi_-$, where $\dim \Pi_+ = \kappa$, $[x, x] > 0$ for $0 \neq x \in \Pi_+$ and $[x, y] = 0$ for $x \in \Pi_+$, $y \in \Pi_-$ (the existence of such decompositions follows from axioms (i)–(v)), the subspace Π_- is complete with respect to the norm $|x| = |[x, x]|^{1/2}$ $(x \in \Pi_-)$.

It can easily be seen that if an indefinite metric $[\cdot, \cdot]$ is given in a Hilbert space $\mathfrak{H}$ by means of the decomposition $\mathfrak{H} = \mathfrak{H}_+ + \mathfrak{H}_-$, where $\kappa = \dim \mathfrak{H}_+ \leq \dim \mathfrak{H}_-$ and $\kappa < \infty$, then $\mathfrak{H}$ satisfies all of the axioms (i)–(vi).

2. Suppose given a space Π_κ and any of its decompositions $\Pi_\kappa = \Pi_+ \dot{+} \Pi_-$ described in axiom (vi). Introduce in Π_κ the (definite) scalar product

$$(x, y) = [x_+, y_+] - [x_-, y_-] \ (x = x_+ + x_-, \ y = y_+ + y_-, \\ x_+, y_+ \in \Pi_+, \ x_-, y_- \in \Pi_-).$$

Show that for these decompositions $\Pi_\kappa = \Pi_+ \dot{+} \Pi_-$ the (Hilbert) topologies defined in Π_κ by the norms $\|x\| = (x, x)^{1/2}$ are equivalent.

3. The above system of axioms (i)–(vi), under the condition that the form $[\cdot, \cdot]$ and the field of scalars over which the linear space Π_κ is given are real, defines a so-called *real space* Π_κ.

Prove that in a real space Π_1 (i.e. when $\kappa = 1$) for any two vectors x, y with $[x, x] \geq 0$ and $[y, y] \geq 0$

$$[x, y]^2 \geqslant [x, x]\,[y, y],$$

the equality sign holding only if x and y are collinear [14b, 21].

4. In a *real* space Π_1 we choose a vector e with $[e, e] = 1$ and consider the set $\mathfrak{L}$ of all vectors x for which

$$[x, x] = 1, \ [x, e] > 0.$$

We introduce a hyperbolic metric in $\mathfrak{L}$ by defining the distance $d(x, y)$ between two points x and y with the formula

$$\operatorname{ch} \frac{d(x, y)}{R} = [x, y],$$

where the constant $R > 0$ (the correctness of this definition follows from the result of problem 3).

The intersection $\mathfrak{L}_n$ of the surface $\mathfrak{L}$ with any $(n+1)$-dimensional linear manifold $\mathfrak{S}$ $(\subset \Pi_1)$ containing positive vectors is an n-dimensional Lobačevskiĭ space with respect to the metric $d(x, y)$ [14b].

$\mathfrak{L}$ itself is naturally called an infinite-dimensional Lobačevskiĭ space (if $\dim \Pi_1 = \infty$).

An axiomatic definition of the space Π_1 (both complex and real) was obtained by the author in connection with an axiomatic definition of an infinite-dimensional Lobačevskiĭ space [18b]. In his thesis [13b] I. S. Iohvidov extended the axiomatic definition of Π_1 to include the case Π_κ with κ any natural number (see also [14a]).

Let $\mathfrak{L}$ be a subspace of $\mathfrak{H}$. The totality of $x \in \mathfrak{H}$ such that $[x, y] = 0$ for all $y \in \mathfrak{L}$ will be called the *J-orthogonal complement* of $\mathfrak{L}$ and be denoted by $\mathfrak{L}^{[\perp]}$. In other words, $\mathfrak{L}^{[\perp]}$ is the set of those $x \in \mathfrak{H}$ for which $Jx \perp \mathfrak{L}$ ($\perp$ is the usual symbol for orthogonality). It is easily seen that $\mathfrak{L}^{[\perp]}$ is a subspace.

We now show that $(\mathfrak{L}^{[\perp]})^{[\perp]} = \mathfrak{L}$. In fact, $\mathfrak{L}^{[\perp]} = (J\mathfrak{L})^{\perp}$ (the symbol $\perp$ in the right side of this equation denotes the orthogonal complement in the usual sense). Therefore

$$((\mathfrak{L})^{[\perp]})^{[\perp]} = [(J\mathfrak{L})^{\perp}]^{[\perp]} = [J\,(J\mathfrak{L})^{\perp}]^{\perp} = J^2\,[\mathfrak{L}^{\perp}]^{\perp} = \mathfrak{L}.$$

Let $\mathfrak{J} = \mathfrak{L} \cap \mathfrak{L}^{[\perp]}$. Clearly, $\mathfrak{J}$ is the totality of elements $x_0 \in \mathfrak{L}$ that are J-orthogonal to $\mathfrak{L}$. In particular, all of the elements of $\mathfrak{J}$ are neutral. The elements of $\mathfrak{J}$ are called the *isotropic vectors of the subspace* $\mathfrak{L}$. It is obvious that $\mathfrak{J}$ is a subspace. This subspace is called the *isotropic subspace* of the space $\mathfrak{L}$.

Suppose $\mathfrak{J} = \mathfrak{L} \cap \mathfrak{L}^{[\perp]} \neq \{0\}$. Then $J\mathfrak{J} \perp \mathfrak{L}$ and $J\mathfrak{J} \perp \mathfrak{L}^{[\perp]}$, and consequently $J\mathfrak{J} \perp \overline{(\mathfrak{L} + \mathfrak{L}^{[\perp]})}$ and $\overline{\mathfrak{L} + \mathfrak{L}^{[\perp]}} \neq \mathfrak{H}$.

It is easily seen that $J(\mathfrak{L} \cap \mathfrak{L}^{[\perp]})$ is in general the orthogonal complement in $\mathfrak{H}$ of the subspace $\overline{\mathfrak{L} + \mathfrak{L}^{[\perp]}}$. Therefore the equality

$$\overline{\mathfrak{L} + \mathfrak{L}^{[\perp]}} = \mathfrak{H}$$

holds if and only if $\mathfrak{L}$ lacks an isotropic subspace: $\mathfrak{J} = \{0\}$.

It is well known that if $\mathfrak{L}_{-}$ is a subspace of $\mathfrak{H}$ then there always exists the decomposition

$$\mathfrak{H} = \mathfrak{L} + \mathfrak{L}^{\perp}.$$

As follows from the arguments just developed, a necessary condition for the validity of the analogous representation

$$\mathfrak{H} = \mathfrak{L}\,[+]\,\mathfrak{L}^{[\perp]} \tag{9}$$

(the symbol $[+]$ means that the sum is direct and the summands are J-orthogonal) is the absence in $\mathfrak{L}$ of an isotropic subspace. There arises the question: what are necessary and sufficient conditions for the validity of the decomposition

$$\mathfrak{H} = \mathfrak{L}\,[+]\,\mathfrak{N}, \tag{10}$$

where $\mathfrak{N}$ is some subspace that is J-orthogonal to $\mathfrak{L}$? An answer to this question will be given for the case when $\mathfrak{L}$ is J-nonnegative. As a preliminary we introduce the following concept.

Definition [13e]. A subspace $\mathfrak{L} \subset \mathfrak{H}$ is said to be *J-projectionally complete* if to any $x \in \mathfrak{H}$ there corresponds an element $x_{\mathfrak{L}} \in \mathfrak{L}$ such that $x - x_{\mathfrak{L}} [\perp] \mathfrak{L}$, i.e. $x - x_{\mathfrak{L}} \in \mathfrak{L}^{[\perp]}$.

If to some $x \in \mathfrak{H}$ there corresponds an element $x_{\mathfrak{L}} \in \mathfrak{L}$ such that $x - x_{\mathfrak{L}} [\perp] \mathfrak{L}$, then $x = x_{\mathfrak{L}} + (x - x_{\mathfrak{L}}) \in \mathfrak{L} + \mathfrak{L}^{[\perp]}$. Consequently, a subspace can be J-projectionally complete only if $\mathfrak{L}$ lacks an isotropic subspace. And in this case it follows from $x_{\mathfrak{L}}, x'_{\mathfrak{L}} \in \mathfrak{L}$ and $\{x - x_{\mathfrak{L}}, x - x'_{\mathfrak{L}}\} \subset \mathfrak{L}^{[\perp]}$ that $x_{\mathfrak{L}} - x'_{\mathfrak{L}} [\perp] \mathfrak{L}$, i.e. $x_{\mathfrak{L}} = x'_{\mathfrak{L}}$. Therefore the assertion that $\mathfrak{L}$ is J-projectionally complete is equivalent to the assertion that

$$\mathfrak{H} = \mathfrak{L} [+] \mathfrak{L}^{[\perp]}.$$

In particular, (10) implies $\mathfrak{N} = \mathfrak{L}^{[\perp]}$ and is therefore equivalent to (9).

Theorem 4 [7b,c,36a]. *In order for a J-nonnegative subspace $\mathfrak{L}$ to be J-projectionally complete it is necessary and sufficient that it be uniformly positive.*

Proof (Ju. L. Daleckiĭ, M. G. Kreĭn). Necessity. Suppose

$$\mathfrak{H} = \mathfrak{L} [+] \mathfrak{N} \quad (\mathfrak{N} = \mathfrak{L}^{[\perp]}).$$

Then any element $y \in \mathfrak{H}$ can be represented in the form

$$y = y_{\mathfrak{L}} + y_{\mathfrak{N}}, \quad y_{\mathfrak{L}} \in \mathfrak{L}, \quad y_{\mathfrak{N}} \in \mathfrak{N},$$

and by a theorem of Banach there exists a constant C such that

$$\| y_{\mathfrak{L}} \| \leqslant C \| y \|.$$

Suppose now x is an arbitrary element of $\mathfrak{L}$. If we put $y = Jx$, we will have

$$(x, x) = (x, J^2 x) = (x, Jy) = [x, y] = [x, y_{\mathfrak{L}}] + [x, y_{\mathfrak{N}}] = [x, y_{\mathfrak{L}}].$$

Consequently,

$$(x, x)^2 = [x, y_{\mathfrak{L}}]^2 \leqslant [x, x] [y_{\mathfrak{L}}, y_{\mathfrak{L}}] \leqslant [x, x] \| y_{\mathfrak{L}} \|^2 \leqslant [x, x] C^2 \| y \|^2 = C^2 [x, x] \| x \|^2,$$

so that

$$[x, x] \geqslant \mu (x, x) \quad (x \in \mathfrak{L}) \quad \text{for} \quad \mu = \frac{1}{C^2}.$$

Sufficiency. Suppose

$$[x, x] \geqslant \mu (x, x) \quad (x \in \mathfrak{L}, \ \mu > 0) \tag{11}$$

and take a fixed element $y \in \mathfrak{H}$. Then $[x, y]$, where x ranges over $\mathfrak{L}$, is a linear functional on $\mathfrak{L}$ that is continuous with respect to the norm $\sqrt{[x, x]}$. By a theorem of Riesz

$$[x, y] = [x, y_{\mathfrak{L}}],$$

where $y_{\mathfrak{L}}$ is a certain vector of $\mathfrak{L}$, i.e. $y_{\mathfrak{N}} = y - y_{\mathfrak{L}} \ [\perp] \mathfrak{L}$. Therefore

$$y = y_{\mathfrak{L}} + y_{\mathfrak{N}}, \quad y_{\mathfrak{N}} [\perp] y_{\mathfrak{L}}.$$

The uniqueness of such a representation follows from (9). Consequently,

$$\mathfrak{H} = \mathfrak{L} [+] \mathfrak{N},$$

Q. E. D.

Consider a subspace $\mathfrak{L} \in \mathfrak{M}_+$, i.e.

$$\mathfrak{L} = \{x : x = x_+ + Kx_+, \, x \in \mathfrak{H}_+\}, \quad K = \operatorname{tg} (\widehat{\mathfrak{L}, \mathfrak{H}_+}), \quad \| K \| < 1.$$

We calculate the angular operator of the subspace $\mathfrak{L}^{[\perp]}$ with respect to $\mathfrak{H}_-$.

For $x \in \mathfrak{L}$ and $y \in \mathfrak{L}^{[\perp]}$

$$\begin{aligned}[y, x] &= [y_+, x_+] - [y_-, x_-] = [y_+, x_+] - [y_-, Kx_+] \\ &= (y_+, x_+) - (K^* y_-, x_+) = 0.\end{aligned}$$

Since $K^* y_- \in \mathfrak{H}_+$ and x_+ ranges over $\mathfrak{H}_+$, we have $y_+ = K^* y_-$. Therefore $\mathfrak{L}^{[\perp]}$ can be written in the form

$$\mathfrak{L}^{[\perp]} = \{y : y = y_- + K^* y_-, \ y_- \in \mathfrak{H}_-\},$$

where $K^* = \operatorname{tg} (\widehat{\mathfrak{L}^{[\perp]}, \mathfrak{H}_-})$ is a continuous operator taking $\mathfrak{H}_-$ into $\mathfrak{H}_+$. From the fact that $\mathfrak{L} \in \mathfrak{M}_+$, and making use of Theorem 3, we conclude that $\mathfrak{L}^{[\perp]}$ is a maximal nonpositive subspace with angular operator K^* (with respect to $\mathfrak{H}_-$).

Let $\mathfrak{L}_1$ and $\mathfrak{L}_2$ be subspaces of a Hilbert space $\mathfrak{H}$ and let $P_{\mathfrak{L}_1}$ and $P_{\mathfrak{L}_2}$ be the orthogonal projections onto these subspaces. The number $\|P_{\mathfrak{L}_1} - P_{\mathfrak{L}_2}\| = d(\mathfrak{L}_1, \mathfrak{L}_2)$ is called the *opening* between the subspaces $\mathfrak{L}_1$ and $\mathfrak{L}_2$. It can easily be verified that the set of all subspaces of $\mathfrak{H}$ with metric $d(\mathfrak{L}_1, \mathfrak{L}_2)$ forms a complete metric space.

Theorem 5. *If* $d(\mathfrak{L}_1, \mathfrak{L}_2) < 1$, *then* $\dim \mathfrak{L}_1 = \dim \mathfrak{L}_2$.

Proof. For arbitrary $x \in \mathfrak{L}_2$

$$y = P_{\mathfrak{L}_1} x = (P_{\mathfrak{L}_2} + P_{\mathfrak{L}_1} - P_{\mathfrak{L}_2}) x = x + Tx,$$

where $T = P_{\mathfrak{L}_1} - P_{\mathfrak{L}_2}$. Since $\|T\| < 1$, the operator $I + T$ has an inverse, i.e. the correspondence $y = P_{\mathfrak{L}_1} x$ ($x \in \mathfrak{L}_2$) is injective. It remains to show that $P_{\mathfrak{L}_1}\mathfrak{L}_2 = \mathfrak{L}_1$. If $P_{\mathfrak{L}_1}\mathfrak{L}_2 \neq \mathfrak{L}_1$, then there exists a nonzero vector y_0 in $\mathfrak{L}_1$ that is orthogonal to $P_{\mathfrak{L}_1}\mathfrak{L}_2$. But then $y_0 \perp \mathfrak{L}_2$, which is impossible by virtue of the injectiveness of the mapping $P_{\mathfrak{L}_1} : \mathfrak{L}_2 \to \mathfrak{L}_1$.

The notion of opening is connected with the notion of *angle of inclination* of one subspace with respect to another. Let $S(\mathfrak{L})$ denote the unit sphere in a space $\mathfrak{L}$. Then the angle of inclination $\alpha(e, \mathfrak{L}_1)$ of a vector $e \in S(\mathfrak{L}_2)$* to $\mathfrak{L}_1$ and the angle of inclination $\theta(\mathfrak{L}_2, \mathfrak{L}_1)$ of $\mathfrak{L}_2$ to $\mathfrak{L}_1$ are defined by the equalities

$$\sin \alpha (e, \mathfrak{L}_1) = \| e - P_{\mathfrak{L}_1} e \|,$$

$$\sin \theta (\mathfrak{L}_2, \mathfrak{L}_1) = \sup_{e \in S(\mathfrak{L}_2)} \sin \alpha (e, \mathfrak{L}_1) = \sup_{e \in S(\mathfrak{L}_2)} \| e - P_{\mathfrak{L}_1} e \|,^{**}$$

$$\left(0 \leqslant \theta (\mathfrak{L}_2, \mathfrak{L}_1) \leqslant \frac{\pi}{2}\right).$$

It turns out that

$$d(\mathfrak{L}_1, \mathfrak{L}_2) = \max (\sin \theta (\mathfrak{L}_2, \mathfrak{L}_1), \sin \theta (\mathfrak{L}_1, \mathfrak{L}_2)).$$

A proof of this well-known fact can be found in the book [2] of N. I. Ahiezer and I. M. Glazman.

It is not difficult to show that for $d(\mathfrak{L}_1, \mathfrak{L}_2) < 1$

$$\theta (\mathfrak{L}_1, \mathfrak{L}_2) = \theta (\mathfrak{L}_2, \mathfrak{L}_1);$$

this equality does not hold in general. The following "triangle inequality" holds:

$$\theta (\mathfrak{L}_3, \mathfrak{L}_1) \leqslant \theta (\mathfrak{L}_3, \mathfrak{L}_2) + \theta (\mathfrak{L}_2, \mathfrak{L}_1), \tag{12}$$

a verification of which is also left to the reader.

We return to the consideration of a Hilbert space with a J-metric.

Let $\mathfrak{L} \subset \mathfrak{P}_+$ and K be its angular operator, i.e.

$$\mathfrak{L} = \{x : x = x_+ + Kx_+, \ x_+ \in \mathfrak{L}_+\}.$$

---

* *Translator's note.* The Russian text has $e \in \mathfrak{L}_2$.

** *Translator's note.* Russian text omits "sin" in the middle term of this equality.

Consider a vector $e \in S(\mathfrak{L})$. It can be written in the form $e = e_+ + Ke_+$, $e_+ \in \mathfrak{L}_+$. Obviously $\sin \alpha(e, \mathfrak{H}_+) = \|Ke_+\|$, while

$$\operatorname{tg} \alpha(e, \mathfrak{H}_+) = \frac{\|Ke_+\|}{\|e_+\|}.$$

Consequently $\operatorname{tg} \theta(\mathfrak{L}, \mathfrak{H}_+) = \|K\|$.

It follows that the subspace $\mathfrak{L}$ is uniformly positive if and only if $\theta(\mathfrak{L}, \mathfrak{H}_+) < \pi/4$. The proof of these equalities is left to the reader.

The following theorem shows that a uniformly positive subspace may be called stably positive.

Theorem 6. *In order for a subspace $\mathfrak{L}$ to have a neighborhood consisting entirely of nonnegative subspaces it is necessary and sufficient that $\mathfrak{L}$ be uniformly positive.*

Proof. If $\mathfrak{L}$ is uniformly positive, then $\theta(\mathfrak{L}, \mathfrak{H}_+) < \pi/4$. By making $d(\mathfrak{L}', \mathfrak{L})$ sufficiently small we can make $\theta(\mathfrak{L}', \mathfrak{L})$ sufficiently small, and by virtue of (12)* we get that $\theta(\mathfrak{L}', \mathfrak{H}_+)$ is also less than $\pi/4$, i.e. $\mathfrak{L}'$ is uniformly positive.

Conversely, if $\|K\| = 1$, we consider the subspace $\mathfrak{L}'$ with angular operator $K' = (1+\epsilon)K$ $(\epsilon > 0)$. Since $\|K'\| > 1$, we see that $\mathfrak{L}'$ is not nonnegative. On the other hand, it is easily seen that $d(\mathfrak{L}', \mathfrak{L})$ can be made arbitrarily small by taking ϵ sufficiently small.

Remark. It can be shown that, if $\mathfrak{L}$ is uniformly positive, every subspace $\mathfrak{L}'$ for which

$$d(\mathfrak{L}, \mathfrak{L}') \leqslant \frac{1}{\sqrt{2}} \frac{1 - \|K\|}{\sqrt{1 + \|K\|^2}}$$

will be nonnegative. The latter estimate is exact in the sense that for any

$$\varepsilon > \frac{1}{\sqrt{2}} \frac{1 - \|K\|}{\sqrt{1 + \|K\|^2}}$$

an $\mathfrak{L}' \notin \mathfrak{P}_+$ can always be found in the neighborhood $d(\mathfrak{L}, \mathfrak{L}') < \epsilon$.

Addenda and problems

1. In a J-space $\mathfrak{H}$ for which $\mathfrak{H}_+$ and $\mathfrak{H}_-$ are both infinite dimensional, there always can be found a pair of nonmaximal definite J-orthogonal subspaces $\mathfrak{L}$ and $\mathfrak{N}$, $\mathfrak{L} \subset \mathfrak{P}_+$, $\mathfrak{N} \subset \mathfrak{P}_-$, with the property that $\mathfrak{L}[+]\mathfrak{N} = \mathfrak{H}$ [27d].

Hint (only for the case when $\mathfrak{L}$ is nonmaximal). Consider a contraction K from $\mathfrak{H}_+$ to $\mathfrak{H}_-$ such that $\|K\| = 1$ and $\|Kx\| < \|x\|$ $(x \in \mathfrak{H}_+, x \neq 0)$, an element

* *Translator's note.* The Russian text has (1).

$x_0 \in \mathfrak{H}_+$, $x_0 \notin (I - K^*K)\mathfrak{H}_+$, and subspaces $\mathfrak{L} = \{x_+ + Kx_+ : x_+ \in \mathfrak{H}_+, x_+ \perp x_0\}$ and $\mathfrak{N} = \{K^*x_- + x_- : x_- \in \mathfrak{H}\}$.

2. Let $\mathfrak{L}$ be a positive subspace. It is a pre-Hilbert space relative to the J-metric $[x, y]$ with norm $|x| = [x, x]^{1/2}$ $(x, y \in \mathfrak{L})$. For each vector $z \in \mathfrak{H}$ consider the functional $f_z(x) = [x, z]$ $(x \in \mathfrak{L})$.

The subspace $\mathfrak{L}$ is uniformly positive if and only if for any $z \in \mathfrak{H}$ the functional $f_z(x)$ is continuous with respect to the norm $|x|$ [8].

Hint. Make use of the Hahn-Banach and Riesz theorems and prove the coincidence of the conjugate spaces of $\mathfrak{L}$ with respect to the norms $\|x\|$ and $|x|$.

3. A neutral subspace $\mathfrak{L}$ is a maximal neutral subspace if and only if $\mathfrak{L}^{[\perp]}$ is semidefinite (i. e. either nonpositive or nonnegative [34]).

4. Every maximal neutral subspace is either a maximal nonnegative subspace or a maximal nonpositive subspace or has both of these properties together (the so-called *hypermaximal neutral* subspace [7c, 8]).

Hint. Take advantage of the result of the preceding problem.

5. A subspace $\mathfrak{L}$ is a hypermaximal neutral subspace if and only if $\mathfrak{L} = \mathfrak{L}^{[\perp]}$.

Hint. Make use of problem 3.

6. Suppose $\dim P_+\mathfrak{H} = \infty$. In order for all of the positive subspaces of $\mathfrak{H}$ to be uniformly positive it is necessary and sufficient that $\dim P_-\mathfrak{H} < \infty$ (the sufficiency was first established by I. S. Iohvidov, and the necessity, by Ju. P. Ginzburg [7c, 14a, 8]).

7. Suppose $\mathfrak{H} = \mathfrak{L} \dot{+} \mathfrak{N}$ is a decomposition of $\mathfrak{H}$ into a direct sum in which $\mathfrak{L}$ is a nonnegative subspace ($\mathfrak{L} \subset \mathfrak{P}_+$) and $\mathfrak{N}$ is a negative subspace. Then $\mathfrak{L}$ is a maximal nonnegative subspace.

8. Suppose $\mathfrak{L}$ $(\subset \mathfrak{H})$ is a subspace. Then there exists a decomposition of $\mathfrak{L}$ into a simultaneously orthogonal and J-orthogonal sum:

$$\mathfrak{L} = \mathfrak{L}^+ \dot{+} \mathfrak{L}^- \dot{+} \mathfrak{L}^0, \qquad (*)$$

in which the summands are respectively positive, negative and neutral subspaces, or reduce to zero [30].

Hint. Consider the spectral decomposition of the Gram operator $G_\mathfrak{L}$ defined in $\mathfrak{L}$ by the equality $(G_\mathfrak{L} x, y) = [x, y]$ $(x, y \in \mathfrak{L})$, and form the decomposition $(*)$ from the invariant subspaces of $G_\mathfrak{L}$.

9. An indefinite subspace $\mathfrak{L}$ $(\subset \mathfrak{H})$ is projectionally complete if and only if it decomposes into a direct sum of two uniformly definite subspaces.

Hint. Take advantage of the decomposition $(*)$ of $\mathfrak{L}$.

10. A subspace $\mathfrak{L}$ $(\subset \mathfrak{H})$ is J-projectionally complete if and only if the Gram

operator $G_{\mathfrak{L}}$ corresponding to it (see problem 8) has a continuous inverse.

Hint. Use Theorems 1 and 4 and the result of problem 9.

11. If an operator E satisfies the conditions

$$EP_{+} = E, \quad P_{+}E = P_{+},$$

then $E\mathfrak{H} \subset^{*} \mathfrak{P}_{+}$ if and only if $\|E\| \leq \sqrt{2}$ [27a].

12. Find a connection between the operator E of problem 11 and the angular operator of the subspace $E\mathfrak{H}$ [8].

13. Suppose given two J-orthogonal definite subspaces. Then it is possible to extend them to two maximal definite subspaces that are J-orthogonal to each other [36b].

§2. Operators in a space with a J-metric

We recall some notation and definitions.

The ring of all bounded linear operators acting in $\mathfrak{H}$ is denoted by $\mathfrak{R}$, while the ideal in $\mathfrak{R}$ consisting of all of the completely continuous operators of $\mathfrak{R}$ is denoted by $\mathfrak{S}_{\infty}$.

A point λ of the complex plane is called a regular point for an operator A if $(A - \lambda I)^{-1} \in \mathfrak{R}$. The set of regular points of an operator A is denoted by $\rho(A)$.

The complement $\sigma(A)$ of $\rho(A)$ is called the spectrum of A.

Suppose $\sigma(A) = \sigma_1 \cup \sigma_2$, where σ_1 and σ_2 are closed nonintersecting sets. Then the space $\mathfrak{H}$ can be uniquely decomposed into a direct sum of invariant (relative to A) subspaces $\mathfrak{H}_1$ and $\mathfrak{H}_2$ such that the spectrum $\sigma(A\,|\,\mathfrak{H}_1)$ of A restricted to $\mathfrak{H}_1$ coincides with σ_1 and $\sigma(A\,|\,\mathfrak{H}_2) = \sigma_2$.

The projection P_1 that projects $\mathfrak{H}$ onto $\mathfrak{H}_1$ and annihilates $\mathfrak{H}_2$ can be written in the form

$$P_1 = -\frac{1}{2\pi i}\oint_{\gamma} (A - \lambda I)^{-1}\, d\lambda,$$

where γ is a simple or complex contour enclosing σ_1. It follows from this formula that if $C (\in \mathfrak{R})$ is any operator that commutes with A then $C\mathfrak{H}_k \subset \mathfrak{H}_k$ $(k = 1, 2)$.

We denote by $\mathfrak{U}(J)$ the set of all J-unitary operators, i.e. those bounded invertible operators in $\mathfrak{H}$ for which $[Ux, Uy] = [x, y]$, $x, y \in \mathfrak{H}$, or, equivalently, $(JUx, Uy) = (Jx, y)$ or

$$U^{*}JU = J. \tag{13}$$

---

* *Translator's note.* The Russian text has $\in$.

We note that if $U \in \mathfrak{U}(J)$ then so does U^*. In fact, (13) can be rewritten in the form $U^*JUJ = I$. Consequently $UJU^*J = I$, i.e.

$$UJU^* = J. \tag{14}$$

And this means that $U^* \in \mathfrak{U}(J)$.

From (13) it follows that $JUJ = U^{*-1}$. Therefore the spectrum of an operator $U \in \mathfrak{U}(J)$ does not vary under inversion with respect to the unit circle. Since $U^{-n} = JU^{*n}J$ $(n = 1, 2, \cdots)$, we have

$$\|U^{-n}\| = \|U^{*n}\| = \|U^n\| \quad (n = 1, 2, \ldots). \tag{15}$$

Definition. An operator $U \in \mathfrak{U}(J)$ is said to be *stable* if

$$\|U^n\| \leqslant C \quad (n = 1, 2, \ldots; \ C = \text{const}).$$

Theorem 7 (R. Phillips [36b]). *An operator $U \in \mathfrak{U}(J)$ is stable if and only if $\mathfrak{H}$ decomposes into a direct sum of two J-orthogonal definite subspaces $\mathfrak{L}_1$ and $\mathfrak{L}_2$:*

$$\mathfrak{H} = \mathfrak{L}_1 [+] \mathfrak{L}_2 \quad (\mathfrak{L}_1 \subset \mathfrak{P}_+, \ \mathfrak{L}_2 \subset \mathfrak{P}_-), \tag{16}$$

that are invariant with respect to U.

Proof. Sufficiency. From the decomposition (16) it follows according to Theorem 4 that $\mathfrak{L}_1$ is uniformly positive while $\mathfrak{L}_2$ is uniformly negative. Since $U\mathfrak{L}_1 = \mathfrak{L}_1$, the operator U is unitary in $\mathfrak{L}_1$ with respect to the scalar product $[x, y]$, which is topologically equivalent to the original scalar product (x, y) by virtue of the uniform positiveness of $\mathfrak{L}_1: (x, x) \geq [x, x] \geq \mu_1(x, x)$. The operators U^n $(n = 1, 2, \cdots)$ are also unitary in $\mathfrak{L}_1$ with respect to the scalar product $[x, y]$. Consequently

$$\mu_1 (U^n x, U^n x) \leqslant [U^n x, U^n x] = [x, x] \leqslant (x, x),$$

and hence

$$\|U^n x\| \leqslant \frac{1}{\sqrt{\mu_1}} \|x\| \quad (x \in \mathfrak{L}_1; \ n = 1, 2, \ldots).$$

The same inequality holds for all $x \in \mathfrak{L}_2$.

For arbitrary $x = x_1 + x_2$ $(x_1 \in \mathfrak{L}_1, x_2 \in \mathfrak{L}_2)$ we have

$$\|U^n x\| = \|U^n x_1\| + \|U^n x_2\| \leqslant \frac{1}{\sqrt{\mu_1}} \|x_1\| + \frac{1}{\sqrt{\mu_2}} \|x_2\|.$$

By a theorem of Banach, $\|x_1\| \leq C_1 \|x\|$ and $\|x_2\| \leq C_2 \|x\|$. Therefore

$$\|U^n x\| \leqslant \left(\frac{C_1}{\sqrt{\mu_1}} + \frac{C_2}{\sqrt{\mu_2}}\right)\|x\| \quad (n = 1, 2, \ldots; x \in \mathfrak{H}).$$

Necessity. If U is stable, then by virtue of (15)

$$\|U^n\| \leqslant C \quad (n = 0, \pm 1, \pm 2, \ldots).$$

By a theorem of B. von Sz.-Nagy, U is similar to a unitary operator V:

$$S^{-1}US = V.$$

Therefore

$$J = U^*JU = S^{*-1}V^*S^*JSVS^{-1}.$$

Setting $G = S^*JS$, we get $G = V^*GV$ or

$$VG = GV.$$

It is obvious that the quadratic form (Gx, x) $(x \in \mathfrak{H})$ defines an indefinite metric in $\mathfrak{H}$. Since G is invertible, one can isolate the positive and negative parts of the spectrum of this operator and decompose $\mathfrak{H}$ into an orthogonal sum of the corresponding invariant subspaces of G, which we denote by $\mathfrak{H}_1$ and $\mathfrak{H}_2$.

The fact that $VG = GV$ implies $V\mathfrak{H}_k = \mathfrak{H}_k$ $(k = 1, 2)$.

Obviously, $(Gx, x) > 0$ for $x \in \mathfrak{H}_1$, $x \neq 0$; $(Gx, x) < 0$ for $x \in \mathfrak{H}_2$, $x \neq 0$; and $(Gx, y) = 0$ for $x \in \mathfrak{H}_1$, $y \in \mathfrak{H}_2$.

Let $\mathfrak{L}_1 = S\mathfrak{H}_1$ and $\mathfrak{L}_2 = S\mathfrak{H}_2$. One can easily see that $U\mathfrak{L}_k = \mathfrak{L}_k$ $(k = 1, 2)$. Since $(Gx, x) = (JSx, Sx)$, we have $(Jx, x) > 0$ for $x \in \mathfrak{L}_1$, $x \neq 0$; $(Jx, x) < 0$ for $x \in \mathfrak{L}_2$, $x \neq 0$; and $(Jx, y) = 0$ for $x \in \mathfrak{L}_1$, $y \in \mathfrak{L}_2$.

The theorem is proved.

Definition. An operator $U \in \mathfrak{U}(J)$ is said to be *strongly stable* if it has a neighborhood $\|U_1 - U\| < \epsilon$ such that all of the J-unitary operators U_1 of this neighborhood are stable.

Theorem 8. *In order for an operator $U \in \mathfrak{U}(J)$ to be strongly stable it is necessary and sufficient that*

$$\mathfrak{H} = \mathfrak{L}_1 [+] \mathfrak{L}_2 \quad (\mathfrak{L}_1 \subset \mathfrak{P}_+, \ \mathfrak{L}_2 \subset \mathfrak{P}_-),$$

where the subspaces $\mathfrak{L}_1$ and $\mathfrak{L}_2$ are invariant with respect to U and

$$\sigma(U \mid \mathfrak{L}_1) \cap \sigma(U \mid \mathfrak{L}_2) = \varnothing. \tag{17}$$

Proof. We will first prove the sufficiency of the condition. Let γ be a contour enclosing $\sigma(U \mid \mathfrak{L}_1)$ but not any part of $\sigma(U \mid \mathfrak{L}_2)$. The projection

$$P_1 = -\frac{1}{2\pi i}\oint_\gamma (U-\lambda I)^{-1}\,d\lambda$$

projects $\mathfrak{H}$ onto $\mathfrak{L}_1$ and vanishes on $\mathfrak{L}_2$. If the norm $\|U' - U\|$ is sufficiently small, then the contour $\gamma \in \rho(U')$ and

$$\max_{\lambda\in\gamma} \|(U-\lambda I)^{-1} - (U'-\lambda I)^{-1}\| < \varepsilon,$$

where $\epsilon > 0$ is arbitrary. Consequently $\|P_1 - P'_1\|$, where

$$P'_1 = -\frac{1}{2\pi i}\oint_\gamma (U'-\lambda I)^{-1}\,d\lambda$$

will also be arbitrarily small. Put $P'_1\mathfrak{H} = \mathfrak{L}'_1$. We will now show that

$$d(\mathfrak{L}_1, \mathfrak{L}'_1) \leqslant \|P_1 - P'_1\|. \tag{18}$$

Let $e \in S(\mathfrak{L}_1)$. Obviously

$$\sin\alpha(e, \mathfrak{L}'_1) = \min_{x\in\mathfrak{L}'_1}\|e - x\| \leqslant \|e - P'_1 e\|$$
$$= \|(P_1 - P'_1)e\| \leqslant \|P_1 - P'_1\|.$$

Consequently

$$\sin\theta(\mathfrak{L}_1, \mathfrak{L}'_1) \leqslant \|P_1 - P'_1\|.$$

Repeating the argument with $\mathfrak{L}_1$ and $\mathfrak{L}'_1$ interchanged, we get that

$$\sin\theta(\mathfrak{L}'_1, \mathfrak{L}_1) \leqslant \|P_1 - P'_1\|.$$

The latter two inequalities imply (18).

Since the subspace $\mathfrak{L}_1(\mathfrak{L}_2)$ is uniformly positive (negative) and $d(\mathfrak{L}_1, \mathfrak{L}'_1)$ $(d(\mathfrak{L}_2, \mathfrak{L}'_2))$ is arbitrarily small, $\mathfrak{L}'_1$ $(\mathfrak{L}'_2)$ is also uniformly positive (negative). Clearly, $\mathfrak{L}'_1$ is invariant with respect to U'^*. It follows that U'^* is stable (note that the proof of sufficiency does not make use of the J-orthogonality of the subspaces, but only of their uniform definiteness [1])).

We pass to a proof of the necessity of the condition of the theorem. Since every strongly stable operator U is stable, Theorem 7 implies that $\mathfrak{H}$ decomposes into a direct sum of J-orthogonal uniformly definite subspaces $\mathfrak{L}_1$ and $\mathfrak{L}_2$:

---

* *Translator's note.* The Russian text has U^1.

1) Though $\mathfrak{L}_1$ and $\mathfrak{L}'_2$ are necessarily J-orthogonal. This can easily be seen by choosing as γ a contour (possibly complex) which goes into itself under inversion with respect to the unit circle. The operator P'_1 will then always be a J-selfadjoint and hence a J-orthogonal projection.

$$\mathfrak{H} = \mathfrak{L}_1 [+] \mathfrak{L}_2 \quad (\mathfrak{L}_1 \subset \mathfrak{P}_+, \ \mathfrak{L}_2 \subset \mathfrak{P}_-), \tag{19}$$

that are invariant with respect to U.

We introduce a new scalar product in $\mathfrak{H}$:

$$(x, y)_1 = [x_1, y_1] - [x_2, y_2],$$

where

$$x = x_1 + x_2, \ y = y_1 + y_2; \quad x_1, x_2 \in \mathfrak{L}_1, \ y_1, y_2 \in \mathfrak{L}_2.$$

One can easily verify that the new norm $\|x\|_1 = (x, x)_1^{1/2}$ is topologically equivalent to the old one $\|x\| = (x, x)^{1/2}$. Here the subspaces $\mathfrak{L}_1$ and $\mathfrak{L}_2$, which are orthogonal with respect to the new scalar product, play the same role for the indefinite scalar product $[x, y]$ as the subspaces $\mathfrak{H}_+$ and $\mathfrak{H}_-$ played for the old scalar product.

It can be directly verified that, with respect to the new scalar product, the operator U is not only J_1-unitary, where $J_1 = P_1 - P_2$ and P_1 and P_2 are the projectors onto $\mathfrak{L}_1$ and $\mathfrak{L}_2$ defined by (19), but also unitary.

If the spectral resolution of the unitary operator U has the form

$$U = \int_0^{2\pi} e^{i\varphi}\, dE(\varphi),$$

then the operator

$$H = \int_0^{2\pi} \varphi\, dE(\varphi)$$

is selfadjoint and

$$U = e^{iH}.$$

It can be easily be seen that H is also J_1-selfadjoint. Since $\mathfrak{L}_1$ and $\mathfrak{L}_2$ are invariant with respect to H, we have (in the matrix form relative to $\mathfrak{L}_1$ and $\mathfrak{L}_2$)

$$H = \begin{pmatrix} H_{11} & 0 \\ 0 & H_{22} \end{pmatrix}.$$

To complete the proof we must show that (17) holds. We assume that this is not the case, i.e. that there exists a number $\lambda_0 = e^{i\omega_0}$ contained in both $\sigma(U|\mathfrak{L}_1)$ and $\sigma(U|\mathfrak{L}_2)$. Then $\omega_0 \in \sigma(H|\mathfrak{L}_1) \cap \sigma(H|\mathfrak{L}_2)$.

In order to prove that under this assumption U is not strongly stable it is sufficient to establish that for any $\epsilon > 0$ there exists a J_1-unitary operator U_1 such that $\|U - U_1\| < \epsilon$ and that at least one point of the spectrum of U_1 does not lie on the unit circle. The latter assertion is obviously equivalent to the following: for any $\epsilon > 0$ there exists a J_1-selfadjoint operator H_1 having at least one nonreal point in its spectrum and such that $\|H - H_1\| < \epsilon$ (see problem 2 on page 136 below).

We pass to a proof of the latter assertion. Since $\omega_0 \in \sigma(H_{11}) \cap \sigma(H_{22})$ it is possible to choose a J_1-selfadjoint operator

$$\tilde{H} = \begin{pmatrix} \tilde{H}_{11} & 0 \\ 0 & \tilde{H}_{22} \end{pmatrix}$$

such that $\|H - \tilde{H}\|_1 < \epsilon/2$ and the number ω_0 is an eigenvalue of the operators $\tilde{H}_{11}$ and $\tilde{H}_{22}$. We denote by $\phi_1 \in \mathfrak{L}_1$ and $\phi_2 \in \mathfrak{L}_2$ the corresponding normed eigenvectors:

$$\tilde{H}_{11}\varphi_1 = \omega_0\varphi_1, \quad \tilde{H}_{22}\varphi_2 = \omega_0\varphi_2, \quad (\varphi_1, \varphi_1) = (\varphi_2, \varphi_2) = 1.$$

Consider the J_1-selfadjoint operator

$$G = \begin{pmatrix} \tilde{H}_{11} & G_{12} \\ -G_{12}^* & \tilde{H}_{22} \end{pmatrix},$$

where $G_{12} = (i\epsilon/2)\,(\cdot, \phi_2)\phi_1$. It is not difficult to see that

$$G(\varphi_1 + \varphi_2) = \left(\omega_0 + i\frac{\varepsilon}{2}\right)(\varphi_1 + \varphi_2),$$

i.e. the nonreal number $\omega_0 + i\epsilon/2$ is an eigenvalue of G. Since $\|G - H_1\| < \epsilon/2$ it follows that $\|G - H_1\| < \epsilon$.

The theorem is proved.

For a finite-dimensional space the sufficient part of Theorem 8 was established by the author in 1950 (see [18d]), while the necessary part was established by I. M. Gel'fand and V. B. Lidskiĭ [6] (see also [37]). A generalization of these results to the case of an infinite-dimensional space was given by V. I. Derguzov [11a, b] (and apparently also by P. Jonas [15] independently) but in a less complete form.

The proof of sufficiency of the conditions of Theorem 8 involves essentially the same arguments as in [16]. The above proof of necessity of these conditions, even when reduced to the case of a finite-dimensional space, is apparently simpler than the others.

Addenda and problems

1. Suppose that for a projection P the subspaces $P\mathfrak{H}$ and $(I-P)\mathfrak{H}$ are J-orthogonal. Then $JP = P^*J$.

2. If an operator U maps all of $\mathfrak{H}$ onto all of $\mathfrak{H}$ and $[Ux, Uy]^* = [x, y]$ $(x, y \in \mathfrak{H})$, then U is J-unitary [13c].

Hint. First show that U is linear and closed and then use a theorem of S. Banach.

3. A bounded operator A is said to be J-hermitian nonnegative if $[Ax, x] \geq 0$ for $x \in \mathfrak{H}$.

The spectrum of a J-hermitian nonnegative operator is real [33, 7a]. 1)

4. If A is a J-hermitian $(JA = A^*J)$ bounded operator with positive spectrum, then there exists a unique operator B with positive spectrum such that $B^2 = A$. Also B is J-hermitian [32, 7a].

Hint. In constructing B make use of the Cauchy-Riesz integral.

5. A bounded operator Y is said to be J-nonexpansive if $[Yx, Yx] \leq [x, x]$ for $x \in \mathfrak{H}$.

If Y and Y^* are J-nonexpansive operators, then the J-hermitian operator JY^*JY has a nonnegative spectrum [33, 7a].

Hint. Use the result of problem 3.

6. If Y and Y^* are J-nonexpansive invertible operators, then $Y = UR$, where U is a J-unitary operator and R is a J-hermitian and J-nonexpansive operator with positive spectrum. This representation is unique [33, 7a].

Hint. Use the results of problems 4 and 5.

7. The null space of a J-nonexpansive operator Y (if it is nontrivial) is uniformly J-positive [7a, b].

8. If Y and Y^* are J-nonexpansive operators, then the null spaces of the operators Y and Y^*JY coincide [7a].

9. If Y is a J-nonexpansive operator and $Y = I + T$, where T is completely continuous, then Y^* is also J-nonexpansive [33, 7a].

Hint. Make use of the Cayley transform of Y.

10. If Y satisfies the conditions of problem 9, then $Y = UR$, where U is a J-unitary operator and R is a J-hermitian operator with positive spectrum (the invertibility of Y is not assumed [33, 7a]).

Hint. Use the results of problems 7–9.

11. Suppose T is a bounded linear operator with domain $\mathfrak{L}$. Then the operator $P^{(J)}_{\mathfrak{L}}$, which associates with each x the set of its J-projections onto

---

* *Translator's note.* The Russian text has $[Ux, Ux]$.

1) The results mentioned in problems 3–6, 9 and 10 are obtained for finite-dimensional operators in [33], and for the general case in [7a].

$\mathfrak{L}(x - P_{\mathfrak{L}}^{(J)} x [\perp] J, P_{\mathfrak{L}}^{(J)} x \in \mathfrak{L})$, can be calculated from the formula

$$P_{\mathfrak{L}}^{(J)} = T (T^* J T)^{-1} T^* J.$$

Here by $(T^* J T)^{-1}$ is meant the operator that associates with each $y \in \mathfrak{H}$ the set of all its preimages under the mapping $T^* J T$ [7b, 11].

12. If a subspace $\mathfrak{L} \subset \mathfrak{H}$ has angular operator K with respect to $\mathfrak{H}_+$, then the usual orthoprojector $P_{\mathfrak{L}}$ and the operator $P_{\mathfrak{L}}^{(J)}$ (see problem 11) can be calculated from the formulas

$$P_{\mathfrak{L}} = (I + K)(I + K^* K)^{-1} (P_+ + K^* P_-); \qquad (*)$$

$$P_{\mathfrak{L}}^{(J)} = (I + K)(I - K^* K)^{-1} (P_+ - K^* P_-). \qquad (**)$$

Formula (**) is in [7b].

§3. Invariant subspaces

This section is devoted to the proof of a basic theorem on the existence of a maximal nonnegative invariant subspace for a certain class of operators. The proof is based on the Schauder-Tihonov fixed point principle.

Tihonov's Theorem. *Suppose* $\mathfrak{S}$ *is a Hausdorff linear topological space,* $\mathfrak{K}$ *is a convex compact subset of* $\mathfrak{S}$ *and* $\Phi(K)$ $(K \in \mathfrak{K})$ *is a continuous function mapping* $\mathfrak{K}$ *into itself.*

Then there exists an element $K_0 \in \mathfrak{K}$ *such that* $\Phi(K_0) = K_0$.

In our case the role of $\mathfrak{S}$ will be played by the linear space $\mathfrak{R}_{21}$ of all bounded linear operators acting from $\mathfrak{H}_+$ into $\mathfrak{H}_-$, and the role of $\mathfrak{K}$ will be played by its hyperball, i. e. the set of all operators $K \in \mathfrak{R}_{21}$ with $\|K\| \leq 1$. The weak operator topology is introduced in $\mathfrak{S}$, i. e. the neighborhood basis of an operator A_0 is composed of all possible sets of the form

$$\mathfrak{U}_{f,g,\epsilon}(A_0) = \{A : |((A - A_0) f, g)| < \varepsilon\},$$

where $f \in \mathfrak{H}_+$, $g \in \mathfrak{H}_-$ and $\epsilon > 0$, and the finite intersections of such sets.

One can easily deduce from A. I. Tihonov's theorem on the compactness of a product of compact spaces that the set $\mathfrak{K}$ is compact in the above topology (see, for example, N. Dunford and J. T. Schwartz [10], Chapter VI).

Let $C(\mathfrak{K})$ denote the set of all continuous functions on $\mathfrak{K}$ with values in $\mathfrak{S}$. We cite some properties of $C(\mathfrak{K})$.

1°. $C(\mathfrak{K})$ is a linear space.

2°. A function $\Phi(K) \in C(\mathfrak{K})$ whenever there exists for arbitrary $\delta > 0$ a function $\Phi_\delta(K) \in C(\mathfrak{K})$ such that

$$\| \Phi(K) - \Phi_\delta(K) \| < \delta \quad (K \in \mathfrak{K}).$$

We denote the set of all bounded linear operators from $\mathfrak{H}_+$ into $\mathfrak{H}_+$, $\mathfrak{H}_-$ into $\mathfrak{H}_-$, and $\mathfrak{H}_-$ into $\mathfrak{H}_+$ by $\mathfrak{R}_{11}$, $\mathfrak{R}_{22}$, and $\mathfrak{R}_{12}$ respectively.

3°. A function $\Phi(K) = A_2KA_1 \in C(\mathfrak{K})$ if $A_1 \in \mathfrak{R}_{11}$ and $A_2 \in \mathfrak{R}_{22}$.

In fact, if one chooses a neighborhood of $\mathfrak{K}_0$* of the form

$$\mathfrak{U} = \{K : |((K - K_0) A_1 f, A_2 g)| < \varepsilon\},$$

then $|((\Phi(K) - \Phi(K_0)) f, g)| < \epsilon$ for all $K \in \mathfrak{U}$.

4°. A function of the form $\Phi(K) = A_2KS_1KS_2K \cdots KS_pKA_1$ belongs to $C(\mathfrak{K})$ if $A_1 \in \mathfrak{R}_{11}$, $A_2 \in \mathfrak{R}_{22}$ and the operators $S_j \in \mathfrak{R}_{12}$ $(j = 1, 2, \cdots \cdots, p)$ are completely continuous.

We will verify this property for $p = 1$, i.e. for $\Phi(K) = A_2KSKA_1$. Let $S_n = \sum_{k=1}^n (\cdot, \psi_k)\phi_k$ $(\phi_k \in \mathfrak{H}_+, \psi_k \in \mathfrak{H}_-)$ be a finite-dimensional operator such that $\|S_n - S\| < \epsilon$, and let $\Phi_n(K) = A_2KS_nKA_1$. Clearly, for $K \in \mathfrak{K}$

$$\| \Phi(K) - \Phi_n(K) \| \leqslant \| A_2 \| \| S - S_n \| \| A_1 \| < \varepsilon.$$

Therefore, by virtue of property 2°, it is sufficient to prove that $\Phi_n(K) \in C(\mathfrak{K})$, and the proof of this fact, according to property 1°, reduces to the case $n = 1$.

In this case $S = (\cdot, \psi)\phi$ and $\Phi(K)f = (KA_1f, \psi)A_2K\phi$. Consequently

$$(\Phi(K)f, g) = (KA_1f, \psi)(K\varphi, A_2^* g) = \xi(K)\eta(K).$$

If the quantities

$$|\xi(K) - \xi(K_0)| \text{ and } |\eta(K) - \eta(K_0)|$$

are sufficiently small, then

$$|\xi(K)\eta(K) - \xi(K_0)\eta(K_0)| = |((\Phi(K) - \Phi(K_0)) f, g)|$$

will be arbitrarily small, since $\xi(K)$ and $\eta(K)$ are bounded functions on $\mathfrak{K}$. The continuity of $\Phi(K)$ now follows.

Theorem 9 (**M. G. Kreĭn** [18d]). *Suppose an operator $A \in \mathfrak{R}$ satisfies the following conditions:*

1) $0 \neq Ax \in \mathfrak{P}_+$ *if* $0 \neq x \in \mathfrak{P}_+$;

2) $A_{12} = P_+AP_- \in \mathfrak{S}_\infty$;

3) *there exists at least one subspace $\mathfrak{L}_0 \in \mathfrak{M}_+$ such that $A\mathfrak{L}_0 \in \mathfrak{M}_+$.*

Then $A\mathfrak{L} \in M_+$ for any $\mathfrak{L} \in \mathfrak{M}_+$. In addition, there exists an $\mathfrak{L}_a \in \mathfrak{M}_+$ such that $A\mathfrak{L}_a = \mathfrak{L}_a$.

---

* *Translator's note.* The Russian text has $\mathfrak{A}_0$.

Proof. The equality $Ax = y$ can be rewritten in the form

$$y_+ = A_{11}x_+ + A_{12}x_-,$$
$$y_- = A_{21}x_+ + A_{22}x_-,$$

where $A_{11} = P_+AP_+$, $A_{21} = P_-AP_+$, $A_{22} = P_-AP_-$.

Suppose now $\mathfrak{L} \in \mathfrak{M}_+$, i.e. $\mathfrak{L} = \{x_+ + Kx_+,\ x_+ \in \mathfrak{H}_+\}$. Then the operator A acts on $\mathfrak{L}$ according to the formulas

$$y_+ = (A_{11} + A_{12}K)\,x_+,$$
$$y_- = (A_{21} + A_{22}K)\,x_+. \tag{1}$$

The operator $W(K) = A_{11} + A_{12}K$ vanishes only at zero. For if it were true that $y_+ = (A_{11} + A_{12}K)\,x_+ = 0$ for some $x_+ \neq 0$, then we would have $y_- = 0$ ($\|y_-\| \leq \|y_+\|$), and hence $Ax = 0$, which is impossible according to condition 1). If the subspace $\mathfrak{L}_0$ from condition 3) is taken as $\mathfrak{L}$, then $\mathfrak{L}'_0 = A\mathfrak{L}_0 \in \mathfrak{M}_+$ and $P_+\mathfrak{L}'_0 = \mathfrak{H}_+$. Since, on the other hand,

$$\mathfrak{L}'_0 = \{(A_{11} + A_{12}K_0)\,x_+ + (A_{21} + A_{22}K_0)\,x_+,\ \ x_+ \in \mathfrak{H}_+\},$$

we see that $W(K_0)\mathfrak{H}_+ = \mathfrak{H}_+$. Thus $W_0 = W(K_0)$ is an invertible operator in $\mathfrak{H}_+$ ($W_0^{-1} \in \mathfrak{R}_{11}$).

Noting that

$$W(K) - W_0 = A_{12}(K - K_0) \in \mathfrak{S}_\infty,$$

we get

$$W(K) = W_0(I + T),$$

where $T = W_0^{-1}A_{12}(K - K_0) \in \mathfrak{S}_\infty$. Since by what has been proved $W(K)$ vanishes only at zero, $I + T$ is invertible by a theorem of Hilbert. Therefore the operator $W(K)$: $W^{-1}(K) = (I + T)^{-1}W_0^{-1}$. From equations (1), for any $\mathfrak{L} \in \mathfrak{M}_+$ with angular operator K, we find that $y_- = \Phi(K)y_+$, where

$$\Phi(K) = (A_{21} + A_{22}K)(A_{11} + A_{12}K)^{-1}.$$

It follows that the nonnegative subspace $\mathfrak{L}' = A\mathfrak{L}$ is maximal and has angular operator $K' = \Phi(K)$. We pass to a proof of the second assertion of the theorem.

It is obvious that $\Phi(\mathfrak{K}) \subseteq \mathfrak{K}$. We will prove that $\Phi(K) \in C(\mathfrak{K})$. The operator $W^{-1}(K) \in \mathfrak{R}_{11}$ for all $K \in \mathfrak{K}$. In particular, $W^{-1}(0) = A_{11}^{-1} \in \mathfrak{R}_{11}$, and therefore

$$W(K) = A_{11}(I - SK),$$

where $S = -A_{11}^{-1}A_{12} \in \mathfrak{S}_\infty$. Let $s = \|S\|$ and a vector ϕ ($\|\phi\| = 1$) be such that $S\phi = s\psi$ ($\|\psi\| = 1$). Put $K_q = q(\cdot, \psi)\phi$. Then for $|q| < 1$

$$(I - SK_q)\psi = \psi - qs\psi = (1 - qs)\psi \neq 0,$$

since the operator $I - SK_q$ is invertible. From the fact that $1 - qs \neq 0$ for all q, $|q| \leq 1$, it follows that $s < 1$. Therefore

$$\Phi(K) = (A_{21} + A_{22}K)(I - SK)^{-1}A_{11}^{-1} = (A_{21} + A_{22}K)\sum_{p=0}^{\infty}(SK)^p A_{11}^{-1}.$$

Put

$$\Phi_n(K) = (A_{21} + A_{22}K)\sum_{p=0}^{n}(SK)^p A_{11}^{-1} \quad (n = 1, 2, \ldots).$$

Clearly the functions $\Phi_n(K)$ uniformly approximate the function $\Phi(K)$ on $\mathfrak{H}$. Since the $\Phi_n(K)$ are continuous by virtue of properties 1° and 4°, $\Phi(K)$ is also continuous by virtue of property 2°. Applying Tihonov's Theorem, we obtain the second assertion of the theorem.

Remark. If $A\mathfrak{P}_+ \subset \mathfrak{P}_+$ and A has an inverse $A^{-1} \in \mathfrak{R}$, then $A^{-1}\mathfrak{P}_- \subset \mathfrak{P}_-$. For suppose $[x, x] < 0$ and $y = A^{-1}x$. If it were then true that $[y, y] \geq 0$, we would have $[Ay, Ay] \geq 0$, i.e. $[x, x] \geq 0$, which is impossible. On the other hand, if y is a neutral element, it can be approximated by negative elements.

Theorem 10 [6d]. *Suppose an operator $A \in \mathfrak{R}$ has the following properties:*

1) $A\mathfrak{P}_+ \subset \mathfrak{P}_+$;
2) *it has an inverse* $A^{-1} \in \mathfrak{R}$;
3) $P_+AP_- \in \mathfrak{S}_\infty$ and $P_-AP_+ \in \mathfrak{S}_\infty$.

Then there exist subspaces $\mathfrak{L}^{\pm} \in \mathfrak{M}_{\pm}$ such that $A\mathfrak{L}^{\pm} = \mathfrak{L}^{\pm}$.

Proof. From conditions 1) and 2) it follows that $A^{-1}\mathfrak{P}_{\pm} \subset \mathfrak{P}_{\pm}$. Let $\mathfrak{L} \in \mathfrak{M}_+$ and $\mathfrak{L}' = A\mathfrak{L}$. Condition 2) implies that $\mathfrak{L}'$ is closed. Suppose that the nonnegative subspace $\mathfrak{L}'$ is not maximal. Then there exists a subspace $\mathfrak{L}'' \in \mathfrak{M}_+$ such that $\mathfrak{L}'' \supset \mathfrak{L}'$. Therefore $A^{-1}\mathfrak{L}'' \supset A^{-1}\mathfrak{L}' = \mathfrak{L}$. Since $\mathfrak{L} \in \mathfrak{M}_+$ it follows that $A^{-1}\mathfrak{L}'' = \mathfrak{L}$, i.e. $\mathfrak{L}'' = A\mathfrak{L} = \mathfrak{L}'$.

Thus it can be asserted that 1)–3) imply the existence of two maximal invariant subspaces, one nonpositive and the other nonnegative.

We give an application of the latter assertion to J-unitary operators.

Theorem 11 [18d]. *If U is a J-unitary operator for which $P_+UP_- \in \mathfrak{S}_\infty$, then U has a maximal nonnegative as well as a maximal nonpositive invariant subspace.*

The proof of this theorem will follow upon verifying that the condition $U_{12} = P_+UP_- \in \mathfrak{S}_\infty$ implies $U_{21} = P_-UP_+ \in \mathfrak{S}_\infty$.

Since U is a J-unitary operator, $UJU^* = J$ (see §3). We write this equality in matrix form:

$$\begin{pmatrix} U_{11} & U_{12} \\ U_{21} & U_{22} \end{pmatrix} \begin{pmatrix} I & 0 \\ 0 & -I \end{pmatrix} \begin{pmatrix} U_{11}^* & U_{21}^* \\ U_{12}^* & U_{22}^* \end{pmatrix} = \begin{pmatrix} I & 0 \\ 0 & -I \end{pmatrix}.$$

Consequently

$$U_{11}U_{21}^* - U_{12}U_{22}^* = 0,$$

and since the operator U_{11} has a continuous inverse (see the proof of Theorem 9) it follows that

$$U_{21}^* = U_{11}^{-1}U_{12}U_{22}^* \in \mathfrak{S}_\infty,$$

and therefore $U_{21} \in \mathfrak{S}_\infty$. The theorem is proved.

In the case when $\mathfrak{H}$ is the space Π_κ ($\kappa = \dim \mathfrak{H}_+ \leq \dim \mathfrak{H}_-$, $\kappa < \infty$) the condition of Theorem 11 is satisfied automatically. This theorem was first obtained for $\kappa = 1$ by the author [21], using geometric methods of the theory of cones (i. e. in effect using the fixed point principle). Then I. S. Iohvidov [13a], using the Cayley-Neumann transform, obtained it for arbitrary finite κ from the corresponding theorem of L. S. Pontrjagin [32] for J-selfadjoint operators (a reverse flow of ideas is employed below in §5; see also [14a], [27a]).

An analogous theorem for a more general class of operators, including J-unitary operators in Π_κ as a special case, was obtained by the author in [18b] (see also [14a]) but again with the use of the fixed point principle. A new method here was the appearance for the first time of subspaces of Π_κ as "points" of a metric space. This method received further development in an ingenious note of M. L. Brodskiĭ [5], in which in particular a proposition is proved for Π_κ that is more complete than Theorem 9.

As in [5], Theorem 9 carries over to operators acting in a direct sum $\mathfrak{B}_+ \dotplus \mathfrak{B}_-$ of two reflexive Banach spaces $\mathfrak{B}_+$ and $\mathfrak{B}_-$. We note that such operators were first considered in a Banach space $\mathfrak{B}$ under the condition $\dim \mathfrak{B} = 1$ by F. Bonsall [4], who also obtained for them an analog of a theorem of the author's in [21].

Further generalizations to the case of operators acting in a direct sum $\mathfrak{G} = \mathfrak{G}_1 + \mathfrak{G}_2$ of two linear topological spaces $\mathfrak{G}_1$ and $\mathfrak{G}_2$ have been obtained by Ky Fan [35a,b]. Instead of the Schauder-Tihonov principle, the arguments of Ky Fan make use of new fixed point theorems.

Addenda and problems

1. Suppose

$$A \in \mathfrak{N}, \ H = J - A^* J A \gg 0$$

and its spectrum $\sigma(A)$ does not cover all of the unit circle. Then

$$\sigma(A) = \sigma_i(A) \cup \sigma_e(A),$$

where $\sigma_i(A)$ lies wholly inside the unit circle and $\sigma_e(A)$ is entirely outside it. Moreover, in the decomposition of $\mathfrak{H}$ into a direct sum of the two invariant (with respect to A) subspaces corresponding to the parts of the spectrum $\sigma_i(A)$ and $\sigma_e(A)$:

$$\mathfrak{H} = \mathfrak{H}_i \dot{+} \mathfrak{H}_e,$$

the first summand is uniformly J-positive while the second is uniformly J-negative [18h].

Hint. First show that by virtue of the two conditions the spectrum $\sigma(A)$ does not intersect the unit circle $|\xi| = 1$, i.e.

$$\sigma(A) = \sigma_i(A) \cup \sigma_e(A), \ \sigma_i(A) \cap \sigma_e(A) = \varnothing.$$

Then, by taking contour integrals of the resolvent of the operator about suitable parts of the spectrum and using the first condition, prove the second assertion.

2. Suppose U is a J-unitary operator and its spectrum $\sigma(U)$ does not intersect the unit circle. Then the space $\mathfrak{H}$ decomposes into a direct sum of two neutral invariant (with respect to U) subspaces, in each of which the spectrum of U is found respectively inside and outside the unit circle.

Hint. Take advantage of the identity

$$J - e^{-i\varphi}(e^{-i\varphi}I - U^*)J - e^{i\varphi}J(e^{i\varphi}I - U) = 0 \quad (0 \leqslant \varphi \leqslant 2\pi)$$

and integrate it with respect to ϕ from 0 to 2π.

We will refine the latter theorem and also establish an analog of this refined theorem for J-selfadjoint operators. As a preliminary, we recall some well-known concepts of the spectral theory of linear operators.

A number λ is said to be an eigenvalue of an operator A if there exists a vector $\phi \neq 0$ (eigenvector) such that $(A - \lambda I)\phi = 0$.

A vector ψ is said to be a *root vector* of A corresponding to λ if there exists a natural number p such that

$$(A - \lambda I)^p \psi = 0.$$

The dimension of the subspace of all eigenvectors corresponding to an eigenvalue λ is said to be the *eigenmultiplicity* of λ, while the dimension of the linear manifold $\mathfrak{L}_\lambda(A)$ of all root vectors (the root linear manifold) is called the *algebraic multiplicity* of λ.

We will say that to an eigenvalue λ there corresponds a *normally* splitting root space if $\dim \mathfrak{L}_\lambda(A) < \infty$ and

$$\mathfrak{H} = \mathfrak{L}_\lambda(A) \dotplus \mathfrak{N}_\lambda(A),$$

where $\mathfrak{N}_\lambda(A)$ is an invariant (for A) subspace in which the operator $A - \lambda I$ is invertible.

We denote by $\tilde{\rho}(A)$ the set of all normal points of an operator A, i.e. the set of all regular points and those eigenvalues to which there correspond normally splitting root spaces.

One can easily show that the normal eigenvalues are isolated points of the spectrum $\sigma(A)$.

If λ is a normal eigenvalue of an operator A, then $\bar{\lambda}$ is a normal eigenvalue of A^*, i.e.

$$\mathfrak{H} = \mathfrak{L}_{\bar{\lambda}}(A^*) \dotplus \mathfrak{N}_{\bar{\lambda}}(A^*).$$

Moreover,

$$\mathfrak{N}_{\bar{\lambda}}(A^*) = \mathfrak{L}_\lambda^\perp(A), \quad \mathfrak{L}_{\bar{\lambda}}(A^*) = \mathfrak{N}_\lambda^\perp(A).$$

Consequently the subspaces $\mathfrak{L}_\lambda(A)$ and $\mathfrak{L}_{\bar{\lambda}}(A^*)$ ($\mathfrak{N}_\lambda(A)$ and $\mathfrak{N}_{\bar{\lambda}}(A^*)$) are *skew-connected*, i.e. neither one has a vector ($\neq 0$) that is orthogonal to the other.

As is well known, the set $\tilde{\rho}(A)$ is open and therefore consists of a finite or countable number of connected components. If M is one of these components, $A - B \in \mathfrak{S}_\infty$ and there is at least one regular point of B in M, then $M \subset \rho(B)$ (a proof of this assertion, as well as of the preceding ones, can be found, for example, in the article [9a] of I. C. Gohberg and M. G. Kreĭn).

We return to the consideration of a J-unitary operator U. The equality $J(U^*)^{-1}J = U$ implies the following simple propositions, the verification of which is left to the reader.

1°. *The spectrum of a J-unitary operator is symmetric with respect to the unit circle.*

2°. *If λ ($|\lambda| \neq 1$) is a normal eigenvalue of an operator U, then* $1/\bar{\lambda}$

is also a normal eigenvalue and the root spaces $\mathfrak{L}_\lambda(U)$ *and* $\mathfrak{L}_{1/\bar\lambda}(U)$ *are J-skew-connected.*

The latter assertion means that in each of the subspaces $L_\lambda(U)$ and $\mathfrak{L}_{1/\bar\lambda}(U)$ there is no vector ($\neq 0$) that is J-orthogonal to the other subspace. The assertion follows from the relation

$$\mathfrak{L}_{1/\bar\lambda}(U) = J\mathfrak{L}_{\bar\lambda}(U^*).$$

The following proposition can also be readily established.

3°. *If* $\lambda\bar\mu \neq 1$, *then*[1] $[\mathfrak{L}_\lambda(U), \mathfrak{L}_\mu(U)] = \{0\}$; *in particular,* $[\mathfrak{L}_\lambda(U), \mathfrak{L}_\lambda(U)] = \{0\}$ *if* $|\lambda| \neq 1$.

It was shown above (Theorem 11) that every J-unitary operator U for which $P_+UP_- \in \mathfrak{S}_\infty$ has a maximal nonnegative as well as a maximal nonpositive invariant subspace.

One can easily see that the set $\mathfrak{G}$ of all J-unitary operators U for which $P_+UP_- \in \mathfrak{S}_\infty$ is a subgroup of the group of J-unitary operators.

Theorem 12 [18d]. *If* $U \in \mathfrak{G}$, *then the nonunitary part (i. e. the part not lying on the unit circle) of the spectrum of U consists of isolated points, viz. normal eigenvalues.*

Proof. Let $T = U - (U^*)^{-1}$. Then

$$T = U - JUJ^* = (P_+ + P_-)U(P_+ + P_-) - (P_+ - P_-)U(P_+ - P_-) = 2(P_-UP_+ + P_+UP_-) \in \mathfrak{S}_\infty.$$

Consequently $U^*U = I + S$, where $S \in \mathfrak{S}_\infty$.

We write down a spectral decomposition of the operator $I + S$:

$$I + S = \sum_j (1 + s_j)(\cdot, \varphi_j)\varphi_j \quad (1 + s_j > 0,\ j = 1, 2, \ldots),$$

where $\{\phi_j\}$ is an orthonormal basis in $\mathfrak{H}$. It is obvious that

$$(I + S)^{-\frac{1}{2}} = \sum_j (1 + s_j)^{-\frac{1}{2}}(\cdot, \varphi_j)\varphi_j = I + \sum[(1 + s_j)^{-\frac{1}{2}} - 1](\cdot, \varphi_j)\varphi_j = I + S_1,$$

where $S_1 \in \mathfrak{S}_\infty$. Therefore $I + S = (I + S_1)^{-2}$, i.e. $U^*U = (I + S_1)^{-2}$, or $(I + S_1)U^*U(I + S_1) = I^*$. Setting $U(I + S_1) = V$, we get that $V^*V = I$. Since V is invertible, it is a unitary operator. From the fact that

$$V - U = US_1 \in \mathfrak{S}_\infty$$

* *Translator's note.* The Russian text has 1.

1) The equality $[\mathfrak{L}_1, \mathfrak{L}_2] = \{0\}$ means that $[x, y] = 0$ for all $x \in \mathfrak{L}_1$ and $y \in \mathfrak{L}_2$.

and the fact that V has regular points both inside and outside the unit circle, it follows that all of the points of the nonunitary spectrum of U are normal and isolated. The theorem is proved.

We are now in a position to formulate the following refinement of Theorem 11 concerning the existence of maximal invariant subspaces of a J-unitary operator.

Theorem 13 [18d]. *Suppose the nonunitary spectrum of an operator* $U \in \mathfrak{G}$ *divides in some way into two nonintersecting parts that are symmetric with respect to the unit circle*:

$$\sigma_{\text{nonun}}(U) = \sigma_{\text{I}} \cup \sigma_{\text{II}}, \quad \sigma_{\text{I}} \cap \sigma_{\text{II}} = \varnothing, \quad \sigma_{\text{I}} = \bar{\sigma}_{\text{II}}^{-1}.$$

Then there exist subspaces $\mathfrak{L}^{\pm} \in \mathfrak{M}_{\pm}$ *such that* $U\mathfrak{L}^{\pm} = \mathfrak{L}^{\pm}$ *and*

$$\sigma_{\text{nonun}}(U \mid \mathfrak{L}^{+}) = \sigma_{\text{I}}, \quad \sigma_{\text{nonun}}(U \mid \mathfrak{L}^{-}) = \sigma_{\text{II}}.$$

Moreover, for each $\mu \in \sigma_{\text{I}}$ (σ_{II}) *the subspace*

$$\mathfrak{L}_{\mu}(U) \subset \mathfrak{L}^{+} \ (\mathfrak{L}^{-}).$$

Here we make use of some standard notation: if $\mathfrak{L}$ is an invariant subspace of an operator T, then the restriction of T to $\mathfrak{L}$ is denoted by $T \mid \mathfrak{L}$.

An operator H densely defined in $\mathfrak{H}$ $(\overline{\mathfrak{D}(H)} = \mathfrak{H})$ is said to be *J-selfadjoint* if $JH = (JH)^*$. We will consider only bounded J-selfadjoint operators. For bounded operators J-selfadjointness simply means that $[Hx, y] = [x, Hy]$.

The Cayley transform of an operator H is the operator

$$U = U_{\zeta} = (H - \zeta I)(H - \bar{\zeta} I)^{-1},$$

where ζ is a normal regular point of H.

All subsequent results on J-selfadjoint operators are based on the following simple proposition, which is well known, as a matter of fact, from linear algebra.

Lemma. *The Cayley transform U of a J-selfadjoint operator H is a J-unitary operator.*

Proof. We have

$$J(H - \zeta I) = (H^* - \zeta I) J; \quad J(H - \zeta I)^{-1} = (H - \bar{\zeta} I)^{-1} J,$$

and therefore

$$U^*JU = (H^* - \zeta I)^{-1} (H^* - \bar{\zeta} I) J (H - \zeta I)(H - \bar{\zeta} I)^{-1} = J.$$

The lemma is proved.

A comparison of this lemma with propositions 1°–3° given at the beginning of the present section leads directly to the following analogs of these propositions.

1°. *The spectrum of a J-selfadjoint operator is symmetric with respect to the real axis.*

2°. *If λ is a normal eigenvalue of a J-selfadjoint operator H, then so is* $\bar{\lambda}$* $(\lambda \neq \bar{\lambda})$. *The subspaces $\mathfrak{L}_\lambda(H)$ and $\mathfrak{L}_{\bar{\lambda}}(H)$ are J-skew-connected.*

3°_1. *If $\lambda \neq \mu$, then $[\mathfrak{L}_\lambda(H), \mathfrak{L}_\mu(H)] = \{0\}$; in particular, $[\mathfrak{L}_\lambda(H), \mathfrak{L}_\lambda(H)] = \{0\}$, if $\lambda \neq \bar{\lambda}$.*

We also require the following lemma.

Lemma. *Suppose H is a J-selfadjoint operator and U is its Cayley transform. If $P_+HP_- \in \mathfrak{S}_\infty$, then $P_+UP_- \in \mathfrak{S}_\infty$.* [1)]

Proof. We rewrite the obvious equality

$$(H - \bar{\zeta} I)\, U = H - \zeta I$$

in the matrix form:

$$\begin{pmatrix} H_{11} - \bar{\zeta} I & H_{12} \\ H_{21} & H_{22} - \bar{\zeta} I \end{pmatrix} \begin{pmatrix} U_{11} & U_{12} \\ U_{21} & U_{22} \end{pmatrix} = \begin{pmatrix} H_{11} - \zeta I & H_{12} \\ H_{21} & H_{22} - \zeta I \end{pmatrix}.$$

In particular,

$$(H_{11} - \bar{\zeta} I)\, U_{12} + H_{12} U_{22} = H_{12},$$

i.e.

$$U_{12} = (H_{11} - \bar{\zeta} I)^{-1} H_{12} (I - U_{22}) \in \mathfrak{S}_\infty .$$

The lemma is proved.

The lemmas just proved and the results established above regarding J-unitary operators directly imply the following theorems.

Theorem 14 [27a, 13d]. *Suppose H is a J-selfadjoint operator for which $P_+HP_- \in \mathfrak{S}_\infty$. Then the nonreal spectrum of H is symmetric with respect to the real axis and consists of normal points.*

Theorem 15 (L. S. Pontrjagin and H. Langer). *If under the conditions of the preceding theorem the nonreal spectrum of H divides in some way into a*

---

* *Translator's note.* The Russian text has λ.

1) And conversely, if $P_+UP_- \in \mathfrak{S}_\infty$, then $P_+HP_- \in \mathfrak{S}_\infty$.

union $\sigma_{\text{nonr}}(H) = \Lambda \cup \overline{\Lambda}$ *of parts* Λ *and* $\overline{\Lambda}$ $(\Lambda \cap \overline{\Lambda} = \emptyset)$ *that are symmetric with respect to the real axis, then there exist subspaces* $\mathfrak{L}^{\pm} \in \mathfrak{M}_{\pm}$ *that are invariant with respect to* H *and are such that*

$$\sigma_{\text{nonr}}(H \mid \mathfrak{L}^{+}) = \Lambda \text{ and } \sigma_{\text{nonr}}(H \mid \mathfrak{L}^{-}) = \overline{\Lambda}.$$

Moreover, if $\lambda \in \Lambda$ $(\overline{\Lambda})$, *then* $\mathfrak{L}_{\lambda}(H) \subset \mathfrak{L}^{+}$ $(\mathfrak{L}^{-})$.

The latter theorem was established by L. S. Pontrjagin for the case $\min(\dim \mathfrak{H}_{+}, \dim \mathfrak{H}_{-}) < \infty$, without an assumption on the boundedness of H.

In the case when $\dim \mathfrak{H}_{+} = \dim \mathfrak{H}_{-} = \infty$ and $\mathfrak{H}$ is separable, the theorem was proved by H. Langer [27a], who also obtained it for unbounded operators H under the additional condition that $\mathfrak{H}_{-} \subset \mathfrak{D}(H)$ (this condition ensures the boundedness of the operator HP_{-}).

As is shown in [18d], a more complete utilization of Theorem 9 permits one to prove Theorem 15 under more general conditions, namely, 1) $\mathfrak{H}_{-} \subset \mathfrak{D}(H)$ and the condition 2) $H_{21}(H_{11} - \zeta I)^{(-1)} \in \mathfrak{S}_{\infty}$ for some point $\zeta \in \rho(H_{11})$, which generalizes the condition $H_{21} \in \mathfrak{S}_{\infty}$.[1)]

Addenda and problems

1. Suppose $A \in \mathfrak{R}$ and $JA + A^{*}J \gg 0$. Then $\sigma(A) = \sigma_{-}(A) \cup \sigma_{+}(A)$, where $\sigma_{-}(A)$ $(\sigma_{+}(A))$ lies wholly within the left (right) half-plane $\operatorname{Re} z \leq 0$ $(\operatorname{Re} z \geq 0)$.

Moreover, in the decomposition of $\mathfrak{H}$ into a direct sum of the invariant (with respect to A) subspaces corresponding to the parts $\sigma_{-}(A)$ and $\sigma_{+}(A)$ of the spectrum:

$$\mathfrak{H} = \mathfrak{H}_{-} \dot{+} \mathfrak{H}_{+}$$

the first summand is uniformly J-positive while the second is uniformly J-negative [18h].

Hint. Show that the Cayley transform of A satisfies the conditions of problem 1 on page 131.

2. If H is a J-symmetric bounded operator, then e^{iH} is a J-unitary operator.

Hint. Show that $d[e^{itH}x, e^{itH}y]/dt = 0$ for all $-\infty < t < \infty$.

3. Suppose U is a J-unitary operator and γ^{*} is the mirror image with respect to the unit circle of a contour γ whose points belong to $\rho(U)$. Then $JP_{\gamma} = (JP_{\gamma^{*}})^{*}$, where

$$P_{\gamma} = -\frac{1}{2\pi i} \int_{(\gamma)} (U - \lambda I)^{-1} \, d\lambda.$$

---

1) If condition 2) is satisfied for some $\zeta \in \rho(H_{11})$, then it is satisfied for all $\zeta \in \rho(H_{11})$; the expression $(H_{11} - \zeta I)^{(-1)}$ denotes the resolvent at ζ of the restriction of the selfadjoint operator $H_{11} = P_{+}HP_{+}$ to $\mathfrak{H}_{+}$.

4. If $\gamma_1 \cap \gamma_2^* = \emptyset$ and U is a J-unitary operator, then

$$[P_{\gamma 1}\mathfrak{H}, P_{\gamma 2}\mathfrak{H}] = \{0\}.$$

5. An operator V is said to be *J-isometric* if it is defined on a linear manifold $\mathfrak{D}(V)$ $(\subset \mathfrak{H})$, maps $\mathfrak{D}(V)$ in a one-to-one manner onto a linear manifold $\mathfrak{R}(V)$ $(\subset \mathfrak{H})$ and is such that

$$[Vx, Vy] = [x, y] \quad (x, y \in \mathfrak{D}(V)).$$

An operator A that is densely defined in $\mathfrak{H}$ is said to be *J-hermitian* if

$$[Ax, y] = [x, Ay] \quad (x, y \in \mathfrak{D}(A)).$$

Show that propositions 3° and 3_1° remain valid for J-isometric and J-hermitian operators respectively ([32], [14a]).

6. If A is a J-hermitian operator and $\mathfrak{C}$ is the linear hull of all of the root manifolds $\mathfrak{L}_\lambda(A)$ for λ with $\operatorname{Im} \lambda > 0$, then $\dim \mathfrak{C} \leq \min (\dim \mathfrak{H}_+, \dim \mathfrak{H}_-)$. In particular, $\dim \mathfrak{C} \leq \kappa$ in the space Π_κ. An analogous assertion holds for the eigenvalues λ with $\operatorname{Im} \lambda < 0$.

Hint. Make use of the results of problem 5.

7. Establish the analog for a J-isometric operator V of the proposition formulated in problem 6.

8. Suppose λ is a real eigenvalue of a J-hermitian operator A in the space Π_κ. Then the root manifold $\mathfrak{L}_\lambda(A)$ can be decomposed into a direct sum: $\mathfrak{L}_\lambda(A) = \mathfrak{L}'_\lambda [+] \mathfrak{L}''_\lambda$, where the subspace $\mathfrak{L}'_\lambda$ is negative and consists entirely of eigenvectors of A, and $\mathfrak{L}''_\lambda$ is a finite-dimensional subspace that is invariant with respect to A.

Thus nonsimple elementary divisors of A are possible only in the finite-dimensional subspace $\mathfrak{L}''_\lambda$. The orders $\{d_j\}$ of these elementary divisors do not depend on the choice of the subspace $\mathfrak{L}'_\lambda$, and

$$\rho(\lambda) = \sum_j \left[\frac{d_j}{2}\right] \leqslant \varkappa,$$

where $[a]$ denotes the integral part of a number a [32].

Analogous decompositions and the estimate of the quantity $\rho(\lambda)$ are also valid for $\mathfrak{L}_\lambda(V)$, where V is a J-isometric operator in Π_κ and $|\lambda| = 1$ [14a].

9. Let λ_k be an arbitrary collection of eigenvalues of a J-hermitian operator A in Π_κ such that $\operatorname{Im} \lambda_k \geq 0$ (≤ 0) for all k. In the case of real λ_k let $\rho(\lambda_k)$ denote the quantity introduced in problem 8 and for nonreal λ_k put $\rho(\lambda_k) = \dim \mathfrak{L}_{\lambda_k}(A)$. Then $\Sigma_k \rho(\lambda_k) \leq \kappa$ [32].

An analogous proposition holds for a J-isometric operator V [14a].

Hint. Make use of the result of problem 8.

§4. A selfadjoint quadratic pencil

1. A linear differential equation for damped small oscillations of a continuum S of bounded dimensions can usually be written in the form

$$T\ddot{g} + R\dot{g} + Vg = 0, \tag{1}$$

where g is a vector function that has values in a Hilbert space [1] $\mathfrak{H}$ and gives the deviation of S from a position of stable equilibrium, T is the kinetic energy operator, V is the potential energy operator and R is the resistence operator.

The operator T is always positive while V is uniformly positive, i. e. $(Vg, g) \geq m(g, g)$, where $g \in \mathfrak{D}(V)$ and $m > 0$.

Usually V, and sometimes also R, turns out to be unbounded. We must therefore, in general, clarify the meaning of equation (1) (see [19c]).

We will carry out formal transformations on equation (1), after which we obtain an equation whose meaning is simple.

To this end we put $u = V^{1/2} g$ and multiply (1) from the left by $V^{-1/2}$; we then obtain the equation

$$C\ddot{u} + B\dot{u} + u = 0, \tag{2}$$

where

$$C = \overline{V^{-\frac{1}{2}} T V^{-\frac{1}{2}}}, \quad B = \overline{V^{-\frac{1}{2}} R V^{-\frac{1}{2}}}$$

(a line above denotes the closure of an operator).

In all concrete problems with a continuum S of bounded dimensions it turns out that the operator C is completely continuous while the operator B is bounded if there is internal resistance and completely continuous if the resistance is merely external.

2. Accordingly, we will consider equation (2), where B and C are bounded and completely continuous operators respectively in a Hilbert space $\mathfrak{H}$.

An elementary solution of equation (2) is any solution of the form

$$u(t) = e^{\mu t} \phi,$$

where $\phi\ (\in \mathfrak{H})$ is a constant vector $(\neq 0)$. The substitution of this expression for $u(t)$ in (2) gives

$$(\mu^2 C + \mu B + I)\varphi = 0.$$

---

1) composed of functions of an appropriate class of spatial coordinates for the continuum.

If $\phi \neq 0$, then $\mu \neq 0$, and upon making the substitution $\mu = 1/\lambda$ we can write

$$L(\lambda)\phi = 0, \tag{3}$$

where $L(\lambda) = \lambda^2 I + \lambda B + C$ is an operator pencil.

A value of λ for which equation (3) has a solution $\phi \neq 0$ is called an *eigenvalue*, and the vector ϕ ($\in \mathfrak{H}$) itself, an *eigenvector*, of the pencil L.

Obviously, the eigenvalues of L always belong to the spectrum $\sigma(L)$ of L, which is defined as the set of all complex λ for which the operator $L(\lambda)$ does not have a bounded inverse.

In many cases it turns out [19b, c] that a successful investigation of the spectral properties of L is achieved by studying the roots of the quadratic operator equation

$$Z^2 + BZ + C = 0, \tag{4}$$

which are sought in the ring $\mathfrak{R}$ of all bounded operators in $\mathfrak{H}$. Let us agree to briefly denote the left side of (4) by $L(Z)$.

One can easily see that if Z_1 is an operator root of (4), i.e. $L(Z_1) = 0$, then any eigenvalue (any eigenvector) of the operator Z_1 will be an eigenvalue (eigenvector) of the pencil L. It is possible to show that, in general, $\sigma(Z_1) \subseteq \sigma(L)$.

We note some properties of the roots Z of the quadratic equation (4).

Every root Z_1 of this equation permits one to decompose the pencil $L(\lambda)$ into linear factors; namely,

$$L(\lambda) = (\lambda I - \hat{Z}_1)(\lambda I - Z_1), \tag{5}$$

where $\hat{Z}_1 = -B - Z_1$.

Generally, $\hat{Z}_1$ will not be a root of equation (4).

But for selfadjoint B and C the roots of equation (4) will include, in addition to Z_1, the operator

$$Z_2 = \hat{Z}_1^* = -B - Z_1^*.$$

For it follows from this equality that

$$B = -Z_1^* - Z_2 = -Z_1 - Z_2^*,$$

and, since $L(Z_1) = 0$,

$$C = -(Z_1 + B) Z_1 = Z_2^* Z_1 = Z_1^* Z_2.$$

Therefore

$$L(Z_2) = Z_2^2 - (Z_1^* + Z_2) Z_2 + Z_1^* Z_2 = 0.$$

The equalities

$$B = -(Z_1 + Z_2^*), \quad C = Z_2^* Z_1$$

obtained for the case $B = B^*$, $C = C^*$ can be considered as analogs of Viète's formulas for a scalar quadratic equation.

The root Z_2 is said to be an *accompanying root* for the root Z_1 of equation (4) (with $B = B^*$, $C = C^*$). Obviously, if Z_2 accompanies Z_1 then, conversely, Z_1 accompanies Z_2.

It can happen that $Z_2 = Z_1$. In this case

$$L(\lambda) = (\lambda I - Z_1^*)(\lambda I - Z_1).$$

If in this connection, for example, Z_1 is a Volterra operator (i. e. a completely continuous operator with spectrum concentrated at zero) then the spectrum $\sigma(\lambda)$ of the pencil L will consist of the one point $\lambda = 0$.

3. A basic theorem for what follows is the next one on the existence of a root Z_1 of (4) with a special property.

Theorem 16 (M. G. Kreĭn and H. Langer [19b,c]). *Suppose B is any selfadjoint bounded operator and C is a positive completely continuous operator. Then equation* (4) *has a root Z_1 such that $Z_1^* Z_1 \leq C$.*

Proof. In the space $\tilde{\mathfrak{H}} = \mathfrak{H} \oplus \mathfrak{H}$, which is the orthogonal sum of two copies of $\mathfrak{H}$, we consider the operator

$$J = \begin{pmatrix} I & 0 \\ 0 & -I \end{pmatrix}. \tag{6}$$

It can be seen by a direct verification that H is a J-selfadjoint operator ($JH = (JH)^*$), where

$$H = \begin{pmatrix} 0 & C^{1/2} \\ -C^{1/2} & -B \end{pmatrix}.$$

According to Theorem 15 the operator H has a nonnegative invariant subspace

$$\mathfrak{L} = \{x_+ + Kx_+;\ x_+ \in \tilde{\mathfrak{H}}_+ = \mathfrak{H}\},$$

where K is a nonexpansive operator acting from $\tilde{\mathfrak{H}}_+$ into $\tilde{\mathfrak{H}}_-$. Writing out the equality $y = Hx$ component by component, we obtain

$$y_+ = C^{1/2}x_-, \quad y_- = -C^{1/2}x_+ - Bx_-,$$

and, taking into account that $x_- = Kx_+$, we find

$$y_- = Ky_+ = KC^{1/2}Kx_+ = -C^{1/2}x_+ - BKx_+.$$

Since this equality is true for any $x_+ \in \mathfrak{H}$, we get

$$-C^{1/2} - BK = KC^{1/2}K,$$

or

$$(KC^{1/2})^2 + B(KC^{1/2}) + C = 0,$$

i.e. $Z_1 = KC^{1/2}$ is a root of (4). By virtue of the fact that $\|K\| \leq 1$ we have

$$(Z_1^*Z_1x, x) \leqslant (K^*KC^{1/2}x, C^{1/2}x) \leqslant (C^{1/2}x, C^{1/2}x) = (Cx, x),$$

and this means that $Z_1^* Z_1 \leq C$.

The theorem is proved.

From the relation $Z_1 = KC^{1/2}$, $\|K\| < 1$ (equivalent to the inequality $Z_1^* Z_1 \leq C$), it follows that $Z_1 \in \mathfrak{S}_\infty$.

4. A pencil L is said to be *strongly damped* if

$$(Bx, x) > 2\sqrt{(Cx, x)(x, x)} \quad \text{for } x \neq 0,$$

i.e. for $x \neq 0$ the quadratic equation

$$(L(\lambda)x, x) = (x, x)\lambda^2 + (Bx, x)\lambda + (Cx, x) = 0$$

always has two distinct negative roots $\lambda_{1,2} = p_\pm(x)$, where

$$p_\pm(x) = \frac{-(Bx, x) \pm \sqrt{(Bx, x)^2 - 4(Cx, x)(x, x)}}{2(x, x)}. \tag{8}$$

If λ_0 is an eigenvalue of L, i.e. $L(\lambda_0)\phi = 0$, then $(L(\lambda_0)\phi, \phi) = 0$. Hence an eigenvalue λ_0 is equal to either $p_+(\phi)$ or $p_-(\phi)$.

Thus all of the eigenvalues of a strongly damped pencil are negative.

It turns out, in addition, that they admit a classification which is quite essential for what follows. In order to establish this classification we introduce some new definitions, that are applicable for a selfadjoint quadratic pencil L in the most general case.

Suppose ϕ_0 is an eigenvector of L, i.e. $L(\lambda_0)\phi_0 = 0$ for some complex λ_0. Then three cases are possible: the quantity $|\lambda_0|^2$ can be either equal to, less than or greater than the ratio $(C\phi_0, \phi_0)/(\phi_0, \phi_0)$. Respectively to these cases, the eigenvector ϕ_0 is said to be either *neutral*, of the *first* or of the *second kind.*

If one associates with the vector $\phi_0 \in \mathfrak{H}$ the vector $\psi_0 = C^{1/2}\phi_0 \oplus \lambda_0\phi_0 \in \tilde{\mathfrak{H}}$ $(= \mathfrak{H} \oplus \mathfrak{H})$, then ϕ_0 will be either a neutral eigenvector of L or an eigenvector of the first or second kind, respectively, depending on whether the eigenvector ψ_0 of the operator H $(H\psi_0 = \lambda_0\psi_0)$ is neutral, positive or negative in $\tilde{\mathfrak{H}}$.

The fact that λ_0 is a root of the quadratic equation $(L(\lambda)\phi_0, \phi_0) = 0$ implies at once that a nonreal eigenvalue λ_0 of L can have only neutral eigenvectors (compare this assertion with proposition 3°_1 on page 135).

If all of the eigenvectors of a pencil L that correspond to one and the same eigenvalue λ_0 are of the first (second) kind, then λ_0 is said to be a *definite* eigenvalue of the *first (second) kind.*

It turns out that the assertion concerning the negativeness of the eigenvalues of a pencil L can be extended in the following way.

1°. *Every eigenvalue of a strongly damped pencil* L *is negative and definite.*

In order to formulate a basic theorem on strongly damped pencils we need the notion of symmetrizability of an operator.

Let $Z \in \mathfrak{R}$, and let S $(\in \mathfrak{R})$ be a selfadjoint operator. We say that Z is symmetrized by S if $SZ = Z^*S$.

If one uses S to introduce a new scalar product

$$(x, y)_S = (Sx, y) \quad (x, y \in \mathfrak{H}),$$

then the condition $SZ = Z^*S$ will mean that Z is symmetric with respect to it:

$$(Zx, y)_S = (x, Zy)_S \quad (x, y \in \mathfrak{H}).$$

If S is uniformly positive, i. e.

$$\inf_{|x|=1} (Sx, x) > 0,$$

then the scalar product $(\cdot, \cdot)_S$ is topologically equivalent to the original one, and the replacement of the original scalar product by it converts the symmetrizable operator Z into an ordinary selfadjoint operator. With respect to the original scalar product it can be asserted in this case that Z will be *similar* to a selfadjoint operator; for in this case $Z = S^{-1/2}HS^{1/2}$.

Starting from Theorem 16, one can prove the following proposition. [1]

Theorem 17 [19b,c]. *If a pencil L is strongly damped, then:*

1) *The equation $L(Z)=0$ has one and only one root $Z_1 (\in \mathfrak{R})$ such that $Z_1^* Z_1 \geq C$, and one and only one root Z_2 such that $Z_2^* Z_2 \leq C$. These roots are accompanying roots, i. e. $Z_2 = -B - Z_1^*$.*

2) *The operator*

$$S = B + Z_1 + Z_1^* = -(B + Z_2 + Z_2^*) = Z_1 - Z_2$$

is uniformly positive.

3) *The roots Z_1, Z_2 are symmetrized by the operator S, with*

$$(Z_2 x, x)_S < (Z_1 x, x)_S < 0 \quad \text{for} \quad x \neq 0 \ (x \in \mathfrak{H}).$$

4) *The spectrum $\sigma(L) = \sigma(Z_1) \cup \sigma(Z_2)$, with*

$$\max\{\lambda : \lambda \in \sigma(Z_2)\} \leqslant \min\{\lambda : \lambda \in \sigma(Z_1)\},$$

i. e. the spectrum $\sigma(Z_2)$ lies to the left of the spectrum $\sigma(Z_1)$.

5) *The eigenvectors of the root Z_1 (Z_2) exhaust all of the eigenvectors of the first (second) kind of the bundle L.*

Since the operator Z_1 is completely continuous and S-symmetrizable, there corresponds to it a complete (in $\mathfrak{H}$) S-orthonormal system of eigenvectors $\{\phi_j\}_1^\infty$, so that

$$Z_1 \varphi_j = \lambda_j^{(1)} \varphi_j \quad (j = 1, 2, \ldots) \tag{9}$$

and

$$(S\varphi_j, \varphi_k) = \delta_{jk} \quad (j, k = 1, 2, \ldots). \tag{10}$$

The continuously invertible operator $S^{1/2}$ transforms the system $\{\phi_j\}_1^\infty$ into an orthonormal basis $\{S^{1/2}\phi_j\}_1^\infty$. Thus assertions 3) and 5) of the theorem imply, in particular,

2°. *A Riesz basis for the space $\mathfrak{H}$ can always be composed from the eigenvectors of the first kind of a strongly damped pencil.*

We will assume below without loss of generality that the numbers $\lambda_j^{(1)}$ $(j = 1, 2, \cdots)$ are arranged in decreasing order with respect to their absolute values, so that

$$\lambda_j^{(1)} \leqslant \lambda_j^{(2)} \leqslant \ldots (<0) \quad (\lambda_j^{(1)} \downarrow 0 \quad \text{for} \quad j \uparrow \infty).$$

---

1) In the lectures the author gave a complete proof from [19c] of Theorem 17. It is omitted here, and instead various extensions and applications of Theorem 17 are communicated.

5. Theorem 17 contains new information even for the elementary case when $\dim \mathfrak{H} < \infty$. Strongly damped pencils $L(\lambda)$ in a finite-dimensional space were investigated by the American engineer R. Duffin [12]. In his original paper this author showed, apparently for the first time, that the eigenvalues of a strongly damped pencil can be partitioned into eigenvalues of the first and second kind and that all eigenvalues of the second kind lie to the left of those of the first kind. Among the other achievements of Duffin we note that he managed to establish distinctive minimaximal and maximinimal properties for eigenvalues of the first and second kinds.

For eigenvalues of the first kind these properties can be expressed with the use of the functional $p_+(x)$ of (8) by the following relations:

$$\lambda_1^{(1)} = \min_{x \in \mathfrak{H}} p_+(x), \ \lambda_j^{(1)} = \min_{\mathfrak{L}_{j-1}} \max_{x \in \mathfrak{L}_{j-1}} p_+(x) \ (j = 1, 2, \dots, n = \dim \mathfrak{H}),$$

where $\mathfrak{L}_{j-1}$ ranges over all subspaces of $\mathfrak{H}$ of codimension $j-1$ ($\dim (\mathfrak{H} \ominus \mathfrak{L}_{j-1}) = j-1$).

In his dissertation R. Kühne [25] extended this result to strongly damped pencils in an infinite-dimensional space $\mathfrak{H}$, in particular, to the pencils being considered by us.[1] In this case one should replace "max" in the above relations by "sup:"

$$\lambda_1^{(1)} = \min_{x \in \mathfrak{H}} p_+(x), \quad \lambda_j^{(1)} = \min_{\mathfrak{L}_j} \sup_{x \in \mathfrak{L}_j} p_+(x) \ (j = 2, 3, \dots). \tag{11}$$

As was shown by H. Langer [27b], the fact that the eigenvalues of the first kind $\lambda_j^{(1)}$ $(j = 1, 2, \cdots)$ of a pencil make up the spectrum $\sigma(Z_1) \setminus 0$ and that a Riesz basis can be formed from the corresponding eigenvectors ϕ_j $(j = 1, 2, \cdots)$ permits one to obtain a very simple proof of (11).

6. As was noted in [19b] and worked out in detail in [27c], it is possible to establish asymptotic formulas for the eigenvalues of a pencil L under various assumptions on the operators B and C.

As was shown in [27c], the following result can be obtained from relations (9)–(11).

3°. *If the operator B in a strongly damped pencil L has the form*

$$B = \beta I + T \quad (\beta > 0), \tag{12}$$

where T is a completely continuous operator, then

---

1) R. Kühne considered pencils $L(\lambda)$ of the form $L(\lambda) = \lambda^2 A + \lambda B + C$.

$$\lambda_j^{(1)}(L) = -\frac{\lambda_j(C)}{\beta}(1 + o(1)) \quad (j \to \infty), \tag{13}$$

where $\lambda_1(C) \geq \lambda_2(C) \geq \cdots$ *are the successive eigenvalues of the operator* C.

If the operator B has the form (12) while the numbers $\lambda_j(C)$ satisfy the condition $\lambda_{j+1}(C)/\lambda_j(C) \longrightarrow 1$ for $j \longrightarrow \infty$, then even without the assumption that L is strongly damped the spectrum of its eigenvalues of the first kind will consist of an infinite nondecreasing sequence of numbers $\{\lambda_j^{(1)}\}_1^\infty$ with the asymptotic property (13) (see problem 8).

7. S. G. Kreĭn [23] recently showed that the problem concerning small oscillations of a viscous fluid situated in a fixed container and having a free surface reduces to the equation

$$y = \mu G y + \frac{1}{\mu} H y, \tag{14}$$

where $G, H \in \mathfrak{S}_\infty$, $G > 0$, $H \geq 0$ and μ is a complex parameter.

In connection with this, equation (14) can easily be transformed to the form $L(\lambda)x = 0$, where $L(\lambda) = \lambda^2 I + \lambda B + C$ is a strongly damped pencil with an operator B of the form (12).

In fact, the substitution $\mu = -\lambda - a$ $(a > 0)$ transforms equation (14) to the form

$$(a^2 G + aI + H) y + \lambda(2aG + I) y + \lambda G y = 0. \tag{15}$$

No matter what the operator $H = H^*$ $(\in \mathfrak{R})$ (neither of the conditions $H \in \mathfrak{S}_\infty$, $H \geq 0$ can be satisfied) the operator $F = a^2 G + H + aI$ will be uniformly positive for sufficiently large a. Taking such an a and carrying out the substitution $x = F^{1/2} y$ in equation (15), we transform it into the equation $L_a(\lambda)x = 0$, where

$$L_a(\lambda) = I + \lambda B_a + \lambda C_a^2,$$

$$B_a = F^{1/2}(2aG + I) F^{-1/2} C_a = F^{-1/2} G F^{-1/2}.$$

If $H \in \mathfrak{S}_\infty$, then $F - aI \in \mathfrak{S}_\infty$, which implies $F^{-1/2} - a^{-1/2} I \in \mathfrak{S}_\infty$ and hence $B_a - a^{-1} I \in \mathfrak{S}_\infty$.

The resultant pencil L_a will be strongly damped if and only if [1)]

$$4(Gx, x)(Hx, x) < (x, x)^2 \quad (x \in \mathfrak{H}, x \neq 0). \tag{16}$$

---

1) We note that condition (16) is automatically satisfied for $H \leq 0$; should $H = H_+ - H_-$ $(H_+ \geq 0)$, condition (16) will be satisfied, for example, if $4\|G\| \|H_+\| < 1$.

In fact, condition (7) for the pencil L means that the quadratic equation $(L(\lambda)f, f) = 0$ has for any $f \neq 0$ two distinct real roots, and this property is invariant under all transformations that return us to the pencil $H - \lambda I + \lambda^2 G$.

In the event condition (16) is satisfied the entire preceding theorem will be applicable to equation (14). If, in addition, $H \in \mathfrak{S}_\infty$, then B_a satisfies condition (12) with $\beta = 1/a$ and $\lambda_n(C_a)/\lambda_n(G) \longrightarrow 1/a$.

All of this permits one to obtain a number of important extensions of the work of N. G. Askerov, S. G. Kreĭn and G. I. Laptev [1], [26]. For example, in the event condition (16) and the conditions $G, H \in \mathfrak{S}_\infty$, $G > 0$ hold it is possible to assert that equation (14) gives rise to two Riesz bases composed respectively of the eigenvectors of (14) of the first and second kind, [1] that the following asymptotic formula holds for the corresponding complete sequence $\mu_1^{(1)} \leq \mu_2^{(1)} \leq \cdots$ of eigenvalues of the first kind of (14):

$$\mu_n^{(1)} = \lambda_n^{-1}(G) \quad (1 + o(1)) \quad (n \to \infty),$$

and that, if in addition $H > 0$, the following analogous formula holds for the corresponding sequence of eigenvalues of the second kind:

$$\mu_n^{(2)} = \lambda_n(H)(1 + o(1)).$$

Thus the asymptotic behavior of the eigenvalues of the first kind is determined by the eigenvalues of the equation $y = \mu G y$ while the asymptotic behavior of the eigenvalues of the second kind is determined by the eigenvalues of the equation $Hy = \mu y$.

In concluding this lecture we should emphasize that we have presented a sufficiently complete description of the properties of the quadratic pencil $L(\lambda)$ only for the case when it is strongly damped.

The polar case is that of a *weakly damped pencil*, when

$$(Bx, x)^2 < 4(x, x)(x, x) \quad \text{for} \quad x \neq 0.$$

In this case, as in a number of "intermediate" cases, a detailed analysis of the properties of a quadratic selfadjoint pencil L can be achieved (see [19b, c]) by means of a systematic utilization of, besides "indefinite" arguments, various methods of the general theory of nonselfadjoint operators (see [16, 17, 9c]).

---

1) It is not difficult to see how the notion of an eigenvector (eigenvalue) of the first or second kind is carried over to equation (14).

Addenda and problems

1. A sequence of nonzero vectors $\psi_0, \psi_1, \cdots, \psi_{p-1}$ is called a *Jordan chain* of a quadratic pencil L corresponding to an eigenvalue λ_0 if

$$L(\lambda_0)\psi_0 = 0, \quad L(\lambda_0)\psi_j + \frac{\partial L(\lambda_0)}{\partial \lambda_0}\psi_{j-1} + \frac{\partial^2 L(\lambda_0)}{2!\,\partial \lambda_0^2}\psi_{j-2} = 0$$
$$(j = 1, 2, \ldots, p-1;\ \psi_{-1} = 0).$$

The linear hull $\mathfrak{L}_{\lambda_0}(L)$ of all Jordan chains of L corresponding to an eigenvalue λ_0 is called a *root manifold* of L. The dimension $\kappa(\lambda_0)$ of this linear manifold is called the *algebraic multiplicity* of the eigenvalue λ_0 of L.

Show that, if $\psi_0, \psi_1, \cdots, \psi_{p-1}$ is a Jordan chain of L corresponding to an eigenvalue λ_0, the vectors

$$\begin{pmatrix} C^{1/2}\psi_0 \\ \lambda_0\psi_0 \end{pmatrix}, \quad \begin{pmatrix} C^{1/2}\psi_1 \\ \lambda_0\psi_1 + \psi_0 \end{pmatrix}, \ldots, \begin{pmatrix} C^{1/2}\psi_{p-1} \\ \lambda_0\psi_{p-1} + \psi_{p-2} \end{pmatrix}$$

form a Jordan chain of the associated operator H (see (6) for its definition) corresponding to the same eigenvalue λ_0.

Conversely, if the vectors

$$\tilde{\psi}_j = \psi_j^{(1)} \oplus \psi_j^{(2)} = \begin{pmatrix} \psi_j^{(1)} \\ \psi_j^{(2)} \end{pmatrix} \quad (j = 0, 1, \ldots, p-1)$$

form a Jordan chain of H corresponding to an eigenvalue λ_0, i.e.

$$H\tilde{\psi}_0 = \lambda_0\tilde{\psi}_0, \quad H\tilde{\psi}_j = \lambda_0\tilde{\psi}_j + \tilde{\psi}_{j-1} \quad (j = 1, 2, \ldots, p-1),$$

then the vectors

$$\psi_0 = \frac{1}{\lambda_0}\psi_0^{(2)}, \quad \psi_j = \frac{1}{\lambda_0}(\psi_j^{(2)} - \psi_{j-1}) \quad (j = 1, 2, \ldots, p-1)$$

form a Jordan chain of L.

Thus a pencil L and an associated operator $\mathfrak{N}$ have the same eigenvalues and, moreover, the same multiplicities.

Show that $\sigma(L) = \sigma(\mathfrak{N})$. To do this one should calculate the elements of the matrix giving the resolvent $(H - \lambda I)^{-1}$ in $\tilde{\mathfrak{H}}$ [19c].

2. Show that 1) each nonreal point λ_0 of the spectrum $\sigma(L)$ of a pencil L is an eigenvalue of finite multiplicity $\kappa(\lambda_0)$ and 2) to each nonreal eigenvalue $\lambda_0 \in \sigma(L)$ there corresponds the eigenvalue $\bar{\lambda}_0 \in \sigma(L)$ of the same multiplicity: $\kappa(\lambda_0) = \kappa(\bar{\lambda}_0)$.

Hint. Make use of the assertion of problem 1.

3. Show that if the operator $B \geq 0$ (is uniformly positive) then the spectrum $\sigma(L)$ lies in (in the interior of) the left half-plane $\operatorname{Re}\lambda \leq 0$.

Hint. Consider the operator $\operatorname{Re}\mathfrak{N} = \frac{1}{2}(\mathfrak{N} + H^*)$.

4. Show that if the condition $B^2 \leq \kappa^2 C$ holds for some κ $(0 < \kappa < 2)$ then the spectrum $\sigma(L)$ lies in the vertical angles

$$\left|\frac{\pi}{2} \pm \arg\lambda\right| \leqslant \arcsin\frac{\varkappa}{2}\left(<\frac{\pi}{2}\right).$$

Hint. Consider the quadratic equation $(L(\lambda)\phi, \phi) = 0$.

Show also that when $0 < \kappa < 1$ the operator $\mathfrak{A} = -\mathfrak{N}^2$ satisfies the inequality

$$|\operatorname{Im}(\mathfrak{A}x, x)| \leqslant \frac{\varkappa}{\sqrt{1-\varkappa^2}} \operatorname{Re}(\mathfrak{A}x, x) \quad (x \in \mathfrak{H}).$$

5. Prove that the completeness in $\tilde{\mathfrak{H}} = \mathfrak{H} \oplus \mathfrak{H}$ of a system of root manifolds of one of the operators $\mathfrak{N}$ and

$$\mathfrak{L} = \begin{pmatrix} 0 & I \\ -C & -B \end{pmatrix}$$

implies the same completeness for the other operator [19c].

Hint. Use the relation $\mathfrak{N}\mathfrak{S} = \mathfrak{S}H$, where

$$\mathfrak{S} = \begin{pmatrix} C^{1/2} & 0 \\ 0 & I \end{pmatrix}.$$

Show that the completeness in $\tilde{\mathfrak{H}}$ of a system of root manifolds of the operator $\mathfrak{L}$ is equivalent to the double completeness of a system of eigen- and associated vectors of the pencil $\Gamma(\lambda) = \lambda^2 L(1/\lambda) = I + \lambda B + \lambda^2 C$ (for the latter concepts see [16, 17] or [9c]).

6. Obtain the following sharpening of Theorem 16.

Let $\sigma_{\text{nonr}}(L)$ (the nonreal part of the spectrum $\sigma(L)$) be broken up in some manner into two disjoint parts Λ and $\sigma_{\text{nonr}}(L)\setminus\Lambda$ that are symmetrically arranged with respect to the real axis. Then the equation $L(Z) = 0$ has a root Z_Λ possessing the following properties: 1) $Z^*_\Lambda . Z_\Lambda \leq C$, 2) the nonreal part of the spectrum $\sigma(Z_\Lambda)$ coincides with Λ and 3) the root manifolds $\mathfrak{L}_\lambda(L)$ and $\mathfrak{L}_\lambda(Z_\Lambda)$ coincide for every $\lambda \in \Lambda$.

Hint. Make more complete use of Theorem 15, following the idea of the proof of Theorem 16.

The operator Z_Λ is always completely continuous by virtue of property 1). It accordingly follows that the origin is the only possible limit point of the nonreal spectrum $\sigma_{\text{nonr}}(L)$.

It is possible to show [19c] that if the system of root manifolds $\mathfrak{L}_\lambda(L)$ $(\lambda \in \Lambda)$ is complete in $\mathfrak{H}$ then the root Z_Λ is uniquely defined by properties 1) and 2). In this case

$$\lambda^2 I + \lambda B + C = (\lambda I - Z_\Lambda^*)(\lambda I - Z_\Lambda),$$

i.e.

$$Z_\Lambda^* Z_\Lambda = C, \quad B = -Z_\Lambda - Z_\Lambda^*,$$

and consequently the operator B is completely continuous.

7. Show that the condition

$$(Bx, x)^2 \leqslant \text{const}\,(x, x)(Cx, x) \quad (x \in \mathfrak{H})$$

implies the complete continuity of the operator B $(B \in \mathfrak{S}_\infty)$.

8. Show that if $B = B^* \in \mathfrak{S}_\infty$ then the condition (17)* for a pencil L to be weakly damped is equivalent to the condition that all of the eigenvalues of L be nonreal [19c].

9. Prove that when the spectrum of condensation of the operator B consists of positive numbers (in particular, when the operator B is uniformly positive) the spectrum $\sigma(L)$ contains only a finite number of nonreal points.

Hint. Consider the operator $S = B + Z_\Lambda + Z_\Lambda^*$, where Z_Λ is the operator of problem 6; make use of its property $SZ_\Lambda = Z_\Lambda^* S = Z_\Lambda^* Z_\Lambda - C \leq 0$ and apply Pontrjagin's theorem of problem 6 on page 137.

§5. Applications to canonical linear differential equations with a periodic Hamiltonian

We will consider the application of our general results J-unitary and related operators to the theory of equations indicated in the title of this section.

1. General properties of linear differential equations. Let us recall some facts from the theory of linear differential equations of the form

$$\frac{dx}{dt} = A(t)\,x, \tag{1}$$

where $A(t)$ $(-\infty < t < \infty)$ is a function with values in the ring $\mathfrak{R}$ of bounded operators acting in a separable Hilbert space $\mathfrak{H}$.

In considering an equation of type (1) below we will always assume that the operator function $A(t)$ satisfies the following conditions of measurability and integrability:

1) $A(t)$ is *weakly measurable*, i.e. the function $(A(t)f, g)$ is locally measurable for any $f, g \in \mathfrak{H}$.

2) The function $\|A(t)\|$ is *locally summable*, i.e. it is summable on any finite interval of the real axis.

A vector function $x(t)$ $(-\infty < t < \infty)$ with values in $\mathfrak{H}$ is said to be

* *Translator's note.* The Russian text has (7).

a solution of equation (1) if it satisfies (1) in the weak sense, i.e. for any $f \in \mathfrak{H}$ the function $(x(t), f)$ is locally absolutely continuous and almost everywhere

$$\frac{d}{dt}(x(t), f) = (A(t)x, f). \tag{2}$$

Under our assumptions 1) and 2) equation (1) possesses the following fundamental property.

For any $x_0 \in \mathfrak{H}$ and $t_0 \in (-\infty, \infty)$ equation (1) has one and only one solution $x(t)$ such that $x(t_0) = x_0$. Moreover, this solution can be obtained in the form of a series

$$x(t) = x_0 + x_1(t) + x_2(t) + \ldots, \tag{3}$$

that converges uniformly in any finite interval and consists of the recursively defined functions

$$x_n(t) = \int_{t_0}^{t} A(t)\, x_{n-1}(t)\, dt \quad (n = 1, 2, \ldots). \tag{4}$$

This assertion implies, in particular, that every solution $x(t)$ of equation (1), being by definition weakly absolutely continuous, is strongly continuous. [1)]

Here we have taken advantage of the fact that if $y(t)$ is a strongly continuous function, then by virtue of 1) and 2) the function $A(t)y(t)$ is weakly locally measurable and weakly locally integrable, and hence the weak integral

$$\int_0^t A(s)\, x(s)\, ds$$

is a strongly continuous function. [2)] The integrals in (4) are understood in

---

1) These assertions, as well as certain subsequent assertions, can be carried over to equations of the form (1) in a Banach space $\mathfrak{B}$. Such equations were first considered by D. L. Kučer [24 a, b] under conditions that go over to conditions 1) and 2) for $\mathfrak{B} = \mathfrak{H}$.

2) The notion of a weakly measurable, weakly integrable abstract function $x(t)$ with values in a Banach space $\mathfrak{B}$ and the theory of the weak integral of such functions were first developed by I. M. Gel'fand in his candidate's dissertation in 1934. A presentation of these matters (without an exact history) can be found in Chapter III of the book of Einar Hille, *Functional analysis and semigroups* (Amer. Math. Soc. Colloq. Publ., Vol. 31, Amer. Math. Soc., Providence, R.I., 1948) or in the same chapter of the 1957 revision, by Hille and R. S. Phillips, of that book. In the case of Hilbert space ($\mathfrak{B} = \mathfrak{H}$) the whole theory of integration of abstract functions is greatly simplified.

the weak sense.

The following majorization is valid for the series composed of the norms of the terms of the series (3):

$$\sum_{n=0}^{\infty} \| x_n(t) \| \ll \sum_{n=0}^{\infty} \frac{1}{n!} \left(\int_{t_0}^{t} \| A(s) \| \, ds \right)^n \| x_0 \|. \tag{5}$$

For any $t, \tau \in (-\infty, \infty)$ we can put

$$x(t) = U(t, \tau) x(\tau).$$

The operator $U(t, \tau)$ will linearly and continuously map all of $\mathfrak{H}$ onto $\mathfrak{H}$; its inverse will be the operator $U(\tau, t)$.

The operator function $U(t) = U(t, 0)$ $(-\infty < t < \infty)$ will be called the "resolving operator function" for equation (1). For any solution $x(t)$ of (1) we will have

$$x(t) = U(t) x(0). \tag{6}$$

Clearly, $U(t, \tau) = U(t) U^{-1}(\tau)$. Since (3) and (5) imply

$$\| x(t) - x(t_0) \| \leqslant \left(\exp \left| \int_{t_0}^{t} \| A(s) \| \, ds \right| - 1\right) \| x(t_0) \|,$$

the operator function $U(t)$ is continuous in the uniform norm (the usual norm in the ring $\mathfrak{R}$). This function is the unique continuous solution of the integral equation

$$U(t) = I + \int_0^t A(s) U(s) \, ds, \tag{7}$$

in which the integral on the right is considered in the weak sense.

In accordance with the expansion (3) the following expansion holds uniformly on any finite interval:

$$U(t) = I + V_1(t) + V_2(t) + \dots, \tag{8}$$

where the $V_n(t)$ are continuous operator functions that are obtainable from the recursion formulas

$$V_n(t) = \int_0^t A(s) V_{n-1}(s) \, ds \quad (V_0 = I).$$

We have the estimates

$$\|V_n(t)\| \leqslant \frac{1}{n!}\left|\int_0^t \|A(s)\|\,ds\right|^n \quad (n = 1, 2, \ldots). \tag{9}$$

The integral equation (7) is equivalent to the system

$$\frac{dU}{dt} = A(t)U, \quad U(0) = I. \tag{10}$$

The first equality (the differential equation) should be understood in the weak sense, i. e. as expressing the following requirement: for any $f, g \in \mathfrak{H}$ the function $(U(t)f, g)$ is locally absolutely continuous, and

$$\frac{d}{dt}(U(t)f, g) = (A(t)U(t)f, g) \tag{11}$$

almost everywhere. The set of those t for which the latter equality holds depends in general on $f, g \in \mathfrak{H}$. By making use of the separability of $\mathfrak{H}$ and condition 2) it is possible to prove the existence of a set $\mathfrak{E}_A$ of full measure (i. e. obtained from the real axis by rejecting a set of measure zero) such that (11) will hold for any $f, g \in \mathfrak{H}$ when $t \in \mathfrak{E}_A$.

According to (6) it can be asserted that every solution $x(t)$ of (1) will have at any point $t \in \mathfrak{E}_A$ a weak derivative dx/dt that, moreover, is equal to $A(t)x$. We have thus invested equation (1) with a new meaning.

2. The monodromy operator of an equation with a periodic coefficient. We will be interested in the case when the operator function $A(t)$ has a certain period T:

$$A(t + T) = A(t) \quad (-\infty < t < \infty). \tag{12}$$

One can easily see that in this case, besides $U(t)$, the operator function $U_1(t) = U(t+T)U^{-1}(T)$ will also be a continuous solution (in the weak sense) of system (10). By virtue of the uniqueness of the solution we will have $U_1(t) = U(t)$, which implies

$$U(t+T) = U(t)\,U(T). \tag{13}$$

The operator $U(T)$ is called the *monodromy operator* of a periodic equation (1) of period T.

We have the following proposition.

1°. *In order for any solution* $x(t)$ *of a periodic equation* (1) *to be bounded on the whole axis it is necessary and sufficient that*

$$\sup_{-\infty < n < \infty} \|U^n(T)\| < \infty. \tag{14}$$

In fact, if any solution $x(t) = U(t)\,x_0$ of (1) is bounded, then

$$\sup_n \| x(nT) \| = \sup_n \| U(nT) x_0 \|$$
$$= \sup_n \| U^n(T) x_0 \| \leqslant \sup_{-\infty < t < \infty} \| x(t) \| < \infty.$$

Thus the sequence of operators $\{U^n(T)\}_{-\infty}^{\infty}$ $(\subset \mathfrak{R})$ takes bounded values at any element $x_0 \in \mathfrak{H}$, and is therefore, according to the well-known Banach-Steinhaus Theorem, bounded in $\mathfrak{R}$.

Conversely, if condition (14) is satisfied then, by putting

$$n_t = \left[\frac{t}{T}\right], \quad r_t = t - n_t T,$$

for any t, according to (13) we will have

$$\| x(t) \| = \| U(t) x(0) \| = \| U(r_t) U^{n_t}(T) x(0) \|,$$

so that

$$\sup_{-\infty < t < \infty} \| x(t) \| \leqslant \max_{0 \leqslant t \leqslant T} \| U(t) \| \sup_n \| U^n(T) \| \| x(0) \|.$$

We recall that, by a well-known theorem of B. von Sz.-Nagy, which we have already made use of once, *condition* (14) *is equivalent to the fact that the operator* $U(T)$ *is similar to a unitary operator.*

In a manner analogous to the way proposition 1° was proved it can prove that all solutions of a periodic equation (1) are bounded on the semi-axis $(0, \infty)$ if and only if

$$\sup_{1 \leqslant n < \infty} \| U^n(t) \| < \infty. \qquad (14')$$

In contrast to condition (14) condition (14′) does not have a simple interpretation.

Let $\mathfrak{A}(T)$ denote the set of all operator functions $A(t)$ $(-\infty < t < \infty)$ with values in $\mathfrak{R}$ that satisfy conditions 1) and 2) of subsection 1 and the condition of periodicity (12).

The set $\mathfrak{A}$ can be regarded as a normed linear space with norm

$$\| A \|_{\mathfrak{A}} = \int_0^T \| A(s) \| \, ds. \qquad (15)$$

We propose as an exercise to prove with the use of equation (7) the assertion

2°. *The mapping* $A(t) \longrightarrow U(t)$ *is a continuous mapping of the space* $\mathfrak{A}$ *into the ring* $\mathfrak{R}$.

3. **Canonical equation.** Equation (1) is said to be *canonical* if it has the form

$$\frac{dx}{dt} = iJH(t)\,x \quad (-\infty < t < \infty), \tag{16}$$

where J is an indefinite [1] signature operator ($J = J^*$, $J^2 = I$) and $H(t)$ is an operator function with selfadjoint values (the Hamiltonian of (16)).

Clearly, the operator function $A(t) = iJH(t)$ will satisfy conditions 1) and 2) if and only if they are satisfied by the operator function $H(t)$.

Assuming these conditions to be satisfied, let us show that

3°. *The resolving operator function* $U(t)$ *of the canonical equation* (16) *is a* J*-unitary function, i. e.*

$$U^*(t)JU(t) = J. \tag{17}$$

In fact, if $x(t)$ is a solution of equation (16) then for any $t \in \mathfrak{E}_H$ ($\equiv \mathfrak{E}_A$) for which $x(t)$ has a weak derivative dx/dt with value (1), we will have

$$\frac{d}{dt}(Jx(t),\ x(t)) = i(H(t)\,x,\ x) - i(Jx,\ JH(t)\,x) = 0.$$

If we show that the function $(Jx(t), x(t))$ is locally absolutely continuous it will follow that

$$(Jx(t),\ x(t)) = (Jx(0),\ x(0)) \quad (-\infty < t < \infty). \tag{18}$$

For $t_1, t_2 \in (a, b)$ we have

$$\begin{gathered} |(Jx(t_2),\ x(t_2)) - (Jx(t_1),\ x(t_1))| \leqslant |(x(t_2) - x(t_1), \\ Jx(t_2))| + |(Jx(t_1),\ x(t_2) - x(t_1))| \leqslant 2C_{a,b}\,\|x(t_2) - x(t_1)\|, \\ (C_{a,b} = \max_{a<t<b} \|x(t)\|), \end{gathered}$$

and since it follows from (1) that

$$\|x(t_2) - x(t_1)\| \leqslant C_{a,b}\int_{t_1}^{t_2} \|H(t)\|\,dt,$$

the local absolute continuity of the function (Jx, x) is established.

From this, by setting $x(t) = U(t)\,x(0)$ in (18) and making use of the arbitrariness in the choice of $x(0) \in \mathfrak{H}$, we deduce relation (17).

---

1) Since we will be interested in conditions for the boundedness of the solutions of (1), the case $J = \pm I$ is trivial. But some of our arguments (subsection 7) furnish new information even for this case.

4. Canonical equations with bounded solutions. Conditions (14) and (14′) for the J-unitary operator $U(t)$ are equivalent (see page 120). According to the terminology introduced on the same page, if they are satisfied the J-unitary operator $U(T)$ is said to be stable. Recalling Theorem 7 (R. Phillips), we arrive at the following proposition.

Theorem. *In order for all of the solutions of a periodic canonical equation* (16) *to be bounded on the real axis* $(-\infty, \infty)$ (*on the semiaxis* $(0, \infty)$) *it is necessary and sufficient that the space* $\mathfrak{H}$ *decompose into a* J-*orthogonal sum of two definite subspaces that are invariant with respect to the monodromy operator* $U(T)$.

Suppose this condition is satisfied, i. e.

$$\mathfrak{H} = \mathfrak{L}_1 [+] \mathfrak{L}_2, \tag{19}$$

where

$$\mathfrak{L}_1 \subset \mathfrak{P}_+, \quad \mathfrak{L}_2 \subset \mathfrak{P}_- \text{ and } U(T)\,\mathfrak{L}_k \subset \mathfrak{L}_k \quad (k = 1, 2). \tag{20}$$

Let U_1 and U_2 denote the operators induced by $U(T)$ in $\mathfrak{L}_1$ and $\mathfrak{L}_2$ respectively. With respect to the scalar product $[x, y]_1 = (Jx, y)$, $x, y \in \mathfrak{L}_1$ (the scalar product $[x, y]_2 = -(Jx, y)$, $x, y \in \mathfrak{L}_2$) the space $\mathfrak{L}_1$ $(\mathfrak{L}_2)$ is a Hilbert space and the operator U_1 (U_2) is an ordinary unitary operator. The following representations will therefore hold:

$$U_k = \exp(iC_k) \quad (k = 1, 2),$$

where C_k $(k = 1, 2)$ is an operator acting in $\mathfrak{L}_k$ $(k = 1, 2)$ that is selfadjoint there.[1] The operators C_k can be chosen so that, for example, they are positive and their spectra lie on the segment $[0, 2\pi]$.

If we set $C = C_1 [+] C_2$ in accordance with the decomposition (19), we will have $U(T) = \exp(iC)$ and C will be a J-selfadjoint operator. Therefore

$$U(T) = \exp(iTJG), \tag{21}$$

where

$$G = \frac{1}{T} JC. \tag{22}$$

We put

$$Q(t) = U(t)\, e^{-iTJG}. \tag{23}$$

According to (13) and (21) we will have

---

1) In essence, we repeat the proof of Theorem 8 (page 121) with a change in notation: C_k in place of H_k $(k = 1, 2)$.

$$Q(t+T) = Q(t) \quad (-\infty < t < \infty).$$

The periodic operator function $Q(t)$ is of interest in that it furnishes a substitution [1)]

$$x = Q(t)y \tag{24}$$

which transforms equation (16) into a canonical equation

$$\frac{dy}{dt} = iJGy \tag{25}$$

with a constant Hamiltonian G.

Under that detailing of the choice of the operators G_k $(k = 1, 2)$ which was indicated above, the Hamiltonian G will be positive: $(Gf, f) > 0$ for $f \neq 0$. In the case of a finite-dimensional space $(\dim \mathfrak{H} < \infty)$ we can also assert that, conversely, the existence of a transformation (24) (with a periodic and absolutely continuous operator function $Q(t)$) that reduces equation (16) to equation (25) with a constant positive G guarantees the boundedness of all of the solutions of equation (16).

This converse assertion is not true in the infinite-dimensional case $(\dim \mathfrak{H} = \infty)$.

5. **A canonical equation with stably bounded solutions.** If the decomposition (19) with properties (20) holds for some J-unitary operator U and, in addition, the spectrum of U in $\mathfrak{L}_1$ does not intersect the spectrum of U in $\mathfrak{L}_2$, i.e.

$$\sigma(U \mid \mathfrak{L}_1) \cap \sigma(U \mid \mathfrak{L}_2) = \varnothing,$$

we will say that the operator U has a *strictly definite spectrum.*

According to Theorem 8 the strict definiteness of the spectrum of a J-unitary operator U is a necessary and sufficient condition for its *strong stability* (we recall that this property means that all of the J-unitary operators in a sufficiently small neighborhood of U are stable).

Let us agree to say that all of the solutions of equation (16) are stably bounded if they remain bounded upon replacing the Hamiltonian $H(t) = H(t+T)$ by a Hamiltonian $H_1(t) = H_1(t+T)$ that is sufficiently close to it. Closeness is understood in the sense of the metric

$$\rho(H, H_1) = \int_0^T \| H(t) - H_1(t) \| \, dt. \tag{26}$$

The following theorem holds.

---

1) In regard to substitutions of this kind see problem 1.

Theorem 18. *In order for all of the solutions of a periodic canonical equation* (16) *to be stably bounded it is necessary and sufficient that the monodromy operator of this equation have a strictly definite spectrum.*

Proof. The sufficiency of the condition follows from the fact that under its fulfillment the monodromy operator is strongly stable and hence will remain stable under a sufficiently small perturbation produced by a sufficiently small variation of the Hamiltonian $H(t)$ (see proposition 2°).

Let us prove the necessity of the condition. Suppose all of the solutions of a periodic equation (16) are stably bounded. Then the monodromy operator $U(t)$ is stable in every case, and we can define an operator G and a continuous periodic operator function $Q(t)$ by formulas (22) and (23).

If we assume that the spectrum of the operator $U(T)$ is not definite, then under an appropriate definition of the operators C_1 and C_2 their spectra will have a nonempty intersection. In this case, as was shown in proving Theorem 8, any neighborhood of the operator C will contain J-selfadjoint operators $\widetilde{C}$ with a nonreal spectrum. For such an operator $\widetilde{C}$ the canonical equation

$$\frac{dy}{dt} = iJ\widetilde{G}y \quad \left(\widetilde{G} = \frac{1}{T}J\widetilde{C}\right) \tag{27}$$

will have unbounded solutions. It can easily be seen that the substitution

$$y = Q^{-1}(t)\,x$$

in equation (27) transforms it into the equation

$$\frac{dx}{dt} = iJ\widetilde{H}(t)\,x, \tag{28}$$

where

$$\widetilde{H}(t) = JQ(t)J\widetilde{G}Q^{-1}(t) + J\frac{dQ}{dt}Q^{-1}(t). \tag{29}$$

Here dQ/dt is the derivative in the weak sense of the operator function $Q(t)$ and exists almost everywhere (for all $t \in \mathfrak{E}_H$). According to (23) and the relation

$$\frac{dU}{dt} = JH(t)U$$

we will have

$$\frac{dQ}{dt} = H(t)Q - iQJG.$$

If we replace $\widetilde{G}$ by G in (29), the operator function $\widetilde{H}_1(t)$ goes over into $H(t)$; therefore

$$\tilde{H}(t) - H(t) = Q^{*-1}(t)(\tilde{G} - G)Q^{-1}(t). \tag{30}$$

Here we have made use of the fact that, together with $U(t)$, the function $Q(t)$ is a J-unitary operator, and hence $JQ(t)J = [Q^*(t)]^{-1}$.

It follows from (30) that equation (28) is also canonical ($\tilde{H} = \tilde{H}^*$) and that for $\tilde{G}$ sufficiently close to G the Hamiltonian $\tilde{H}(t)$ will be arbitrarily close to the Hamiltonian $H(t)$ in the sense of the metric (26).

On the other hand, if equation (27) has unbounded solutions, then, obviously, so will equation (28).

We have arrived at a contradiction with the original assumption concerning the stable boundedness of all of the solutions of equation (16). The theorem is proved.

6. The central zone of stability of the canonical equation with a parameter. We consider the canonical equation with a real parameter

$$\frac{dx}{dt} = \lambda iJH(t)x. \tag{31}$$

Let $U(t; \lambda)$ denote the resolving operator of this equation defined by the system

$$\frac{dU}{dt} = i\lambda JH(t)U, \quad U(0) = I. \tag{32}$$

According to the results of subsection 1 (see (8) and (9)) it admits the expansion

$$U(t; \lambda) = I + \sum (i\lambda)^n U_n(t),$$

where

$$U_n(t) = \int_0^t JH(s)U_{n-1}(s)\,ds \quad (n = 1, 2, \ldots)$$

and the following estimates hold:

$$\|U_n(t)\| \leqslant \frac{1}{n!}\left|\int_0^t \|H(s)\|\,ds\right|^n.$$

In particular, for the monodromy operator $U(T; \lambda)$ we will have

$$U(T; \lambda) = I + i\lambda J\int_0^T H(s)\,ds + O(\lambda^2) \quad (\lambda \to 0). \tag{33}$$

A real point $\lambda = \lambda_0$ is called a *point of strong stability* of equation (31) if for $\lambda = \lambda_0$ all of the solutions of equation (31) are stably bounded, or, what is equivalent, if the operator $U(T;\ \lambda_0)$ is strongly stable.

Clearly, the set of points of strong stability, unless empty, is an open set, and hence decomposes into a system of open intervals (the *stability intervals* of equation (31)).

Theorem 19. *If the operator*

$$H_{\mathrm{av}} = \frac{1}{T}\int_0^T H(s)\,ds$$

is uniformly positive ($H_{\mathrm{av}} \gg 0$), *then there exists a* $\delta > 0$ *such that all of the points of the "punctured" neighborhood* $-\delta < \lambda < \delta$, $\lambda \neq 0$, *are points of strong stability of equation* (31).

Proof. For $\lambda \neq 0$ we put

$$W_\lambda = \frac{U(T;\ \lambda) - I}{i\lambda T} = JH_{\mathrm{av}} + O(|\lambda|).$$

Clearly

$$JW_\lambda + W_\lambda^* J = 2H_{\mathrm{av}} + O(|\lambda|).$$

Since $H_{\mathrm{av}} \gg 0$ by assumption, for sufficiently small $|\lambda|$ we will have

$$JW_\lambda + W_\lambda^* J \gg 0.$$

On the other hand (see problem 1 on page 136), if the latter condition is satisfied, then the space $\mathfrak{H}$ decomposes into a direct sum of two definite subspaces $\mathfrak{L}_1$ and $\mathfrak{L}_2$ invariant with respect to W_λ and in addition the spectra $\sigma(W_\lambda \mid \mathfrak{L}_k)$ ($k = 1, 2$) are disjoint (they will lie on different sides of the real axis). It is obvious that $U(T;\ \lambda)\mathfrak{L}_k \subset \mathfrak{L}_k$ ($k = 1, 2$) and the spectra of the monodromy operator $U(T;\ \lambda)$ in $\mathfrak{L}_k$ ($k = 1, 2$) will be disjoint. Therefore the operator $U(T;\ \lambda)$ is strongly stable for the indicated λ.

The theorem is proved.

This theorem implies the existence of a maximal open interval $(\lambda_{-I}, \lambda_I)$ that contains the origin and all other points which are points of strong stability of equation (31).

The interval $(\lambda_{-I}, \lambda_I)$ is called the *central zone of stability* of equation (31).

7. Theorem on the central zone of a canonical equation with a Hamiltonian of positive type. A Hamiltonian $H(t) = H(t + T)$ is said to be a Hamiltonian of *positive type* if

$$H(t) \geqslant 0 \quad (0 \leqslant t \leqslant T) \text{ and } H_{av} \gg 0. \tag{34}$$

We consider the boundary problem

$$\frac{dx}{dt} = i\lambda JH(t)x, \quad x(0) + x(T) = 0. \tag{35}$$

In a finite-dimensional space $\mathfrak{H}$ its spectrum would be determined from the equation

$$\det(U(T;\ \lambda) + I) = 0.$$

By the spectrum of problem (35) we mean the closed set Ω of those λ for which the operator $U(T;\ \lambda) + I$ does not have a continuous inverse.

Since according to (33) the spectrum of $U(T;\ \lambda)$ is contained for sufficiently small $|\lambda|$ in a sufficiently small neighborhood of the point 1 (and hence $U(T;\ \lambda)$ has a continuous inverse) there exists a neighborhood of the origin that does not contain any points of Ω. Let λ_1 (λ_{-1}) denote the least positive (negative) number in absolute value of Ω.

Theorem 20. *All of the solutions of a periodic canonical equation* (35) *with a Hamiltonian of positive type are stably bounded whenever*

$$\lambda_{-1} < \lambda < \lambda_1, \quad \lambda \neq 0. \tag{36}$$

Thus $\lambda_{-l} \leq \lambda_{-1} < 0 < \lambda_1 \leq \lambda_l$, where $(\lambda_{-l},\ \lambda_l)$ is the central zone of stability of equation (31).

Proof. For a λ from the interval (36) the operator $(U(T;\ \lambda) + I)^{-1}$ $(\in \mathfrak{R})$ exists, and hence also the operator

$$V(\lambda) = -i\,[U(T;\ \lambda) - I]\,[U(T;\ \lambda) + I]^{-1}.$$

If we can show that the operator $JV(\lambda)$ is uniformly positive for the values of λ in question, then we will obtain as a direct consequence (see the proof of Theorem 19) the strong stability of $U(T;\ \lambda)$ and thereby the proof of the theorem.

To this end we consider the derivative

$$\frac{dV}{d\lambda} = -2i\,[U(T;\ \lambda) + I]^{-1}\,\frac{d\dot{U}(T;\ \lambda)}{d\lambda}\,[U(T;\ \lambda) + I]^{-1}. \tag{37}$$

To calculate the derivative $dU(T;\ \lambda)/\,d\lambda$ we differentiate the equalities of (32) with respect to λ. Putting $\dot{U}(t;\ \lambda) = dU(t;\ \lambda)/\,d\lambda$,* we obtain

$$\begin{cases} \dfrac{d\dot{U}(t;\ \lambda)}{dt} = i\lambda JH(t)\,\dot{U}(t;\ \lambda) + iJH(t)\,U(t;\ \lambda); \\ \dot{U}(0;\ \lambda) = 0. \end{cases}$$

* *Translator's note.* The Russian text has dt.

From here we find that

$$\dot{U}(t;\ \lambda) = i\lambda \int_0^t U(t;\ \lambda) U^{-1}(s;\ \lambda) JH(s) U(s;\ \lambda)\, ds$$
$$= i\lambda U(t;\ \lambda) J \int_0^t U^*(s;\ \lambda) H(s) U(s;\ \lambda)\, ds.$$

Thus

$$\frac{dU(T;\ \lambda)}{d\lambda} = i\lambda U(T;\ \lambda) JR(\lambda),$$

where

$$R(\lambda) = \int_0^T U^*(s;\ \lambda) H(s) U(s;\ \lambda)\, ds. \tag{38}$$

Inserting the expression obtained for $U(T;\ \lambda)$ into (37) and taking into account that

$$(U + I)^{-1} UJ = (U^{-1} + I)^{-1} J = J(JU^{-1}J + I)^{-1}$$
$$= J(U^* + I)^{-1},$$

we get

$$\frac{dV}{d\lambda} = 2\lambda J [U^*(T;\ \lambda) + I]^{-1} R(\lambda) [U(T;\ \lambda) + I]^{-1}.$$

Inasmuch as $V(0) = 0$,

$$JV(\lambda) = 2\int_0^\lambda \mu [U^*(T;\ \mu) + I]^{-1} R(\mu) [U(T;\ \mu) + I]^{-1} d\mu.$$

The operator under the integral sign is nonnegative (nonpositive) for $\mu > 0$ $(\mu < 0)$ and is uniformly positive (negative) for $\mu > 0$ $(\mu < 0)$ and sufficiently close to zero. For it follows from (34) and (38) that $R(\mu) \geq 0$; and if μ $(\neq 0)$ is sufficiently close to zero then $U(s;\ \mu)$ is uniformly (in $s \in (0,\ T)$) sufficiently close to I while $R(\mu)$ is sufficiently close to the operator $TH_{av} \gg 0$ and is therefore also $\gg 0$.

Thus $JV(\lambda) \gg 0$ for $\lambda_{-1} < \lambda < \lambda_1$, $\lambda \neq 0$, and the theorem is proved.

Remark. Theorem 20 admits both a generalization, which can be proved with the use of analogous arguments, and a sharpening that requires some minor additional considerations.

Let $\Omega(\theta)$ denote for any real θ the closed set of those λ for which the operator $U(T;\ \lambda) - e^{i\theta} I$ does not have a continuous inverse. The set $\Omega(\theta) = \Omega(\theta + 2\pi)$ can be interpreted as the spectrum of the boundary problem

$$\frac{dx}{dt} = i\lambda JH(t)x, \quad x(T) - e^{i\theta}x(0) = 0. \tag{39}$$

Let $\lambda_{-1}(\theta)$ and $\lambda_1(\theta)$ ($\theta \not\equiv 0 \pmod{2\pi}$) be the least in absolute value of the positive and negative numbers respectively of the spectrum $\Omega(\theta)$. Then, for a canonical equation (35) with a Hamiltonian of positive type, it is possible to assert in the generalization of (36) that

$$\lambda_{-I} \leqslant \lambda_{-1}(\theta) < 0 < \lambda_1(\theta) \leqslant \lambda_I.$$

Moreover, it turns out that

$$\lambda_I = \max_{0<\theta<2\pi} \lambda_1(\theta), \quad \lambda_{-I} = \min_{0<\theta<2\pi} \lambda_{-1}(\theta). \tag{40}$$

It can happen (see below) that in the Hilbert space $\mathfrak{H}$ there is defined an involution of complex conjugation ($f \longrightarrow \bar{f}$) with respect to which the operator $U(T;\lambda)$ ($-\infty < \lambda < \infty$) is real. In this case the spectrum of a J-unitary operator $U(T;\lambda)$ is arranged symmetrically not only with respect to the unit circle but also the real axis, and the functions $\lambda_{\pm}(\theta)$ will be even: $\lambda_k(-\theta) = \lambda_k(\theta)$ ($k = 1, 2,$). Moreover, in this case $\lambda_1(\theta)$ is nondecreasing while $\lambda_{-1}(\theta)$ is nonincreasing in the interval $0 < \theta \leq \pi$, so that

$$\begin{aligned} \lambda_1 &= \lambda_1(\pi) = \max_{0<\theta\leqslant\pi} \lambda_1(\theta) = \lambda_I, \\ \lambda_{-1} &= \lambda_{-1}(\pi) = \min_{0<\theta\leqslant\pi} \lambda_{-1}(\theta) = \lambda_{-I}. \end{aligned} \tag{41}$$

For the case $\dim \mathfrak{H} < \infty$ all of these facts are well known from the papers [18e, c, g] and [20].

The case of a real operator $U(T;\lambda)$ is encountered in considering a "classical Hamilton" system.

Suppose $\mathfrak{H}$ is a real space (an infinite-dimensional Hilbert space or a finite-dimensional Euclidean space of even dimension) and suppose it decomposes into an orthogonal sum $\mathfrak{H} = \mathfrak{H}_1 \oplus \mathfrak{H}_2$ of two subspaces of equal dimension: $\dim \mathfrak{H}_1 = \dim \mathfrak{H}_2$.

We consider in $\mathfrak{H}$ a Hamilton equation with real parameter

$$\frac{dx}{dt} = \lambda J_\Gamma H(t)x, \tag{42}$$

where $H(t) = H^*(t)$ is an operator function satisfying conditions 1) and 2) of subsection 1 and the operator J_Γ is defined with respect to the decomposition $\mathfrak{H} = \mathfrak{H}_1 \oplus \mathfrak{H}_2$ by the operator matrix

$$J_\Gamma = \begin{pmatrix} 0 & I \\ -I & 0 \end{pmatrix}.$$

Since $J_\Gamma^* = -J_\Gamma$, $J_\Gamma^2 = -I$, if we put $J_\Gamma = iJ$, we will have $J = J^*$, $J^2 = I$, so that equation (42) is a particular case of a canonical equation.

We now consider the resolving operator $U(t;\lambda)$ $(-\infty < \lambda < \infty)$ of equation (42) as an operator acting in the complex Hilbert space $\mathfrak{H}_c$ which is the complex hull of the space $\mathfrak{H}$. For any t and, in particular, for $t = T$ the spectrum of $U(t;\lambda)$ will be symmetrically distributed with respect to the real axis.

If the Hamiltonian $H(t) = H(t+T)$ is of positive type, we will have relation (41).

8. **Infinite-dimensional analog of Hill's equation.** We now consider the differential equation of second order in $\mathfrak{H}$

$$\frac{d^2y}{dt^2} + \mu P(t)\, y = 0, \tag{43}$$

where $P(t) = P(t+T)$ is a periodic operator function whose values are selfadjoint operators: $P(t) = P^*(t)$ $(\in \mathfrak{R})$.

For $\mu \in (-\infty, \infty)$, $\mu \neq 0$, the formal substitutions

$$\lambda = \sqrt{|\mu|}, \quad y_1 = \lambda y, \quad y_2 = \frac{dy}{dt}; \quad x = \begin{pmatrix} y_1 \\ y_2 \end{pmatrix} = y_1 \oplus y_2$$

transform equation (43) into a canonical equation in the double Hilbert space $\tilde{\mathfrak{H}} = \mathfrak{H} \oplus \mathfrak{H}$, viz.

$$\frac{dx}{dt} = \lambda \begin{pmatrix} 0 & I \\ -I & 0 \end{pmatrix} \begin{pmatrix} \pm P(t) & 0 \\ 0 & I \end{pmatrix} x, \tag{44}$$

which can also be written in the form of (42) with

$$H(t) = \begin{pmatrix} \pm P(t) & 0 \\ 0 & I \end{pmatrix}. \tag{45}$$

The + or − sign is placed in front of $P(t)$, depending on whether $\mu > 0$ $(\mu = \lambda^2)$ or < 0 $(\mu = -\lambda^2)$ respectively.

We see that if the operator function $P(t)$ is required to satisfy the conditions of measurability and integrability, 1) and 2) of subsection 1, then equation (43) can be regarded as an equation in the "weak sense" (see subsection 1).

Moreover, under these assumptions on the operator function $P(t)$ $(\equiv P(t+T) \equiv P^*(t))$ a number of results concerning the boundedness or stable boundedness of the solutions of equation (43) can be obtained as a direct consequence of the corresponding results for the solutions of canonical differential equations.

The manner in which the notion of stable boundedness of all of the solutions of the equation of second order

$$\frac{d^2y}{dt^2} + P(t)\,y = 0 \tag{46}$$

is introduced is completely analogous to the way it was done for canonical equations. Here, in place of $H(t)$ we vary the operator function $P(t)$ with preservation of the properties $P(t) = P(t+T) = P^*(t)$ (as well as properties 1)–2)), the smallness of the variation being determined by means of the metric

$$\rho(P, \widetilde{P}) = \int_0^T \| \widetilde{P}(t) - P(t) \| \, dt.$$

Clearly, if a point $\lambda = \lambda_0$ is a point of strong stability of (44), then the corresponding point $\mu_0 = \pm\lambda_0^2$ will be a point of strong stability of (43) (i.e. all of the solutions of (43) will be stably bounded for $\mu = \mu_0$).

Such a theorem, by virtue of the indicated transformations, follows directly from Theorem 19.

Theorem 21. *If the coefficient* $P(t) = P(t+T)$ *in equation* (43) *is an operator function of positive type* ($P(t) \geq 0$, $P_{av} \gg 0$), *then there exists a* μ_1 *such that all of the solutions of equation* (43) *will be stably bounded for* $0 < \mu < \mu_1$.

In fact, the $+$ sign belongs in front of $P(t)$ in (44) and (45) for $\mu > 0$, and hence, besides $P(t)$, the Hamiltonian (45) will also be of positive type.

In the finite-dimensional case ($\dim \mathfrak{H} < \infty$) the author showed [18g] that one can take as μ_1 the first positive eigenvalue of the boundary problem

$$\frac{d^2y}{dt^2} + \mu P(t)\,y = 0, \;\; y(0) + y(T) = 0, \;\; y'(0) + y'(T) = 0. \tag{47}$$

When the matrix function defining $P(t)$ is real, this result is "exact" (the indicated value for μ_1 is maximal). In the scalar case this exact result was first obtained in 1899 by A. M. Ljapunov [28, pages 401 and 409].

Apparently, an analogous result also holds in the infinite-dimensional case.

In every case, when $P(t) \geq 0$ the corresponding result follows directly from Theorem 20.

Theorem 22. *All of the solutions of equation* (43) *with coefficient* $P(t) = P(t+T)$ *of positive type are stably bounded as long as* $0 < \mu < \mu_1$, *where* μ_1 *is the least positive number of the spectrum of the boundary problem* (47).

By the spectrum of the boundary problem (47) we mean the spectrum of the boundary problem (35) with Hamiltonian (45)* (with the + sign in front of $P(t)$ and $J = -iJ_\Gamma$).

9. **Generalization of Ljapunov's test.** In 1892 A. M. Ljapunov [28, page 203], while studying the scalar Hill equation $y'' + p(t)y = 0$ ($p(t) = p(t+T) \geq 0$, $p_{av} > 0$), showed that its solutions are stably bounded whenever

$$T\int_0^T p(s)\,ds < 4.$$

Theorem 22 permits one to obtain an analogous result for the vector equations (46) in an n-dimensional or Hilbert space $\mathfrak{H}$.

Theorem 23. *All of the solutions of the equation*

$$\frac{d^2y}{dt^2} + P(t)\,y = 0 \tag{43}$$

with a coefficient $P(t)$ *of positive type are stably bounded whenever*

$$I - \frac{T}{4}\int_0^T P(s)\,ds \gg 0. \tag{48}$$

Proof. Condition (48) is equivalent to the inequality

$$T\rho < 4, \tag{49}$$

where ρ is the upper point of the spectrum of the operator TP_{av}. Since the latter operator is positive, the number ρ is also its norm:

$$\rho = \left\| \int_0^T P(s)\,ds \right\| = \| TP_{av} \|. \tag{50}$$

If we show that the number μ_1 from Theorem 22 satisfies the estimate

$$\mu_1 > \frac{4}{T_\rho}, \tag{51}$$

then (49) will accordingly imply that $\mu = 1 < \mu_1$ and the validity of Theorem 23 will follow from Theorem 22.

* *Translator's note.* The Russian text has (4).

We consider the scalar boundary problem

$$-\frac{d^2y}{dt^2}=f(t),\quad y(0)+y(T)=y'(0)+y'(T)=0, \tag{52}$$

where $f(t)\in L_1(0, T)$.

This problem always has a unique solution which together with its derivative is absolutely continuous and is given by the formula

$$y(t)=\int_0^T g(t-s)f(s)\,ds \quad (0\leqslant t\leqslant T),$$

where $g(t-s)$ is the Green's function for problem (52):

$$g(t)=\frac{T}{\pi^2}\sum_{-\infty}^{\infty}\frac{1}{(2k+1)^2}\exp\left(\frac{i(2j+1)\pi t}{T}\right)=-\frac{|t|}{2}+\frac{T}{4}, \tag{53}$$
$$(-T<t<T).$$

In the finite-dimensional case, the boundary problem (47) is clearly equivalent to the weighted integral equation

$$y(t)=\mu\int_0^T g(t-s)P(s)y(s)\,ds, \tag{54}$$

considered in the class $C_P(\mathfrak{H})$ of all continuous vector functions $y(t)$ $(0\leq t\leq T)$ with values in $\mathfrak{H}$.

The class $C_P(\mathfrak{H})$ will be regarded as a pre-Hilbert space with scalar product

$$[y_1, y_2]_P=\int_0^T (P(s)y_1(s), y_2(s))\,ds \quad (y_1,\ y_2\in C_P(\mathfrak{H})). \tag{55}$$

One can easily see that the integral operator K of equation (54)

$$(Ky)(t)=\int_0^T g(t-s)P(s)y(s)\,ds \quad (y\in C_P(\mathfrak{H})) \tag{56}$$

transforms $C_P(\mathfrak{H})$ into itself and is symmetric, nonnegative and completely continuous on $C_P(\mathfrak{H})$.[1)]

---

1) In essence, $C_P(\mathfrak{H})$ is a space with two norms: $\|y\|_P=[y, y]_P^{1/2}$ and $\|y\|_B=\sup_{0\leq t\leq T}\|y(t)\|$. It is a Banach space with respect to the second one. One can easily verify that K is a completely continuous operator with respect to the norm $\|\cdot\|_B$, and this implies, on the basis of a general theorem of the author's [18a],

The situation is somewhat more complicated when $\mathfrak{H}$ is infinite dimensional. It is easy to see that the definition of a pre-Hilbert space $C_P(\mathfrak{H})$ with scalar product (55) and formula (56) for the linear operator K transforming $C_P(\mathfrak{H})$ into itself still makes sense in this case.

But K will no longer be completely continuous in this case, although it will be bounded and we will presently obtain an estimate for its norm.

By making use of the expansion (53), we can formally write

$$Ky = \sum_{j=-\infty}^{\infty} c_j K_j y, \tag{57}$$

where

$$(K_j y)(t) = \exp\left(\frac{i(2j+1)\pi t}{T}\right) \int_0^T \exp\left(-\frac{i(2j+1)\pi s}{T}\right) P(s)\, y(s)\, ds$$
$$(j = 0, \pm 1, \pm 2, \ldots),$$

and

$$c_j = \frac{T}{\pi^2} \sum_{-\infty}^{\infty} \frac{1}{(2j+1)^2} \quad (j = 0, \pm 1, \pm 2, \ldots).$$

We note that

$$\sum_{-\infty}^{\infty} c_j = g(0) = \frac{T}{4}. \tag{58}$$

We define unitary operators V_j $(j = 0, \pm 1, \pm 2, \cdots)$ in $\mathfrak{H}_P$ by putting

$$(V_j y)(t) = \exp\left(-\frac{i(2j+1)\pi t}{T}\right) y(t) \quad (y \in C_P(\mathfrak{H}),$$
$$j = 0, \pm 1, \pm 2, \ldots).$$

Clearly,

$$K_j = V_j K_0 V_j^{-1} \quad (j = 0, \pm 1, \pm 2, \ldots)$$

and therefore

$$\| K_j \| = \| K_0 \| \quad (j = 0, \pm 1, \pm 2, \ldots). \tag{59}$$

Since

---

the complete continuity of K with respect to the norm $\|\cdot\|_P$.

By means of the same general theorem, we can conclude in the case $\dim \mathfrak{H} = \infty$ that the boundedness of K with respect to $\|\cdot\|_B$ implies its boundedness with respect to $\|\cdot\|_P$.

$$(K_0 y)(t) = \int_0^T P(s)\, y(s)\, ds \quad (y \in C_P(\mathfrak{H})),$$

the operator K_0 is symmetric and nonnegative, and it maps $C_P(\mathfrak{H})$ into the space of vector functions taking a constant value from $\mathfrak{H}$.

Since any symmetric operator A with the property $\mathfrak{R}(A) \subset \mathfrak{D}(A)$ has norm $\|A\| = \sup\,[(Af, f)/(f, f)]$, where f ranges over $\mathfrak{D}(A)$, in the present case

$$\|K_0\| = \sup_{f \in \mathfrak{H}} \{[K_0 f, f]_P / [f, f]_P\}. \tag{60}$$

By taking into account that for a function from $C_P(\mathfrak{H})$ taking a constant value $f \in \mathfrak{H}$ we have

$$[f, f]_P = \int_0^T (P(s) f, f)\, ds = T\,(P_{\mathrm{av}} f, f),$$

while

$$[K_0 f, f]_P = T\,[P_{\mathrm{av}} f, f]_P = T^2 (P_{\mathrm{av}}^2 f, f),$$

we get, according to (60), that

$$\|K_0\| = T \sup_{f \in \mathfrak{H}} [(P_{\mathrm{av}}^2 f, f)/(P_{\mathrm{av}} f, f)] = T\, \|P_{\mathrm{av}}\| = \rho.$$

It is not difficult to conclude from this and from (58) and (59) that the expansion (57) always converges in the sense of the metric of the scalar product (55), and by virtue of this expansion and (50)

$$\|K\| \leqslant \sum_{j=-\infty}^{\infty} c_j \|K_j\| = \rho \sum_{-\infty}^{\infty} c_j = \rho \frac{T}{4}. \tag{61}$$

We now consider a nonhomogeneous integral equation, which we write in operator form: $y = \lambda K y - K f$ where f is an arbitrarily given vector function from $C_P(\mathfrak{H})$ and y is an unknown vector function from $C_P(\mathfrak{H})$.

This integral equation is equivalent to the boundary problem

$$\frac{d^2 y}{dt^2} + \mu P(t)\, y = P(t) f; \quad y(0) + y(T) = y'(0) + y'(T) = 0.$$

When μ does not belong to the spectrum of problem (47) (in particular, when $0 < \mu < \mu_1$), one can easily obtain in explicit form [1] the unique solution of this boundary problem, which in operator notation can be written in the form

$$y = (I - \mu K)^{-1} Kf = -\frac{1}{\mu} f + \frac{1}{\mu^2}\left(\frac{1}{\mu} - K\right)^{-1} f.$$

It can be seen from the explicit expression for this solution that the point μ_1 is a singular point of the solution y as a function of μ.

Since the operator K is nonnegative, it follows that $1/\mu_1$ is the least upper bound of the spectrum of K, i.e. $\mu_1 = 1/\|K\|$.

Recalling estimate (61), we get, finally, inequality (51).

The theorem is proved.

In conclusion we mention that a large part of the above results for the finite-dimensional case were obtained in the author's papers [18e, c, h] (see also [6, 37a, 9c]).

The first investigations of canonical systems in an infinite-dimensional Hilbert space were made by V. I. Derguzov [11]. He investigated from the point of view of stability certain classes of canonical equations with a Hamiltonian $H(t)$ whose values are unbounded nonselfadjoint operators. It is to just such canonical systems that particular problems on the parametric resonance of elastic systems with an infinite number of degrees of freedom reduce.

The papers of V. I. Derguzov gave rise to an elucidation by the author of the possibility of transferring the methods available to him (some of them had not been published) to the infinite-dimensional case. [2] The above propositions were also a result of this.

The contents of this lecture, in a less complete and polished form, were presented in Chapter VI of the mimeographed "Lectures" of the author [18h].

---

1) An expression for $y(t)$ in explicit form will involve the operator $(U(T; \lambda)+I)^{-1}$ $(\lambda = \sqrt{\mu})$, where $U(t; \lambda)$ is the resolving operator for the canonical equation (35) with Hamiltonian $H(t)$ from (45) (with the + sign in front of $P(t)$).

2) P. Jonas partially performed such a transfer independently of V. I. Derguzov and the author in his thesis work (under the supervision of H. Langer) [15].

Addenda and problems

1. Generalizing the definition of A. M. Ljapunov, we will say that equation (1) with a periodic coefficient $A(t) = A(t+T)$ is reducible if there exists a continuous weakly differentiable periodic operator function $Q(t) = Q(t+T)$ such that the substitution $x = Q(t)y$ transforms equation (1) into the equation

$$\frac{dy}{dt} = \Gamma y$$

with a constant operator $\Gamma\,(\in \mathfrak{R})$.

Show that a periodic equation (1) is reducible if the spectrum of its monodromy operator $U(T)$ does not enclose the origin [1] (i.e. the origin can be connected with the point at infinity by a continuous line which does not intersect the spectrum of the monodromy operator) [18h].

Show that a J-selfadjoint operator can be obtained as Γ if equation (1) is canonical $(A(t) = JH(t))$ [37b].

Hint. As Γ one can choose the product of T^{-1} and any $\ln U(T)$, for example, the one obtained by the formula

$$\ln U(T) = -\frac{1}{2\pi i}\oint_{\gamma} \ln\zeta\,(U(T) - \zeta I)^{-1}\,d\zeta,$$

where γ is a simple or complex contour enclosing the spectrum of $U(T)$ but not the origin.

In the case of a canonical equation the contour γ should be chosen so that it goes over into itself under inversion with respect to the unit circle.

2. Suppose that in the equation for an n-dimensional vector function y

$$\frac{d^2y}{dt^2} + \mu P(t)\,y = 0 \quad (P(t) = P(t+T) = P^*(t)) \tag{$*$}$$

the coefficient $P(t)$ satisfies the following conditions:

$$\int_0^T P(s)\,ds = 0$$

and for any $f \in \mathfrak{H}$ $(f \neq 0)$

$$\int_0^T \| P(s) f \|\,ds > 0.$$

Then all of the solutions of equation $(*)$ are stably bounded whenever $\mu_{-1} < \mu < \mu_1$, $\mu \neq 0$, where μ_1 and μ_{-1} are the least in absolute value of the positive and negative numbers respectively of the spectrum of problem (47).

---

1) This condition is obviously always satisfied in the finite-dimensional case. The reducibility of a periodic equation (1) in a finite-dimensional $\mathfrak{H}$ was established by A. M. Ljapunov (see [28], pages 194–195).

This result was established for a scalar equation by A. M. Ljapunov (see [28], pages 404–406), and for a vector equation in a finite-dimensional space by the author [18f].

Hint. Transform equation (*) to the canonical form (35) by means of the substitutions

$$y_1 = y,\ y_2 = \frac{1}{\mu}\frac{dy}{dt} + Q(t)\,y,$$

where

$$Q(t) = \int_0^t P(s)\,ds.$$

It may be necessary to add a bit more in order to generalize this result to the case of equation (*) in an infinite-dimensional Hilbert space $\mathfrak{H}$.

In completing the lectures it should be noted that for lack of time, and perhaps for lack of suitable skill in "compacting" his material, the author was unable to present all of the, even at the present time, motley gamut of problems and methods of the foregoing theory. Many of the more recent achievements in this theory were ignored. In particular, the lectures in no way reflect the investigations of Langer and the author [19a] in the theory of spectral functions of J-selfadjoint operators, the series of recent investigations of M. A. Naĭmark in the theory of algebras of operators in Π_κ spaces (see, for example, [29a, b]), the results in the theory of spectral representations of indefinite functions and kernels [18g, 14b, 31] and others.

The lectures also in no way reflect the applications of the theory of J-selfadjoint operators to questions in quantum field theory (for mathematical references on this subject see [3]).

A self-contained discussion of all of these questions would require a separate monograph.

BIBLIOGRAPHY

[1] N. G. Askerov, S. G. Kreĭn and G. I. Laptev, *On a class of non-selfadjoint boundary-value problems*, Dokl. Akad. Nauk SSSR 155 (1964), 499–502 = Soviet Math. Dokl. 5 (1964), 424–427. MR 28 #3347.

[2] N. I. Ahiezer and I. M. Glazman, *The theory of linear operators in Hilbert space*, GITTL, Moscow, 1950; English transl., Ungar, New York, 1961. MR 13, 358.

[3] F. A. Berezin, *On the Lee model*, Mat. Sb. 60 (102) (1963), 425–446; English transl., Amer. Math. Soc. Transl. (2) 56 (1966), 249–272. MR 27 #3306.

[4] F. F. Bonsall, *Indefinitely isometric linear operators in a reflexive Banach space*, Quart. J. Math. Oxford Ser. (2) 6 (1955), 179–187. MR 17, 512.

[5] M. L. Brodskiĭ, *On properties of an operator mapping the nonnegative part of a space with indefinite metric into itself,* Uspehi Mat. Nauk 14 (1959), no. 1 (85), 147–152. (Russian) MR 21 #5145.

[6] I. M. Gel'fand and V. B. Lidskiĭ, *On the structure of the regions of stability of linear canonical systems of differential equations with periodic coefficients,* Uspehi Mat. Nauk 10 (1955), no. 1 (63), 3–40; English transl., Amer. Math. Soc. Transl. (2) 8 (1958), 143–181. MR 17, 482; 19, 960.

[7] Ju. P. Ginzburg, a) *On J-nonexpansive operators in Hilbert space,* Naučn. Zap. Fiz.-Mat. Fak. Odessa Gos. Ped. Inst. 22 (1958), no. 1, 13–20. (Russian)
b) *Projection in a Hilbert space with bilinear metric,* Dokl. Akad. Nauk SSSR 139 (1961), 775–778 = Soviet Math. Dokl. 2 (1961), 980–983. MR 25 #4336.
c) *On subspaces of a Hilbert space with an indefinite metric,* Naučn. Zap. Kaf. Mat. Fiz. Estest. Odessa Gos. Ped. Inst. 25 (1961), no. 2, 3–9. (Russian)

[8] Ju. P. Ginzburg and I. S. Iohvidov, *A study of the geometry of infinite-dimensional spaces with bilinear metric,* Uspehi Mat. Nauk 17 (1962), no. 4 (106), 3–56 = Russian Math. Surveys 17 (1962), no. 4, 1–51. MR 26 #2850.

[9] I. C. Gohberg and M. G. Kreĭn, a) *The basic propositions on defect numbers, root numbers and indices of linear operators,* Uspehi Mat. Nauk 12 (1957), no. 2 (74), 43–118; English transl., Amer. Math. Soc. Transl. (2) 13 (1960), 185–264. MR 20 #3459; 22 #3984.
b) *Transformations of kernels of integral equations and their influence on the spectra of these equations,* Ukrain. Mat. Ž. 13 (1961), no. 3, 12–38; English transl., Amer. Math. Soc. Transl. (2) 35 (1964), 263–295. MR 27 #1788.
c) *Introduction to the theory of linear non-selfadjoint operators in Hilbert space,* "Nauka", Moscow, 1965; English transl., Transl. Math. Monographs, vol. 18, Amer. Math. Soc., Providence, R.I., 1969. MR 36 #3137.

[10] N. Dunford and J. T. Schwartz, *Linear operators.* I: *General theory,* Pure and Appl. Math., vol. 7, Interscience, New York, 1958; Russian transl., IL, Moscow, 1962. MR 22 #8302.

[11] V. I. Derguzov, a) *On the stability of the solutions of the Hamilton equations with unbounded periodic operator coefficients,* Mat. Sb. 63 (105) (1964), 591–619. (Russian) MR 29 #325.
b) *On the stability of the solutions of Hamiltonian equations in Hilbert space with unbounded periodic operator coefficients,* Dokl. Akad. Nauk SSSR 152 (1963), 1294–1296 = Soviet Math. Dokl. 4 (1963), 1543–1546. MR 28 #314.

[12] R. J. Duffin, *A minimax theory for overdamped networks*, J. Rational Mech. Anal. 4 (1955), 221–233. MR 16, 979.

[13] I. S. Iohvidov, a) *Unitary operators in a space with an indefinite metric*, Har'kov. Gos. Univ. Uč. Zap. 29 = Zap. Mat. Otd. Fiz.-Mat. Fak. i Har'kov. Mat. Obšč. (4) 21 (1949), 79–86. (Russian) MR 18, 405.
b) *Unitary and selfadjoint operators in a space with an indefinite metric*, Candidate's Dissertation, Odessa, 1950. (Russian)
c) *Boundedness of J-isometric operators*, Uspehi Mat. Nauk 16 (1961), no. 4 (100), 167–170; English transl., Amer. Math. Soc. Transl. (2) 47 (1965), 67–71. MR 24 #A1628.
d) *On some classes of operators in a space with a general indefinite metric*, Trudy Fifth Conf. Functional Analysis and Probability, Izdat. Akad. Nauk Azerbaĭdžan. SSR, Baku, 1961. (Russian) MR 26 #6790.
e) *Regular and projectively complete linear manifolds in spaces with a general Hermitian bilinear metric*, Dokl. Akad. Nauk SSSR 139 (1961), 791–794 = Soviet Math. Dokl. 2 (1961), 994–997. MR 25 #4337.

[14] I. S. Iohvidov and M. G. Kreĭn, a) *Spectral theory of operators in spaces with an indefinite metric*. I, Trudy Moskov. Mat. Obšč. 5 (1956), 367–432; English transl., Amer. Math. Soc. Transl. (2) 13 (1960), 105–175. MR 18, 320; 22 #3983.
b) *Spectral theory of operators in spaces with an indefinite metric*. II, Trudy Moskov. Mat. Obšč.8 (1959), 413–496; English transl., Amer. Math. Soc. Transl. (2) 34 (1963), 283–373. MR 21 #6543.

[15] P. Jonas, *Einige Betrachtungen zur Stabilität kanonischer Differentialgleichungen im Hilbertraum*, Dissertation, Technische Universität Dresden, 1964.

[16] M. V. Keldyš, *On the characteristic values and characteristic functions of certain classes of non-selfadjoint equations*, Dokl. Akad. Nauk SSSR 77 (1951), 11–14. (Russian) MR 12, 835.

[17] M. V. Keldyš and V. B. Lidskiĭ, *On the spectral theory of non-selfadjoint operators*, Proc. Fourth All-Union Math. Congr. (Leningrad, 1961) vol. 1, Izdat. Akad. Nauk SSSR, Leningrad, 1963, pp. 101–120. (Russian) MR 30 #1414.

[18] M. G. Kreĭn, a) *On completely continuous linear operators in function spaces with two norms*, Ž. Inst. Mat. Acad. Nauk Ukrain. SSR 9 (1947), 104–129. (Ukrainian)
b) *On some problems connected with the circle of ideas of Ljapunov in stability theory*, Uspehi Mat. Nauk 3 (1948), no. 3 (25), 166–169. (Russian) MR 10, 128.
c) *A generalization of some investigations of A. M. Ljapunov on linear differential equations with periodic coefficients*, Dokl. Akad. Nauk SSSR 73 (1950), 445–448. (Russian) MR 12, 100.

d) *On an application of the fixed-point principle in the theory of linear transformations of spaces with an indefinite metric,* Uspehi Mat. Nauk 5 (1950), no. 2 (36), 180–190; English transl., Amer. Math. Soc. Transl. (2) 1 (1955), 27–35. MR 14, 56; 17, 514.

e) *The basic propositions of the theory of λ-zones of stability of a canonical system of linear differential equations with periodic coefficients,* In Memory of A. A. Andronov, Izdat. Akad. Nauk SSSR, Moscow, 1955, pp. 413–498. (Russian) MR 17, 738.

f) *On criteria of stable boundedness of solutions of periodic canonical systems,* Prikl. Mat. Meh. 19 (1955), 641–680. (Russian) MR 17, 1088.

g) *Integral representation of a continuous Hermitian-indefinite function with a finite number of negative squares,* Dokl. Akad. Nauk SSSR 125 (1959), 31–34. (Russian) MR 22 #877.

h) *Lectures on the theory of stability of solutions of differential equations in a Banach space,* Izdat. Akad. Nauk Ukrain. SSR, Kiev, 1964. [1] (Russian) MR 32 #1428.

i) *A new application of the fixed-point principle in the theory of operators in a space with an indefinite metric,* Dokl. Akad. Nauk SSSR 154 (1964), 1023–1026 = Soviet Math. Dokl. 5 (1964), 224–228. MR 29 #6314.

[19] M. G. Kreĭn and Heinz Langer, a) *On the spectral function of a self-adjoint operator in a space with indefinite metric,* Dokl. Akad. Nauk SSSR 152 (1963), 39–42 = Soviet Math. Dokl. 4 (1963), 1236–1239. MR 29 #2650.

b) *On the theory of quadratic pencils of self-adjoint operators,* Dokl. Akad. Nauk SSSR 154 (1964), 1258–1261 = Soviet Math. Dokl. 5 (1964), 266–269. MR 29 #6315.

c) *Certain mathematical principles of the linear theory of damped vibrations of continua,* Proc. Internat. Sympos. Appl. Theory of Functions in Continuum Mechanics (Tbilisi, 1963) vol. II: Fluid and Gas Mechanics, Math. Methods, "Nauka", Moscow, 1965, pp. 283–322. (Russian) MR 33 #3493.

[20] M. G. Kreĭn and G. Ja. Ljubarskiĭ, *Analytic properties of the multipliers of periodic canonical differential systems of positive type,* Izv. Akad. Nauk SSSR Ser. Mat. 26 (1962), 549–572; English transl., Amer. Math. Soc. Transl. (2) 89 (1969), 1–28. MR 26 #400.

[21] M. G. Kreĭn and M. A. Rutman, *Linear operators leaving invariant a cone in a Banach space,* Uspehi Mat. Nauk 3 (1948), no. 1 (23), 3–95; English transl., Amer. Math. Soc. Transl. (1) 10 (1962), 199–325. MR 10, 256; 12, 341.

---

1) *Editor's note.* The subject matter of this monograph will be updated in a forthcoming book by M. G. Kreĭn and Ju. L. Daleckiĭ, *Stability of solutions of differential equations in Banach space,* "Nauka", Moscow, late 1970. (Russian)

[22] M. G. Kreĭn and V. A. Jakubovič, *Hamiltonian systems of linear differential equations with periodic coefficients. Analytic methods in the theory of non-linear vibrations*, Proc. Internat. Sympos. Non-Linear Vibrations (1961) vol. 1, Izdat. Akad. Nauk Ukrain. SSR, Kiev, 1963, pp. 277–305. (Russian) MR 28 #312.

[23] S. G. Kreĭn, *Oscillations of a viscous fluid in a container*, Dokl. Akad. Nauk SSSR 159 (1964), 262–265 = Soviet Math. Dokl. 5 (1964), 1467–1471. MR 31 #6461.

[24] D. L. Kučer, a) *On the stability and boundedness of the solutions of certain types of differential equations in normed linear spaces*, Candidate's Dissertation, Odessa, 1950. (Russian)
b) *Some questions on the integration of normal differential equations in Banach spaces*, Naučn. Zap. Kaf. Mat. Fiz. Estest. Odessa Gos. Ped. Inst. 25 (1961), no. 2, 25–37. (Russian)

[25] R. Kühne, *Eine Klasse J-selbstadjungierter Operatoren und stark gedämpfte Scharen*, Dissertation, Dresden, 1964.

[26] G. I. Laptev, *Boundary problems for differential equations of second order in a Banach space and their applications*, Candidate's Dissertation, Voronezh, 1964. (Russian)

[27] H. Langer, a) *Zur Spektraltheorie J-selbstadjungierter Operatoren*, Math. Ann. 146 (1962), 60–85. MR 25 #1450.
b) *Eine Verallgemeinerung eines Satzes von L. S. Pontrjagin*, Math. Ann. 152 (1963), 434–436. MR 28 #1492.
c) *Spektraltheorie linearer Operatoren in J-Räumen und einige Anwendungen auf die Schar* $L(\lambda) = \lambda^2 I + \lambda B + C$, Wiss. Z. Techn. Univ. Dresden (to appear).
d) *On maximal dual pairs of invariant subspaces of J-selfadjoint operators*, Mat. Zametki 7 (1970), 443–447. (Russian)

[28] A. M. Ljapunov, *Collected works*. Vol. II, Izdat. Akad. Nauk SSSR, Moscow, 1956. (Russian) MR 27 #5980.

[29] M. A. Naĭmark, a) *Unitary representations of solvable groups in spaces with an indefinite metric*, Izv. Akad. Nauk SSSR Ser. Mat. 27 (1963), 1181–1185; English transl., Amer. Math. Soc. Transl. (2) 49 (1965), 86–91. MR 27 #5866.
b) *Commutative algebras of operators in the space* π_1, Rev. Roumaine Math. Pures Appl. 9 (1964), 499–528. (Russian) MR 31 #6134.

[30] R. Nevanlinna, *Über metrische lineare Räume. IV: Zur Theorie der Unterräume*, Ann. Acad. Sci. Fenn. Ser. A. I. Math.-Phys. No. 163 (1954). MR 15, 717.

[31] V. I. Pljuščeva, a) *An integral representation for Hermite indefinite matrices with k-negative squares*, Ukrain. Mat. Ž. 14 (1962), 30–39. (Russian) MR 25 #3946.

b) *An integral representation of continuous Hermite-indefinite kernels*, Dokl. Akad. Nauk SSSR 145 (1962), 534–537 = Soviet Math. Dokl. 3 (1962), 1061–1064. MR 25 #2414.

c) V. I. Gorbačuk, [1] *An integral representation of Hermitian indefinite kernels (the case of several variables)*, Ukrain. Mat. Ž. 16 (1964), 232–236. (Russian) MR 29 #3849.

[32] L. S. Pontrjagin, *Hermitian operators in spaces with indefinite metric*, Izv. Akad. Nauk SSSR Ser. Mat. 8 (1944), 243–280. (Russian) MR 6, 273.

[33] V. P. Potapov, *The multiplicative structure of J-contractive matrix functions*, Trudy Moskov. Mat. Obšč. 4 (1955), 125–236; English transl., Amer. Math. Soc. Transl. (2) 15 (1960), 131–243. MR 17, 958; 22 #5733.

[34] E. Scheibe, *Über hermitische Formen in topologischen Vektorräumen*, I, Ann. Acad. Sci. Fenn. Ser. A I No. 294 (1960). MR 25 #4339.

[35] Ky Fan, a) *Invariant subspaces of certain linear operators*, Bull. Amer. Math. Soc. 69 (1963), 773–777. MR 28 #1494.

b) *Invariant cross-sections and invariant linear subspaces*, Israel J. Math. 2 (1964), 19–26. MR 30 #1382.

[36] R. S. Phillips, a) *Dissipative operators and hyperbolic systems of partial differential equations*, Trans. Amer. Math. Soc. 90 (1959), 193–254; Russian transl., Matematiki 6 (1962), no. 4, 11–70. MR 21 #3669.

b) *The extension of dual subspaces invariant under an algebra*, Proc. Internat. Sympos. Linear Spaces (Jerusalem, 1960) Jerusalem Academic Press, Jerusalem and Pergamon Press, Oxford, 1961; Russian transl., Matematika 8 (1964), no. 6, 81–108. MR 24 #A3512.

[37] V. A. Jakubovič, a) *Structure of the group of simplectic matrices and of the set of unstable canonical systems with periodic coefficients*, Mat. Sb. 44 (86) (1958), 313–352. (Russian) MR 20 #3325.

b) *The small parameter method for canonical systems with periodic coefficients*, Prikl. Mat. Meh. 23 (1959), 15–34 = J. Appl. Math. Mech. 23 (1959), 17–43. MR 21 #5774.

Translated by:

S. Smith

---

1) Née Pljuščeva

Amer. Math. Soc. Transl.
(2) Vol. 93, 1970

ON SOME NEW BANACH ALGEBRAS AND WIENER-LÉVY TYPE THEOREMS FOR FOURIER SERIES AND INTEGRALS*

M. G. KREĬN

Introduction

1. In the fall of 1964 the author discussed with V. M. Adamjan, a participant of the author's seminar, the conditions under which an infinite Hankel matrix $\|c_{j+k}\|_0^\infty$ defines a nuclear operator in the space l_2. In the course of the discussion Adamjan observed that there is a quite simple formulation of a necessary and sufficient condition for such a matrix to define a Hilbert-Schmidt operator, viz.

$$\sum_0^\infty (n+1)|c_n|^2 < \infty.$$

In fact, the infinite sum here coincides with the sum $\sum_{j,k=0}^\infty |c_{j+k}|^2$. This simple observation will be essentially used in establishing Theorems I–III below.

Let $M = M(0, 2\pi)$ be the commutative Banach algebra of all measurable complex functions $f(t)$ $(0 \le t \le 2\pi)$ with the usual definition of the product and norm

$$\|f\| = \operatorname{ess} \sup_{0 \le t \le 2\pi} |f(t)|. \tag{0.1}$$

Let $C = C\{0, 2\pi\}$ denote the subalgebra of all continuous functions $f(t)$ on $(0, 2\pi)$ such that $f(0) = f(2\pi)$.

For $f \in M$ we will let $\sigma(f)$ denote the *value spectrum* of f, i.e. the complement in the complex plane C^1 of the set $\rho(f)$ of all points $z \in C^1$ for which $(f(t) - z)^{-1} \in M$.

Let

$$c_n(f) = \frac{1}{2\pi} \int_0^{2\pi} e^{-int} f(t)\, dt \quad (n = 0, \pm 1, \pm 2, \cdots), \tag{0.2}$$

---

* Translation of Mat. Issled. 1 (1966), no. 1, 82–109.

$$\mu_\pm(f) = \left\{ \sum_{n=1}^{\infty} n |c_{\pm n}(f)|^2 \right\}^{1/2} \tag{0.3}$$

and

$$\mu(f) = \left\{ \sum_{n=-\infty}^{\infty} |n|\, |c_n(f)|^2 \right\}^{1/2} (= (\mu_+^2(f) + \mu_-^2(f))^{1/2}). \tag{0.4}$$

We let M_+ (C_+) denote the set of all $f \in M$ $(f \in C)$ for which $\mu_+(f) < \infty$ and put

$$\|f\|_+ = \operatorname{ess} \sup_{0 \le t \le 2\pi} |f(t)| + \sum_{n=1}^{\infty} n |c_n(f)|^2. \tag{0.5}$$

The sets M_-, C_- and the norm $\|\cdot\|_-$ are defined analogously with the replacement everywhere of $\mu_+(f)$ by $\mu_-(f)$. And if in the definition of M_+, C_+ and $\|\cdot\|_+$ we replace $\mu_+(f)$ everywhere by simply $\mu(f)$, we obtain

$$M_\cap = M_+ \cap M_-, \quad C_\cap = C_+ \cap C_- \tag{0.6}$$

with the definition of the norm in $M_\cap$ being

$$\|f\|_\cap = \operatorname{ess} \sup_{0 \le t \le 2\pi} |f(t)| + \mu(f) \quad (f \in M_\cap). \tag{0.7}$$

It is easily seen that $C_\pm$ and $C_\cap$ are Banach spaces; at the same time they are elementwise linear manifolds in the space C that are dense in this space with respect to its norm.

It remains to note that the supply of real functions in each of the three spaces $M_\pm$, $M_\cap$ $(C_\pm, C_\cap)$ is one and the same.

The following proposition holds.

Theorem I. *Let $\mathfrak{A}$ be one of the Banach spaces $M_\pm$, $M_\cap$, $C_\pm$, $C_\cap$, and let $\|\cdot\|_{\mathfrak{A}}$ be the corresponding norm. Then $\mathfrak{A}$ is a Banach algebra of functions with the usual multiplication:*

$$\|f_1 f_2\|_{\mathfrak{A}} \le \|f_1\|_{\mathfrak{A}} \|f_2\|_{\mathfrak{A}} \quad (f_1, f_2 \in \mathfrak{A}, \ \|1\|_{\mathfrak{A}} = 1).$$

This algebra has the property that together with any of its elements f it also contains the element $\Phi(f)$, where $\Phi(z)$ is any function that is holomorphic in a domain G (which is simply connected if $\mathfrak{A} = M_\pm$) containing the spectrum $\sigma(f)$.

This theorem will be obtained as a corollary of certain general theorems (Theorem 4 and Corollary 1 of Theorem 2), which can be established by pure operator methods.

By drawing on the methods of the Gel'fand theory of commutative rings (Banach algebras), we can sharpen the second assertion of Theorem I for the algebras $C_{\pm}$ and $C_{\cap}$.

Theorem II.[1] *Each algebra* $\mathfrak{A} = C_{\pm}, C_{\cap}$ *has the following property: suppose* $f \in \mathfrak{A}$ *and the closed curve* $z = f(t)$ $(0 \leq t \leq 2\pi)$ *is continuously inscribed in a domain* G *of a many sheeted Riemann surface over the complex plane* C^1; *then* $\Phi(f) \in \mathfrak{A}$ *as long as the function* $\Phi(z)$ *is holomorphic in* G.

Theorem II will be obtained as a corollary of the general Theorem 5.

2. Theorems I and II turn out to be useful in studying the asymptotic behavior of the determinant

$$D_n(f) = \det \| c_{j-k}(f) \|_0^n \quad (n = 0, 1, 2, \cdots). \tag{0.8}$$

As is well known [1, 2], G. Szegö in 1952 managed to show that the following relation holds for a periodic positive function f having a derivative belonging to the class of periodic functions $\mathrm{Lip}\,\alpha$ $(0 < \alpha \leq 1)$:

$$\lim (D_n(f)/G^{n+1}(f)) = \sum_{n=1}^{\infty} n |c_n(\ln f)|^2, \tag{0.9}$$

where $G(f)$ is the geometric mean of the values of f, i.e.

$$G(f) = \exp\left\{\frac{1}{2\pi} \int_0^{2\pi} \ln f(t)\, dt\right\}. \tag{0.10}$$

Later M. Kac [3, 4], using probability theoretic arguments, established relation (0.9) for the case when the condition $f'(t) \in \mathrm{Lip}\,\alpha\, (0 < \alpha \leq 1)$ is replaced by the condition $\sum_{n=1}^{\infty} n^{1/2} |c_n(f)| < \infty$.

Still later G. Baxter [5, 6] obtained a sharpening of the latter result by proving that relation (0.9) is valid for a positive function $f \in C$ if

$$\sum_{n=1}^{\infty} n^{1/2} |c_n(f)| < \infty.$$

The elegant methods of G. Baxter [5, 6] are based on an extensive use of Wiener-Lévy type theorems. Using these theorems, Baxter also obtained a generalization of relation (0.9) for the case of complex functions $f \in C$.

The new Theorems I and II of Wiener-Lévy type permit one to obtain a sharpening of several of Baxter's results. In particular, it is possible to

1) See the footnote on page 189.

establish the following proposition for strictly positive[1] functions $f \in M$:

In order for relation (0.9) *to hold for a strictly positive function* $f \in M$ *it is necessary, and if* $f \in C$ *then also sufficient, that* $\mu(f) < \infty$.

Clearly, this proposition contains in itself both of the above-formulated overlapping results of G. Szegö and G. Baxter. The necessity of the condition $\mu(f) < \infty$ follows at once from Theorem I. Its sufficiency is proved with the use of Theorem I and some arguments from the theory of perturbations of selfadjoint operators.

A complete proof of this proposition and some related ones will be given in another place.

It may turn out that the formulated proposition, at least in the part concerning a positive function $f \in C$, can be sharpened: viz. *for such a function the limit* $\lim_{n\to\infty} [D_n(f)/G^{n+1}(f)]$ *exists finitely if and only if* $\mu(f) < \infty$.

3. In the same way as we selected the subalgebras $M_\pm$, $M_\cap$ ($C_\pm$, $C_\cap$) from the algebra M (C), we can select three analogous subalgebras from the Banach algebra $M(-\infty, \infty)$ of all bounded measurable functions $f(t)$ $(-\infty < t < \infty)$ on an infinite axis (the Banach algebra $C\{-\infty, \infty\}$ of all continuous functions $f(t)$ $(-\infty < t < \infty)$ on the closed real axis: $f(\infty) = \lim_{t\to-\infty} f(t) = \lim_{t\to\infty} f(t) = f(\infty)$). But in order to obtain the analytic character of these subalgebras it turns out to be necessary to have recourse to the use of the generalized Fourier-Bochner transform of bounded measurable functions.

We remain within the limits of the usual Fourier transform if we consider the problem of selecting the subalgebras analogous to $C_\pm$ and $C_\cap$ from the well-known Wiener algebra W, viz. the algebra of all functions $f(t)$ $(-\infty \leq t \leq \infty)$ that are representable in the form

$$f(t) = f(\infty) + \int_{-\infty}^{\infty} c_f(\lambda)\, e^{i\lambda t}\, d\lambda \quad (-\infty \leq t \leq \infty) \tag{0.11}$$

where $f(\infty)$ is any complex number and $c_f(\lambda) \in L_1(-\infty, \infty)$.[2]

The "analog"[3] of C_+ will now be the space W_+ of all functions $f(t)$ of the form (0.11), where $c_f(t)$ $(\in L_1(-\infty, \infty))$ has the additional property that

---

1) A function $f(t)$ is said to be *strictly positive* if it is positive almost everywhere and, in addition, ess $\inf_t f'(t) > 0$.

2) If $t \to -\infty$ then by Lebesgue's theorem the integral in (0.11) tends to zero; we therefore consider it to be equal to zero for $t = \pm\infty$.

3) The exact analog of C_+ will be another space; the space W_+ is in reality the complete analog of another algebra, viz. the algebra of all functions $f \in C$ which have an absolutely convergent Fourier series and for which $\mu_+(f) < \infty$.

$$\int_0^\infty \lambda|c_f(\lambda)|^2\,d\lambda < \infty.$$

In this connection we will naturally define the norm $\|f\|_+$ for $f \in W_+$ by the equality

$$\|f\|_+ = |f(\infty)| + \int_{-\infty}^{\infty} |c_f(\lambda)|\,d\lambda + \int_0^\infty \lambda|c_f(\lambda)|^2\,d\lambda \quad (f \in W_+). \tag{0.12}$$

The Banach space W_- with norm $\|\cdot\|_-$ is defined analogously, after which we define the space $W_\cap$ as the intersection of the spaces $W_\pm$: $W_\cap = W_+ \cap W_-$, and with the norm

$$\|f\|_\cap = |f(\infty)| + \int_{-\infty}^{\infty} |c_f(\lambda)|\,d\lambda + \int_{-\infty}^{\infty} |\lambda|\,|c_f(\lambda)|^2\,d\lambda. \tag{0.13}$$

We have

Theorem III.[1] *Each of the subspaces $W_\pm$, $W_\cap$ is a Banach algebra of functions. Any one of these algebras $\mathfrak{A}(= W_\pm, W_\cap)$ has the following property: if $f \in \mathfrak{A}$ and the curve $z = f(t)$ $(-\infty < t < \infty)$ is continuously inscribed in a domain G of a many sheeted Riemann surface, then $\Phi(f) \in \mathfrak{A}$ as long as the function $\Phi(z)$ is holomorphic in G.*

A continual analog of the theorem of G. Szegö was obtained in the above-mentioned paper of M. Kac [3] (see also [4]). This result of Kac later received further development in a paper of N. Ahiezer [7], and was recently significantly sharpened in a paper of Hirschman [8].

Theorem III permits one to make some additions to all of these investigations.

We note, finally, that the methods developed in papers [9-11] in combination with our new theorems of Wiener-Lévy type permit one to give in new classes of functions a solution of the problem concerning the factorization of functions and matrix functions on a circle and on the real axis, and hence of the Riemann-Hilbert problem of linear conjugation on these curves.

§1. General theorems on the Banach algebra $\mathfrak{A}_{12}$

1. Let $\mathfrak{A}$ be a Banach algebra over the field of complex numbers with unit E and norm $\|\cdot\|$:

$$\|XY\| \leq \|Y\| \cdot \|X\|. \tag{1.1}$$

---

1) See the footnote on page 189.

Let $\mathfrak{J}$ $(\subset \mathfrak{A})$ be a two-sided ideal of $\mathfrak{A}$ in which there is given a norm $\|\cdot\|_{\mathfrak{J}}$ converting $\mathfrak{J}$ into a Banach space and such that

1) $$\|X\| \le \|X\|_{\mathfrak{J}} \quad (X \in \mathfrak{J}), \tag{1.2}$$

2) $$\max(\|AX\|_{\mathfrak{J}}, \|XA\|_{\mathfrak{J}}) \le \|X\|, \|A\| \quad (A \in \mathfrak{A}, X \in \mathfrak{J}). \tag{1.3}$$

Let $P_1 (\in \mathfrak{A})$ be an idempotent $(P_1^2 = P_1)$, and $P_2 = E - P_1$ the complementary idempotent.

We consider the linear manifold

$$\mathfrak{A}_{12} = \{X;\ X \in \mathfrak{A},\ P_1 X P_2 \in \mathfrak{J}\}.$$

Since $P_1 P_j P_2 = 0$, we have $P_j \in \mathfrak{A}_{12}$ $(j = 1, 2)$, and hence $E = P_1 + P_2 \in \mathfrak{A}_{1,2}$.

Theorem 1. *The set* $\mathfrak{A}_{12}$ *is a subalgebra of the algebra* $\mathfrak{A}$. *It is a Banach algebra with respect to the norm*

$$\|A\|_{12} = (\|P_1 A P_1\|^2 + \|P_2 A P_1\|^2 + \|P_2 A P_2\|^2 + \|P_1 A P_2\|_{\mathfrak{J}}^2)^{1/2} \quad (A \in \mathfrak{A}_{12}). \tag{1.4}$$

Proof. If $A, B \in \mathfrak{A}_{12}$ then $P_1 A P_2 \in \mathfrak{J}$ and $P_1 B P_2 \in \mathfrak{J}$, and hence

$$P_1 A B P_2 = P_1 A (P_1 + P_2) B P_2 = P_1 A P_1 B P_2 + P_1 A P_2 B P_2 \in \mathfrak{J},$$

since, by definition, $\mathfrak{J}$ is a two-sided ideal.

From $A = (P_1 + P_2) A (P_1 + P_2) = \Sigma P_j A P_k$ it follows that

$$\|A\| \le \sum_{j,k=1}^{2} \|P_j A P_k\| \le 2 \left\{ \sum_{j,k=1}^{2} \|P_j A P_k\|^2 \right\}^{1/2}.$$

According to (1.2), $\|P_1 A P_2\| \le \|P_1 A P_2\|_{\mathfrak{J}}$, and hence

$$\|A\| \le 2\, \|A\|_{12}. \tag{1.5}$$

Suppose now $\{A_n\} \subset \mathfrak{A}_{12}$ and

$$\|A_n - A_m\|_{12} \to 0 \text{ for } n, m \to \infty. \tag{1.6}$$

Then, according to (1.5), $\|A_n - A_m\| \to 0$ for $n, m \to \infty$, and hence an element $A \in \mathfrak{A}$ can be found such that $\|A - A_n\| \to 0$ for $n \to \infty$. By virtue of (1.1) we will have

$$\|P_j A P_k - P_j A_n P_k\| \to 0 \text{ for } n \to \infty \quad (j, k = 1, 2). \tag{1.7}$$

From (1.6) and (1.3) it also follows that

$$\|P_1 A_n P_2 - P_1 A_m P_2\|_{\mathfrak{J}} \longrightarrow 0 \text{ for } n, m \longrightarrow \infty .$$

Therefore an element C can be found in the ideal $\mathfrak{J}$ such that

$$\|C - P_1 A_n P_2\|_{\mathfrak{J}} \longrightarrow 0 \text{ for } n \longrightarrow \infty . \tag{1.8}$$

By virtue of (1.2) the element C will also be the limit of the sequence $\{P_1 A_n P_2\}$ with respect to the norm $\|\cdot\|$, and therefore

$$P_1 AP_2 = C \in \mathfrak{J}. \tag{1.9}$$

Thus $A \in \mathfrak{A}_{12}$, and according to (1.7)–(1.9) the element A will be the limit of the sequence $\{A_n\}$ in the metric of the algebra $\mathfrak{A}_{12}$.

We also show that

$$\|AB\|_{12} \leq \|A\|_{12} \|B\|_{12} \quad (A, B \in \mathfrak{A}_{12}). \tag{1.10}$$

If we put $C = AB$ and

$$A_{jk} = P_j AP_k, \; B_{jk} = P_j BP_k, \; C_{jk} = P_j CP_k \quad (j, k = 1, 2),$$

we will have

$$C_{jk} = A_{j1} B_{1k} + A_{j2} B_{2k} \quad (j, k = 1, 2).$$

Therefore, according to (1.1) and (1.3),

$$\|C_{11}\| \leq \|A_{11}\| \, \|B_{11}\| + \|A_{12}\|_{\mathfrak{J}} \|B_{21}\|,$$

$$\|C_{12}\|_{\mathfrak{J}} \leq \|A_{11}\| \, \|B_{12}\|_{\mathfrak{J}} + \|A_{12}\|_{\mathfrak{J}} \|B_{22}\|,$$

which implies that

$$\|C_{11}\|^2 + \|C_{12}\|_{\mathfrak{J}}^2 \leq$$
$$(\|A_{11}\|^2 + \|A_{12}\|_{\mathfrak{J}}^2) (\|B_{11}\|^2 + \|B_{21}\|^2 + \|B_{22}\|^2 + \|B_{12}\|_{\mathfrak{J}}^2).$$

It is shown analogously that

$$\|C_{21}\|^2 + \|C_{22}\|^2 \leq (\|A_{21}\|^2 + \|A_{22}\|^2)(\|B_{11}\|^2 + \|B_{21}\|^2 + \|B_{22}\|^2 + \|B_{12}\|_{\mathfrak{J}}^2).$$

Adding the last two inequalities, we get that $\|C\|_{12} \leq \|A_{12}\| \, \|B\|_{12}$, i.e. (1.10).

The theorem is proved.

Remark 1.1. Incidentally, it can shown that the convergence of a sequence $\{A_n\} \subset \mathfrak{A}_{12}$ in the metric of $\mathfrak{A}_{12}$ is equivalent to the convergence of $\{A_n\}$ and $\{P_1 A_n P_2\}$ in the metrics of $\mathfrak{A}$ and $\mathfrak{J}$ respectively.

Therefore the norm $\|\cdot\|_{12}$ is topologically equivalent to the simpler norm

$$\|A\|^1_{12} = \|A\| + \|P_1 AP_2\|_{\mathfrak{J}}.$$

It is easily seen that this norm will have the "ring" property (1.1) if $\|P_1\| = \|P_2\| \leq 1$.

2. Let $\rho(A)$ denote the open set of all regular points z of an element $A \in \mathfrak{A}$ (i.e. the set of all complex z for which $(A - zE)^{-1}$ exists) and let $\rho_\infty(A)$ denote the connected component of $\rho(A)$ that includes the point at infinity.

Theorem 2. *If* $A \in \mathfrak{A}_{12}$ *then* $(A - zE)^{-1} \in \mathfrak{A}_{12}$ *for all* $z \in \rho_\infty(A)$.

Proof. Let $R(z) = R(A;\, z) = (A - zE)^{-1}$ $(z \in \rho_\infty(A))$. Multiplying the equality $(A - zE)\, R(z) = E$ from the left by P_1 and from the right by P_2, we get

$$0 = P_1(A - zE)\, R(z)P_2 = P_1(A - zE)\, P_1P_1R(z)P_2 + P_1(A - zE)\, P_2R(z)P_2. \tag{1.12}$$

The second term of the right side (which we denote by $C(z)$) belongs to $\mathfrak{J}$. Thus from (1.12) we deduce that

$$(P_1AP_1 - zE)P_1R(z)P_2 = -P_1(A - zE)P_1P_1R(z)P_2 = C(z) \in \mathfrak{J}.$$

The element $(P_1AP_1 - zE)^{-1}$ will exist for $|z| > r = \|P_1AP_2\|$ and hence

$$P_1R(z)P_2 = (P_2AP_1 - zE)^{-1}C(z) \in \mathfrak{J} \text{ for } |z| > r. \tag{1.13}$$

We must prove that for all $z \in \rho_\infty(A)$

$$P_1R(z)P_2 \in \mathfrak{J}. \tag{1.14}$$

To this end we make use of the relation

$$R(z) + (\zeta - z)\, R(\zeta)\, R(z) = R(\zeta),$$

which is valid for any $z,\, \zeta \in \rho(A)$.

Multiplying it from the left by P_1 and from the right by P_2, we get

$$\begin{aligned} &P_1R(z)P_2 + (\zeta - z)P_1R(\zeta)P_1P_1R(z)P_2 \\ &= P_1R(\zeta)P_2 + (z - \zeta)P_1R(\zeta)P_2R(z)P_2. \end{aligned} \tag{1.15}$$

We denote the sum on the right by $F(z, \zeta)$. If the first term $P_1R(\zeta)P_2$ of this sum belongs to $\mathfrak{J}$ then $\mathfrak{J}$ will also contain the second term and hence the sum $F(z, \zeta)$ itself. Therefore, if $P_1R(\zeta)P_2 \in \mathfrak{J}$ and

$$|\zeta - z| \, \|P_1 R(z) P_2\| < 1$$

we also have

$$P_1 R(z) P_2 = [1 + (\zeta - z) P_1 R(z) P_1]^{-1} F(z, \zeta) \in \mathfrak{J}.$$

Suppose now z_1 is an arbitrary point of $\rho_\infty(A)$ and $z_0 \in \rho_\infty(A)$ is such that $|z_0| > r$ (and hence $P_1 R(z_0) P_2 \in \mathfrak{J}$). We connect the points z_0 and z_1 by a continuous curve γ and put

$$\delta = \max_{z \in \gamma} \|P_1 R(z) P_1\|.$$

Then by virtue of what has been proved the validity of the relation $P_1 R(\zeta) P_2 \in \mathfrak{J}$ for some point $\zeta \in \gamma$ will imply the validity of (1.14) for all z from the neighborhood $|z - \zeta| < 1/\delta$. Since (1.14) is satisfied for $z = z_0$, we can show with the use of well-known arguments that it also holds for $z = z_1$.

Thus the theorem is proved.

Corollary 1. *Let $\Phi(z)$ be a function that is holomorphic in a simply connected domain G, and suppose the spectrum of an element $A \in \mathfrak{A}_{12}$ is contained in G. Then $\Phi(A) \in \mathfrak{A}_{12}$.*

In fact, the function $\Phi(A)$ can be defined by the equality

$$\Phi(A) = -\frac{1}{2\pi i} \int_\gamma \Phi(z) R(A; z) dz, \tag{1.16}$$

where γ is a positively directed smooth contour in G which encloses the spectrum $\sigma(A)$ of A.

The chosen properties of γ imply that it lies in $\rho_\infty(A)$, and therefore $R(A; z)$ will be a function which, when restricted to γ, has its values in $\mathfrak{A}_{12}$ and is continuous in the metric of $\mathfrak{A}_{12}$.

Remark 1.2. The corollary can be easily extended to the case when the function $\Phi(z)$ is piecewise holomorphic in an open set $G(\supset \sigma(A))$ which is the set-theoretic sum of a finite number of nonintersecting simply connected domains.

But in the applications the requirement of simple connectedness is very restrictive. In the following section we get rid of this requirement (and in one important case, of the requirement that $\Phi(z)$ be single valued) at the expense of specializing the algebra $\mathfrak{A}$ and constricting the algebra $\mathfrak{A}_{12}$.

§2. Banach algebras of operators with completely continuous nondiagonal cells

1. Let $\mathfrak{B}$ be a complex Banach space, let $\mathfrak{A}(\mathfrak{B})$ be the Banach algebra of all bounded linear operators A acting in $\mathfrak{B}$ with the usual norm

$$\|A\| = \sup_{x \in \mathfrak{B}} (\|Ax\|/\|x\|),$$

and let $\mathfrak{J}_\infty$ be the ideal consisting of all of the completely continuous operators $A \in \mathfrak{A}$.

Let $P_1 (\in \mathfrak{A})$ and $P_2 = I - P_1$ be two complementary projectors and let $\mathfrak{J}_j \subset \mathfrak{J}_\infty (j = 1, 2)$ be two two-sided ideals of $\mathfrak{A}$. In addition, we will assume that a norm $\|\cdot\|_j$ $(j = 1, 2)$ is given in $\mathfrak{J}_j$ which converts $\mathfrak{J}_j$ into a Banach space and is such that

$$\begin{aligned} &1) \qquad \|A\| \le \|A\|_j \quad (A \in \mathfrak{J}_j), \\ &2) \qquad \max(\|XA\|_j, \|AX\|_j) \le \|X\| \, \|A\|_j. \end{aligned} \qquad (j = 1, 2)$$

According to Theorem I it will be possible to assert that each of the linear manifolds

$$\mathfrak{A}_{12} = \{X; X \in \mathfrak{A}, P_1 X P_2 \in \mathfrak{J}_1\}, \ \mathfrak{A}_{21} = \{X; X \in \mathfrak{A}, P_2 X P_1 \in \mathfrak{J}_2\}$$

is (under a suitable definition of a norm in it) a Banach algebra with unit. Of interest to us will be the algebra $\mathfrak{A}_\cap = \mathfrak{A}_\cap (P_1, P_2; \mathfrak{J}_1, \mathfrak{J}_2)$, which is their intersection:

$$\mathfrak{A}_\cap = \mathfrak{A}_{12} \cap \mathfrak{A}_{21}.$$

Thus the algebra $\mathfrak{A}_\cap$ consists of all $A \in \mathfrak{A}(\mathfrak{B})$ for which

$$P_1 A P_2 \in \mathfrak{J}_1, \ P_2 A P_1 \in \mathfrak{J}_2.$$

By arguments similar to those used in the proof of Theorem 1, it can be shown that $\mathfrak{A}_\cap$ becomes a Banach algebra if we introduce in $\mathfrak{A}_\cap$ the norm

$$\|A\|_\cap = \sqrt{\|P_1 A P_1\|^2 + \|P_2 A P_2\|^2 + \|P_1 A P_2\|_1^2 + \|P_2 A P_1\|_2^2} \quad (A \in \mathfrak{J}_\cap).$$

Theorem 3. *If an operator $A \in \mathfrak{A}_\cap$ has an inverse $B = A^{-1}$ in $\mathfrak{A}(\mathfrak{B})$ then $B \in \mathfrak{A}_\cap$.*

Proof. Let

$$A_1 = P_1 A P_1 + P_2 A P_2.$$

Since $A - A_1 = P_1AP_2 + P_2AP_1 \in \mathfrak{J}_\infty$ and A is invertible, the operator A_1 will have the following properties (see, for example, [12], §2): $\mathfrak{R}(A_1)$ will be closed and the codimension of $\mathfrak{R}(A_1)$ will be finite and will coincide with the dimension of the set of zeros $\mathfrak{Z}(A_1)$ of A_1: $\operatorname{codim}\mathfrak{R}(A_1) = \dim \mathfrak{Z}(A_1) < \infty$. Therefore we can always find two subspaces $\mathfrak{L}$ and $\mathfrak{N}$ in $\mathfrak{B}$ such that the following direct sums will occur:

$$\mathfrak{L} = \mathfrak{Z}(A_1) \dotplus \mathfrak{L},\ \mathfrak{L} = \mathfrak{R}(A_1) \dotplus \mathfrak{N}.$$

Let $\hat{A}_1$ denote the restriction of A_1 to $\mathfrak{L}$. The operator $\hat{A}_1$ is a continuous and one-to-one mapping of $\mathfrak{L}$ onto $\mathfrak{R}(A_1)$; consequently, by Banach's theorem its inverse $\hat{A}_1^{-1}$ will also be continuous. Let B_1 denote the operator coinciding with $\hat{A}_1^{-1}$ on $\mathfrak{R}(A_1)$ and equal to zero on $\mathfrak{N}$. Clearly $B_1 \in \mathfrak{A}(\mathfrak{B})$ and for any $x \in \mathfrak{B}$

$$B_1A_1x = B_1A_1(x - Qx) = B_1\hat{A}_1(x - Qx) = x - Qx,$$

where Q is the operator that projects $\mathfrak{B}$ onto $\mathfrak{Z}(A_1)$ and is equal to zero on $\mathfrak{L}$ (and hence $I - Q$ projects $\mathfrak{B}$ onto $\mathfrak{L}$). Thus

$$B_1A_1 = I - Q, \tag{2.1}$$

Let $A_{jk} = P_jAP_k$, $B_{jk} = P_jBP_k$ $(j, k = 1, 2)$ and $B = A^{-1}$. From $AB = I$ it follows, in particular, that

$$A_{11}B_{12} + A_{12}B_{22} = 0,$$

and since $A_1 = A_{11} + A_{22} = A_{11} + A_{22}P_2$ and $P_2B_{12} = 0$, we have

$$A_1B_{12} + A_{12}B_{22} = 0.$$

Multiplying this equality from the left by B_1 and taking (2.1) into account, we find that

$$B_{12} = QB_{12} + B_1A_{12}B_{22}.$$

Since $\mathfrak{J}_1$ is a two-sided ideal in $\mathfrak{A}(\mathfrak{B})$ and, by assumption, $A_{12} \in \mathfrak{J}_1$, the second term of the right side belongs to $\mathfrak{J}_1$. The first term is a finite-dimensional operator: $\dim(QB_{12}\mathfrak{B}) \le \dim(Q\mathfrak{B}) = \dim\ \mathfrak{Z}(A_1)$, and hence it also belongs to $\mathfrak{J}_1$.[1)]

Thus $B_{12} \in \mathfrak{J}_1$; it is proved analogously that $B_{21} \in \mathfrak{J}_2$. The theorem is proved.

---

1) As is well known, any two-sided ideal $\mathfrak{J}$ of the algebra $\mathfrak{A}(\mathfrak{B})$ contains all of the finite-dimensional operators of this algebra.

2. In order to avoid any misunderstanding we emphasize that $\rho(A)$ for $A \in \mathfrak{A}(\mathfrak{B})$ will now denote the set of all of the regular points z of an operator A in $\mathfrak{B}$, i.e. the set of all complex z for which $A - zI$ has an inverse $(A - zI)^{-1} \in \mathfrak{A}(\mathfrak{B})$. By $\sigma(A)$ for $A \in \mathfrak{A}(\mathfrak{B})$ we will denote the spectrum of an operator A in $\mathfrak{B}$, i.e. the complement of $\rho(A)$ in the whole complex plane.

Since the algebra $\mathfrak{A}_\cap$ contains I and hence, together with an element A, the element $A - zI$, Theorem 3 is equivalent to the following theorem (which is formally more general).

Theorem 3′. *If* $A \in \mathfrak{A}_\cap$ *then* $(A - zI)^{-1} \in \mathfrak{A}_\cap$ *for all* $z \in \rho(A)$.

We leave it to the reader to show, by slightly generalizing the arguments in the proof of the second assertion of Theorem 2, that for $A \in \mathfrak{A}$ the resolvent $R(A; z) = (A - zI)^{-1}$ will already be a holomorphic operator function in any domain contained in $\rho(A)$. From here we at once obtain

Theorem 4. *Suppose* $A \in \mathfrak{A}_\cap$ *and* $\Phi(z)$ *is a piecewise holomorphic function in an open set* G *containing the spectrum* $\sigma(A)$. *Then* $\Phi(A) \in \mathfrak{A}_\cap$.

Proof. By definition, the operator $\Phi(A)$ can be obtained by formula (1.16), where now γ is possibly a complicated contour lying wholly in G and enclosing the spectrum $\sigma(A)$. Since the resolvent $R(A; z)$ is a continuour vector function on γ with values in the Banach space $\mathfrak{A}$, the integral (1.16) will belong to $\mathfrak{A}_\cap$.

§3. Theorem on a commutative algebra of operators with completely continuous nonnegative cells

1. Let $\mathfrak{R}$ be a commutative Banach algebra, let $\mathfrak{M}$ be the structure space of its maximal ideals, and let $x \longrightarrow x(M)$ $(M \in \mathfrak{M})$ be the Gel′fand homomorphism of $\mathfrak{R}$ into the algebra of complex functions on $\mathfrak{M}$ [13].

Let G be a polydomain, viz. an open set consisting of a finite number of connected components and situated on a many sheeted Riemann surface over the complex plane C^1. We agree to say that the spectrum (in $\mathfrak{R}$) of an element $x \in \mathfrak{R}$ is continuously inscribed in G if there is given a mapping of the graph $(x(M), M)$ $(M \in \mathfrak{M})$ into G: $(x(M), M) \longrightarrow x_G(M)$, that induces a continuous function $x_G(M)$ on M with values in G situated directly over the corresponding values of $x(M)$ $(M \in \mathfrak{M})$.

An important role will be played below by the following

Šilov's Theorem [13, 14].[1] *Let* x *be an element in a commutative Banach algebra* $\mathfrak{R}$ *whose spectrum in* $\mathfrak{R}$ *is continuously inscribed in a polydomain* G: $x(M) \to x_G(M)$. *Then, for any function* Φ *that is piecewise holomorphic in* G, *an element* $y \in \mathfrak{R}$ *can be found such that* $\Phi(x_G(M)) = y(M)$.

Clearly, the element y will be uniquely determined by the element $x \in \mathfrak{R}$, the mapping of its spectrum into G and the function Φ if and only if the algebra $\mathfrak{R}$ is semisimple (i.e. $x(M) \neq 0$ for $x \neq 0$).

2. We now return to a consideration of the objects $\mathfrak{A}(\mathfrak{B})$ and $\mathfrak{A}_\cap = \mathfrak{A}_\cap(P_1, P_2; I_1, I_2)$ studied in §2. We have

Theorem 5.[1] *Suppose* $\mathfrak{R}$ *is a commutative Banach algebra composed of operators* $X \in \mathfrak{A}(\mathfrak{B})$ *and suppose the intersection* $\mathfrak{R}_\cap = \mathfrak{R} \cap \mathfrak{A}_\cap$ *is dense (with respect to the norm of* $\mathfrak{R}$) *in* $\mathfrak{R}$ *and is also a Banach algebra under a suitable choice of norm.*

Then, for any function Φ *that is piecewise holomorphic in a polydomain* G *and any element* $X \in \mathfrak{R}$ *whose spectrum in* $\mathfrak{R}$ *is continuously inscribed in* G, *an element* $Y \in \mathfrak{R}$ *can be found such that* $Y(M) = \Phi(X(M))$ $(M \in \mathfrak{M})$.

Proof. If an element A has an inverse $B = A^{-1} \in \mathfrak{R}(\subset \mathfrak{A}(\mathfrak{B}))$, then by Theorem 3 the operator $B \in \mathfrak{A}_\cap$ and hence is also contained in the Banach algebra $\mathfrak{R}_\cap$. This fact and the fact that $\mathfrak{R}_\cap$ is dense in $\mathfrak{R}$ imply (see [15], Lemma 2 and the corollary to it) that the compact spaces $\mathfrak{M}$ and $\mathfrak{M}_\cap$ of the algebras $\mathfrak{R}$ and $\mathfrak{R}_\cap$ are homeomorphic and, moreover, by virtue of this homeomorphism the Gel'fand homomorphism $X \to X(M)$ of $\mathfrak{R}$ into the algebra of functions induces in $\mathfrak{R}_\cap$ a Gel'fand homomorphism of $\mathfrak{R}_\cap$. After this the assertion of Theorem 5 is obtained by directly applying Šilov's Theorem to the algebra $\mathfrak{R}_\cap$.

§4. Proof of Theorems I and II

We choose as $\mathfrak{B}$ the Hilbert space $L_2 = L_2(0, 2\pi)$. Here, in place of $\mathfrak{A}(\mathfrak{B})$, we will write $\mathfrak{A}(L_2)$. With a function $f \in M$ (see the Introduction) we associate an operator $A_f \in \mathfrak{A}(L_2)$ by putting $A_f\phi = f\phi(\phi \in L_2)$. The correspondence $f \to A_f$ is an isometric and isomorphic mapping of the commutative

---

1) *Added in proof.* It recently became known to the author that the final formulation of Šilov's Theorem has the additional requirement that the function $\Phi(z)$ be normal. This condition means that $\Phi(z)$ always takes different values at different points of its polydomain G lying over one and the same point z (as, for example, $\ln z$ and $\sqrt[n]{z}$). As a result, the requirement that $\Phi(z)$ be normal must be added in Theorems II and III (of the Introduction) and Theorem 5 (of § 3).

algebra M onto the subalgebra $\mathfrak{R}\ (\subset \mathfrak{A}(L_2))$ of operators $A_f (f \in M)$. Under this mapping the subspaces $M_\pm$ and $M_\cap\ (\subset M)$ go over into the subspaces $\mathfrak{R}_\pm$ and $\mathfrak{R}_\cap\ (\subset \mathfrak{R})$.

For an element

$$\phi = \sum_{n=-\infty}^{\infty} c_n(\phi) e_n \in L_2 \left\{ e_n = \frac{1}{\sqrt{2\pi}} e^{nit} \in L_2,\ n = 0, \pm 1, \pm 2, \cdots \right\}$$

we put

$$P_1 \phi = \sum_{n=1}^{\infty} c_n(\phi) e_n,\ P_2 \phi = \sum_{n=-\infty}^{0} c_n(\phi) e_n.$$

Clearly, P_1 and P_2 are mutually complementary orthoprojectors $(P_1^* = P_1 = P_1^2,\ P_2 + P_1 = 1)$.

Let $\mathfrak{S}_2\ (\mathfrak{S}_\infty)$ denote the ideal in $\mathfrak{A}(L_2)$ of all Hilbert-Schmidt (completely continuous) operators acting in L_2.

Let us find conditions on $f \in M$ under which $P_1 A_f P_2 \in \mathfrak{S}_2$, i.e.

$$\|P_1 A_f P_2\|^2 = \sum_{n=-\infty}^{\infty} \|P_1 A_f P_2 e_n\|^2 = \sum_{n=-\infty}^{\infty} \|P_1 A_f e_n\|^2 < \infty. \tag{4.1}$$

Since

$$(A_f e_n, e_m) = \frac{1}{2\pi} \int_0^{2\pi} f(t) e^{(n-m)it}\, dt = c_{m-n}(f) (n, m = 0, \pm 1, \pm 2, \cdots),$$

we get

$$A_f e_n = \sum_{m=-\infty}^{\infty} c_{m-n}(f) e_m,\ P_1 A_f e_n = \sum_{m=1}^{\infty} c_{m-n}(f) e_m,$$

and hence the left side of (4.1) is equal to

$$\sum_{n=-\infty}^{0} \sum_{m=1}^{\infty} |c_{m-n}(f)|^2 = \sum_{n=0}^{\infty} \sum_{m=1}^{\infty} |c_{m+n}(f)|^2 = \sum_{n=0}^{\infty} n |c_n(f)|^2.$$

Thus the condition $P_1 A_f P_2 \in \mathfrak{S}_2$ is equivalent to the condition $\mu_+(f) < \infty$, i.e. the condition $f \in M_+$. At the same time we note that (see (0.5))

$$\|f\|_+ = \|A_f\| + \|P_1 A_f P_2\|_2 \quad (f \in M_+).$$

It is shown analogously that the condition $P_2 A_f P_1 \in \mathfrak{S}_2$ is equivalent to the condition $f \in M_-$ and that

$$\|f\|_{-} = \|A_f\| + \|P_2 A_f P_1\|_2.$$

On the basis of Theorem 1 (and Remark 1.1) we conclude that $\mathfrak{R}_{\pm}$ and $\mathfrak{R}_{\cap}$ are Banach subalgebras of $\mathfrak{R}$, and hence $M_{\pm}$ and $M_{\cap}$ are Banach subalgebras of M.

Since a product of continuous functions is a continuous function, we have also established that $C_{\pm} (\subset M_{\pm})$ and $C_{\cap} (\subset M_{\cap})$ are Banach algebras of functions.

Thus the first assertion of Theorem I is proved.

The second assertion of Theorem I is obtained for the algebras $M_{\pm}$ on the basis of Corollary 1 (of Theorem 1) and for the algebra $M_{\cap}$ on the basis of Theorem 4 (applied to the algebra $\mathfrak{A}_{\cap}$ extracted from $\mathfrak{A}(L_2)$ by means of the ideals $\mathfrak{J}_1 = \mathfrak{J}_2 = \mathfrak{S}_2$).

Let $\mathfrak{L}$ denote the set of all trigonometric polynomials of C, viz. the set of all $f \in C$ which have only a finite number of nonzero Fourier coefficients $c_n(f)$. Clearly, $\mathfrak{L} \subset C_{\cap}$ and is dense in C with respect to its norm. Consequently the algebras $C_{\pm}$ and $C_{\cap}$ are dense in C with respect to the uniform norm.

Thus Theorem II is obtained for $\mathfrak{A} = C_{\cap}$ on the basis of Theorem 5. It is also obtained on the basis of Theorem 5 for $\mathfrak{A} = C_{\pm}$ if we show that $P_j A_f P_k \in \mathfrak{S}_{\infty} (j \neq k)$ whenever $f \in C$. Clearly, it suffices to show this for $j = 1, k = 2$.

For $f \in C$ and any $\epsilon > 0$ there exists a trigonometric polynomial g such that $\|f - g\| < \epsilon$.

Then $\|A_f - A_g\| = \|f - g\| < \epsilon$, and hence

$$\|P_1 A_f P_2 - P_1 A_g P_2\| = \|P_1 (A_f - A_g) P_2\| \leq \|A_f - A_g\| < \epsilon.$$

Since $P_1 A_g P_2 \in \mathfrak{S}_2 \subset \mathfrak{S}_{\infty}$, we get $P_1 A_f P_2 \in \mathfrak{S}_{\infty}$.

Theorem II is completely proved.

§5. Proof of Theorem III

Theorem III is obtained by directly applying Theorem 5 to some special algebras $\mathfrak{R}_{\cap}$.

We choose as $\mathfrak{B}$ the Hilbert space $L_2 = L_2(-\infty, \infty)$, as $\mathfrak{R}$ the commutative Banach algebra of all operators $A_f \in \mathfrak{A}(L_2)$ induced by the functions f of the Wiener algebra W (see the Introduction) according to the rule

$$A_f\phi = f\phi \quad (\phi \in L_2),$$

and as P_1 the orthogonal projector in L_2 defined by the equality[1)]

$$P_1\phi = \frac{1}{\sqrt{2\pi}} \int_0^\infty e^{it\lambda}\, \tilde{\phi}(\lambda)\, d\lambda,$$

where $\tilde{\phi}(\lambda)$ $(-\infty < \lambda < \infty)$ is the Fourier-Plancherel transform of $\phi(t)$, i.e.

$$\tilde{\phi} = \frac{1}{\sqrt{2\pi}} \int_{-\infty}^\infty e^{it\lambda}\, \phi(t)\, dt.$$

We prove Theorem III for the algebra W_+; the proof is completely analogous for W_- and is even simpler in one respect for $W_\cap$.

To this end we choose as $\mathfrak{I}_1$ the Banach algebra (with the norm $\|\cdot\|_2$) of all Hilbert-Schmidt operators acting in L_2, and as $\mathfrak{I}_2$ the Banach algebra $\mathfrak{S}_\infty$ (with the uniform norm $\|\cdot\|$) of all completely continuous operators acting in L_2.

From the representation (0.11) for $f \in W$ we find on the basis of well-known properties of the Fourier transform that for $\psi = A_f\phi = f\phi\,(\phi \in L_2)$

$$\tilde{\psi}(\lambda) = f(\infty)\,\tilde{\phi}(\lambda) + \int_{-\infty}^\infty c_f(\lambda - \mu)\,\tilde{\phi}(\mu)\, d\mu.$$

Hence for $\chi_1 = P_1 A_f P_2 \phi$ we have

$$\tilde{\chi}_1(\lambda) = \begin{cases} \int_{-\infty}^0 c_f(\lambda - \mu)\,\tilde{\psi}(\mu)\, d\mu & (0 < \lambda < \infty) \\ 0 & (-\infty < \lambda < 0) \end{cases} \tag{5.1}$$

and, analogously, for $\chi_2 = P_2 A_f P_1 \phi$

$$\tilde{\chi}_2(\lambda) = \begin{cases} \int_0^\infty c_f(\lambda - \mu)\,\tilde{\psi}(\mu)\, d\mu & (-\infty < \lambda < 0) \\ 0 & (0 < \lambda < \infty). \end{cases} \tag{5.2}$$

As is well known, the Fourier-Plancherel transform unitarily maps $L_2(-\infty, \infty)$ onto $L_2(-\infty, \infty)$. Therefore the operator $P_1 A_f P_2$ $(P_2 A_f P_1)$ will be a Hilbert-Schmidt (completely continuous) operator if the integral operator given by (5.1) ((5.2)) is a Hilbert-Schmidt (completely continuous) operator. The proof of the complete continuity of the integral operator given by (5.2) for

1) All improper integrals below are understood in the sense of convergence in the space $L_2(-\infty, \infty)$.

$c_f(\lambda) \in L_1(-\infty, \infty)$ can be found, for example, in [12] or [10]. As to the integral operator given by (5.1), it will obviously be a Hilbert-Schmidt operator if and only if its kernel belongs to $L_2(E_2)$ (E_2 is the two-dimensional Euclidean plane), i.e.

$$\int_0^\infty \int_{-\infty}^0 |c_f(\lambda - \mu)|^2 d\mu\, d\lambda = \int_0^\infty \lambda |c_f(\lambda)|^2\, d\lambda < \infty. \tag{5.3}$$

Hence the condition $P_2 A_f P_1 \in \mathfrak{S}_\infty$ is satisfied for any function $f \in W$, while the condition $P_1 A_f P_2 \in \mathfrak{S}_2$ is equivalent to the condition $f \in W_+$.

Thus in the case in question the algebra $\mathfrak{R}_\cap = \mathfrak{R} \cap \mathfrak{A}_\cap(L_2)$ consists of those and only those operators A_f $(\in \mathfrak{A}(L_2))$ for which $f \in W_+$, and, as is easily seen, this algebra becomes a Banach algebra if one puts $\|A_f\|_\cap = \|f\|_+$, where $\|f\|_+$ is defined by (0.12). Since the mapping $f \to A_f$ is an algebraic isomorphism of W onto $\mathfrak{R}$ while the inverse mapping $A_f \to f$ maps $\mathfrak{R}_\cap$ into W_+, the space W_+ is a subalgebra of W. With the introduction of the norm $\|\cdot\|_+$ into it, W_+ becomes a Banach algebra.

The set $\mathfrak{L}$ of all $f \in W$ to which there correspond finite functions $c_f(\lambda)$ is dense in W.

On the other hand, it is obvious that $\mathfrak{L} \subset W_+$; consequently $\mathfrak{R}_\cap$ is dense in $\mathfrak{R}$. Thus Theorem 5 can be applied to $\mathfrak{R}_\cap$.

Since the Wiener algebra W can be considered as an algebra of functions on the set of its maximal ideals ([13], Chapter III), the isomorphism $A_f \to f$ mapping $\mathfrak{R}$ onto W is also a Gel'fand isomorphism. As has been shown, this isomorphism maps the ring $\mathfrak{R}_\cap$ onto W_+.

Thus Theorem III for $\mathfrak{A} = W_+$ is actually a corollary of Theorem 5.

Interchanging the roles of the ideals $\mathfrak{S}_2$ and $\mathfrak{S}_\infty$ (or the projectors P_1 and P_2), we obtain Theorem III for $\mathfrak{A} = W_-$.

And if we extract $\mathfrak{R}_\cap$ from $\mathfrak{R}$ with the use of the ideals $\mathfrak{J}_1 = \mathfrak{J}_2 = \mathfrak{S}_2$, then Theorem 5 applied to $\mathfrak{R}_\cap$ implies Theorem III for $\mathfrak{A} = W_\cap$.

Theorem III is proved.

Remark 5.1. Condition (5.3), as we explained, means that the operator $P_1 A_f P_2$ belongs to $\mathfrak{S}_2$. Other variants of Theorem III can be formulated by replacing the starting ideal $\mathfrak{S}_2$ with some other symmetrically normed ideal of $\mathfrak{A}(L_2)$ (see concerning this idea [16]), for example, $\mathfrak{S}_p$ $(p \geq 1)$ (viz. the ideal of all operators $A \in \mathfrak{A}(L_2)$ for which $\mathrm{Tr}[(A^* A)^{p/2}] < \infty$). But the corresponding condition $P_1 A_f P_2 \in \mathfrak{S}_p$ can be written in the form of an integral inequality only when p is an even natural number. As p increases this condition becomes more complicated: it has its simplest form for $p = 2$;

for $p = 4$ it is already written in the form

$$\int_0^\infty \lambda\left[\int_\lambda^\infty |c_f(\mu)|^2\, d\mu\right]^2 d\lambda < \infty .$$

Clearly, analogous remarks can be made in regard to Theorems I and II.

In conclusion we note that we could generalize the general abstract part of the paper (the results of §§1–3) by considering the algebras $\mathfrak{A}_\cap$ extracted from the given Banach algebra $\mathfrak{A}$ by means of the conditions

$$P_j A P_k \in \mathfrak{J}_1,\ P_k A P_j \in \mathfrak{J}_2 \quad (1 \leq j < k \leq n),$$

where $P_1, P_2, \cdots, P_n$ is a system of mutually orthogonal idempotents of $\mathfrak{A}$, the sum of which is equal to the unit E, and $\mathfrak{J}_1$ and $\mathfrak{J}_2$ are two-sided ideals of $\mathfrak{A}$ with suitable properties (in this connection one of these ideals can be improper, i.e. coincide with the whole algebra $\mathfrak{A}$).

Supplement

I. C. Gohberg acquainted himself with the manuscript of the present article and obtained a number of generalizations of the results presented here (see [17]). In response the author reexamined Theorems I–III formulated in the Introduction and discovered that in the part concerning the rings $M_\cap$, $C_\cap$ and $W_\cap$ the assertions of these theorems can be obtained in an elementary way (and under a more general formulation) without drawing on general theorems on Banach algebras.

We explain this with the example of the space $L_2(0, 2\pi)$.

Let $\omega(n)$ $(n = 0, 1, \cdots)$ be a sequence of positive numbers admitting the representation

$$\omega(n) = \int_0^\pi \frac{\sin^2 ns}{s^2}\, d\sigma(s), \tag{S.1}$$

where $\sigma(s)$ $(0 \leq s \leq \pi,\ \sigma(0) = 0)$ is a nondecreasing function. Of interest to us will be the case when

$$\int_0^\pi \frac{d\sigma(s)}{s^2} = \infty . \tag{S.2}$$

This condition is equivalent to the requirement that $\omega(n) \to \infty$ for $n \to \infty$ (see, for example, [18], §21).

For any function $f \in L_2(0, 2\pi)$ we put

$$\Omega(f) = \left[\sum_{n=-\infty}^{\infty} \omega(n) |c_n(f)|^2 \right]^{1/2} (\leq \infty)\ (\omega(-n) = \omega(n);\ n = 1, 2, \cdots).$$

It is easily seen that for the periodic extension of any function $f \in L_2(0, 2\pi)$ we have

$$\Omega^2(f) = \frac{1}{8\pi} \int_0^{\pi} \frac{d\sigma(s)}{s^2} \int_0^{2\pi} |f(t+s) - f(t-s)|^2\, dt. \tag{S.3}$$

The following proposition is easily proved with the use of this identity.

Theorem A. *If the condition $\Omega(f) < \infty$ is satisfied for some function $f \in L_2(0, 2\pi)$, then this condition will also be satisfied for $g = \Phi(f)$, where $\Phi(z)$ is any function that is holomorphic in a domain G containing the spectrum $\rho(f)$ of f.*

In fact, if $z \notin \rho(f)$, i.e. $\delta(z_1 f) = \text{ess inf}\, |f(t) - z| > 0$, then, using (S.3), we easily obtain for the function $g_z(t) = (f(t) - z)^{-1}$ the inequality

$$\Omega(g_z) \leq \Omega(f)/\delta(z, f).$$

Next, representing $\Phi(f)$ by means of the Cauchy integral

$$\Phi(f(t)) = -\frac{1}{2\pi i} \int_{\gamma} \Phi(z) g_z(t)\, dz,$$

where γ is a suitably chosen directed contour, we show that

$$\Omega(\Phi(f)) \leq \frac{\Omega(f)}{2\pi} \int_{\gamma} \frac{|dz|}{\delta(z, f)}.$$

We note that the sequence $\{\omega(n)\}$ always satisfies the condition ([18], §21)

$$\lim_{n\to\infty} \frac{\omega(n)}{n^2} = \sigma(+0).$$

If in (S.1) we put $\sigma(0) = 0$ and $\sigma(s) = 1$ for $0 \leq s \leq \pi$, we get $\omega(n) = n^2$.

If in (S.1) we put

$$\sigma(s) = 4 \int_0^t \frac{r^2}{\sin^2 r}\, dr, \quad \text{where } r = \min(\tfrac{\pi}{2}, s)\ (0 \leq t \leq \pi).$$

we get

$$\sigma(n) = 4 \int_0^{\pi/4} \frac{\sin^2 ns}{\sin^2 s}\, ds = \int_0^{2\pi} \frac{\sin^2 ns}{\sin^2 s}\, ds = n.$$

Thus in this case $\omega(n) = n$ $(n = 0, 1, \cdots)$ and hence $\Omega(f) = \mu(f)$.

It can be shown that *a sequence of numbers* $\{\omega(n)\}_0^\infty$ *with* $\omega(0) = 0$ *will admit the representation* (S.1) *if and only if* (i) *the form*

$$\sum_{n,m=0}^{\infty} [\omega(n+m) - \omega(n-m)]\, \xi_n\, \xi_m$$

is nonnegative for any finite real sequence $\{\xi_n\}_0^\infty$, *and* (ii) $\omega(n) = O(n^2)$ *for* $n \longrightarrow \infty$.

Other characteristics of the sequence $\{\omega(n)\}$ can be found in [18], §24, and [19].

A good idea of the variety of possible seminorms $\Omega(f)$ (to within topological equivalence) can be obtained from an interesting paper of P. L. Ul'janov [20] (see also [21], Chapter V, §7).

We note further that the set $\mathfrak{A}_\Omega$ of all $f \in M(0, 2\pi)$ for which $\Omega(f) < \infty$ forms a Banach algebra with the norm

$$|f|_\Omega = \text{ess sup}\,|f| + \Omega(f).$$

This remark together with Theorem A obviously contains in itself all of the assertions of Theorem I concerning the space $M_\cap$.

For a function $f \in C$ for which $\Omega(f) < \infty$ it is possible to formulate a a Theorem B generalizing the corresponding assertion of the theorem concerning the algebra $C_\cap$. We leave this for the reader to do.

In order to obtain a corresponding generalization of the assertion of the theorem in regard to the algebra $W_\cap$ we introduce into consideration an even function $\omega(\lambda)$ $(-\infty < \lambda < \infty)$ admitting the representation

$$\omega(\lambda) = \int_0^\infty \frac{\sin^2 s\lambda}{s^2}\, d\sigma(s) \quad (-\infty < \lambda < \infty), \tag{S.4}$$

where $\sigma(s)$ $(0 \le s < \infty, \sigma(0) = 0)$ is a nondecreasing function such that

$$\int_0^\infty \frac{d\sigma(s)}{s^2} = \infty. \tag{S.5}$$

For $f \in W_1(-\infty, \infty)$ we now put

$$\Omega(f) = \left[\int_{-\infty}^{\infty} \omega(\lambda)|c_f(\lambda)|^2 \, d\lambda\right]^{1/2}.$$

Then for any $f \in L_1(-\infty, \infty)$ we will have the identity

$$\Omega^2(f) = \frac{1}{8\pi}\int_0^{\infty} \frac{d\sigma(s)}{s^2} \int_{-\infty}^{\infty} |f(t+s) - f(t-s)|^2 \, dt.$$

By using this identity, we can establish

Theorem C. *Suppose $f \in W$ and the curve $z = f(t)$ $(-\infty \leq t \leq \infty)$ is continuously inscribed in a domain G of a many sheeted Riemann surface. Then $\Omega(\Phi(f)) < \infty$ if $\Omega(f) < \infty$ and Φ is any function that is holomorphic in G.*

This theorem contains in itself the corresponding assertion of Theorem III in regard to W_Ω. It admits further generalizations, which we cannot go into here.

We note that a continuous even function $\omega(\lambda)$ $(-\infty < \lambda < \infty)$ admits the representation (S.4) if and only if (i) the integral form

$$\int_0^{\infty}\int_0^{\infty} [\omega(\lambda+\mu) - \omega(\lambda-\mu)]\, \xi(\lambda)\, \xi(\mu)\, d\lambda\, d\mu$$

is nonnegative for any finite continuous function $\xi(\lambda)$ $(-\infty < \lambda < \infty)$, (ii) $\omega(\lambda) = O(\lambda^2)$ for $\lambda \to \infty$, and (iii) $\omega(\lambda) \to \infty$ for $\lambda \to \infty$.

Condition (iii) can be dropped if one does not require the fulfilment of equality (S.5).[1)]

In regard to these assertions see [18, 22].

I wish to express my sincere gratitude to I. C. Gohberg for a valuable discussion of the results of this paper.

BIBLIOGRAPHY

[1] G. Szegö, *On certain Hermitian forms associated with the Fourier series of a positive function,* Comm. Sém. Math. Univ. Lund 1952, Tome supplémentaire, 228–238. MR 14, 553.

[2] Ulf Grenander and G. Szegö, *Toeplitz forms and their applications,* California Monographs in Math. Sciences, Univ. of California Press, Los Angeles, Calif., 1958; Russian transl., IL, Moscow, 1961. MR 20 #1349; 24 #A349.

[3] M. Kac, *Toeplitz matrices, translation kernels and a related problem in probability theory,* Duke Math. J. 21 (1954), 501–509. MR 16, 31.

1) The equivalence of conditions (S.5) and (iii) was apparently first established in [23].

[4] ———, *Probability and related topics in physical sciences,* Lectures in Appl. Math., Proc. Summer Sem. (Boulder, Colo., 1957) vol. 1, Interscience, New York, 1957; reprint, Amer. Math. Soc., Providence, R.I., 1970; Russian transl., "Mir", Moscow, 1963. MR 21 #1635.

[5] G. Baxter, *A convergence equivalence related to polynomials orthogonal on the unit circle,* Trans. Amer. Math. Soc. 99 (1961), 471–487. MR 23 #A3422.

[6] ———, *Polynomials defined by a difference system,* J. Math. Anal. Appl. 2 (1961), 223–263. MR 23 #A3421.

[7] N. Ahiezer, *A functional analogue of some theorems on Toeplitz matrices,* Ukrain. Mat. Ž. 16 (1964), 445–462; English transl., Amer. Math. Soc. Transl. (2) 50 (1966), 295–316. MR 30 #411.

[8] I. I. Hirschman, Jr.,[1] *On a formula of Kac and Ahiezer,* J. Math. Mech. 16 (1966), 167–196. MR 34 #8089.

[9] M. G. Kreĭn, *Integral equations on the half-line with a kernel depending on the difference of the arguments,* Uspehi Mat. Nauk 13 (1958), no. 5 (83), 3–120; English transl., Amer. Math. Soc. Transl. (2) 22 (1962), 163–288. MR 21 #1507.

[10] I. C. Gohberg and M. G. Kreĭn, *Systems of integral equations on a half-line with kernels depending on the difference of the arguments,* Uspehi Mat. Nauk 13 (1958), no. 2 (80), 3–72; English transl., Amer. Math. Soc. Transl. (2) 14 (1960), 217–287. MR 21 #1506; 22 #3954.

[11] I. C. Gohberg, *The factorization problem in normed rings, functions of isometric and symmetric operators, and singular integral equations,* Uspehi Mat. Nauk 19 (1964), no. 1 (115), 71–124 = Russian Math. Surveys 19 (1964), no. 1, 63–114. MR 29 #487.

[12] I. C. Gohberg and M. G. Kreĭn, *The basic propositions on defect numbers, root numbers and indices of linear operators,* Uspehi Mat. Nauk 12 (1957), no. 2 (74), 43–118; English transl., Amer. Math. Soc. Transl. (2) 13 (1960), 185–264. MR 20 #3459; 22 #3984.

[13] I. M. Gel'fand, D. A. Raĭkov and G. E. Šilov, *Commutative normed rings,* Fizmatgiz, Moscow, 1960; English transl., Chelsea, New York, 1964. MR 23 #A1242; 34 #4940.

[14] G. E. Šilov, *Letter to the editor: On locally analytic functions,* Uspehi Mat. Nauk 21 (1966), no. 6 (132), 177–182. (Russian) MR 36 #1696.

[15] M. G. Sonis, *On the Wiener relation in commutative rings,* First Republican Mathematical Conference of Young Researchers (Kiev, 1964) part II, Izdat. Akad. Nauk Ukrain. SSR Inst. Mat., Kiev, 1965. MR 32 #7368.

---

1) The author is grateful to Professor Hirschman for sending the first three pages (the Introduction) of a preprint of this paper.

[16] I. C. Gohberg and M. G. Kreĭn, *Introduction to the theory of linear non-selfadjoint operators in Hilbert space,* "Nauka", Moscow, 1965; English transl., Transl. Math. Monographs, vol. 18, Amer. Math. Soc., Providence, R.I., 1969. MR 36 #3137.

[17] I. C. Gohberg, *A generalization of theorems of M. G. Kreĭn of the type of the Wiener-Levy theorems,* Mat. Issled. 1 (1966), no. 1, 110–130. (Russian) MR 34 #3366.

[18] I. S. Iohvidov and M. G. Kreĭn, *Spectral theory of operators in spaces with an indefinite metric.* II, Trudy Moskov. Mat. Obšč. 8 (1959), 413–496; English transl., Amer. Math. Soc. Transl. (2) 34 (1963), 283–373. MR 21 #6543.

[19] ———, *Letter to the editor,* Trudy Moskov. Mat. Obšč. 15 (1966), 452–454 = Trans. Moscow Math. Soc. 1966, 505–508. MR 34 #4906.

[20] P. L. Ul'janov, *On some equivalent conditions of convergence of series and integrals,* Uspehi Mat. Nauk 8 (1953), no. 6 (58), 133–141. (Russian) MR 16, 242.

[21] N. K. Bari, *Trigonometric series,* Fizmatgiz, Moscow, 1961; English transl., Macmillan, New York and Pergamon Press, Oxford, 1964. MR 23 #A3411; 30 #1347.

[22] M. G. Kreĭn, *On a general method of decomposing Hermite-positive nuclei into elementary products,* C. R. (Dokl.) Acad. Sci. URSS 53 (1946), 3–6. MR 8, 277.

[23] J. von Neumann and J. Schoenberg, *Fourier integrals and metric geometry,* Trans. Amer. Math. Soc. 50 (1941), 226–251. MR 3, 37.

Translated by:
S. Smith

Amer. Math. Soc. Transl.
(2) Vol. 93, 1970

EXPONENTIAL LAW FOR SPACES OF CONTINUOUS LINEAR MAPPINGS*

UDC 513.836

D. A. RAĬKOV

The Cantor exponential notation F^E for the set of all the mappings of E into F is well suited for the expression of such "natural equivalences" as the equivalence $G^{E\times F} \sim (G^F)^E$ which to each $u: E \times F \to G$ associates $\tilde{u}: E \to G^F$ defined by the formula

$$\tilde{u}_x (= \tilde{u}(x)) = u(x, \cdot), \quad \text{i.e.} \quad \tilde{u}_x(y) = u(x, y) \text{ for all } \quad x \in E, y \in F.$$

An analogous "exponential law" is valid for some categories whose objects are equipped with a structure of a specific type. Obviously, the set $H(E, F)$ of all the morphisms of E into F then plays the role of F^E and a suitably defined product $E \cdot F$ plays the role of $E \times F$. (One equips $H(E, F)$ and $E \cdot F$ with appropriate structure in order that they may be considered as objects of a category.) Thus in the case of the category of vector spaces, $H(E, F)$ is the vector space $\mathfrak{L}(E, F)$ of all linear mappings of E into F and $E \cdot F$ is the tensor product $E \otimes F$. Then

$$\mathcal{L}(E \otimes F, G) \sim \mathcal{L}(E, \mathcal{L}(F, G)), \tag{1}$$

where the isomorphism $\sim$ is defined by the formula $\tilde{u}_x(y) = u(x \otimes y)$. In the category of Banach spaces this isomorphism induces the exponential law

$$L_b(E \hat{\otimes} F, G) \sim L_b(E, L_b(F, G)), \tag{2}$$

where $L_b(X, Y)$ is the Banach space of all continuous linear mappings of X into Y and $E \hat{\otimes} F$ is the completion of $E \otimes F$ with respect to the projective tensor norm.

However, the extension of the exponential law (2) to larger classes of locally convex spaces encounters essential difficulties. First of them is the fact that, for $E = L(F, G)$, in the right side of (1) the identity mapping of $L(F, G)$ into itself corresponds to the canonical mapping $\omega: L(F, G) \otimes F \to G$ defined by $\omega(v \otimes y) = v(y)$. This identity should be continuous under any circumstances, but if F is non-normable, and $L(F, G) \otimes F$ is equipped with the projective tensor topology, then ω cannot be continuous for any topology

* Translation of Mat. Sb. 67 (109) (1965), 279–302.

on $L(F, G)$ (cf. [6]). The second difficulty arises from the fact that spaces $L(X, Y)$ are traditionally equipped with the topology of bounded convergence (i.e. the topology of uniform convergence on bounded sets) or with a weaker one. But even in the case when G is one-dimensional and E and D are spaces of type (F) it is not known whether (2) holds, i.e. whether $(E\hat{\otimes}F)'_b \sim L_b(E, F'_b)$.[1)]

Analogous problems also arise in topology when one considers spaces C of continuous mappings: the exponential law $C(S \times T, U) \sim C(S, C(T, U))$ holds if T is locally bicompact and the spaces C are equipped with the topology of compact convergence,[2)] but it does not hold if T is not locally bicompact (for any topologization of the space C). Spanier [10] has shown that these difficulties can be avoided if one replaces topology by a more general structure which is called by him a quasi-topology. An analogous theory also can be constructed for vector spaces.

However, it turned out to be possible to extend significantly the domain of applicability of the exponential law within the category of locally convex spaces. Of course, to do this it was necessary to reject projective topologization for for tensor products as well as bounded convergence (which also has an essentially projective character) for the spaces of continuous linear mappings. Inductive tensor products subjected not to the usual, but to "local" completion are considered here. (The existence of a local completion is established for arbitrary locally convex separable spaces.) Spaces of continuous linear mappings are equipped with a new "inductive" topology (which majorizes the topology of uniform convergence on bounded sets), and a considerable part of the paper deals with this.

The basic result of this paper is the fact that if we consider $E \tilde{\otimes} F$ as the local completion of the inductive tensor product $E \otimes_i F$ and $L_i(X, Y)$ as the space $L(X, Y)$ of continuous linear mappings of X into Y equipped with the inductive topology, then the canonical isomorphism

$$L_i(E \tilde{\otimes} F, G) \sim L_i(E, L_i(F, G)),$$

holds for arbitrary locally complete bornological spaces E, F and G. (All other spaces occurring in the last formula are locally complete and bornological.) Therefore locally complete bornological spaces form a D-category in

1) Cf. [3], question 7.

2) Cf. [1], § 3, Corollary 2 of Theorem 3.

the sense of A. S. Švarc [11] with the functors Ω_A and Σ_A given by the formulas $\Omega_A(X) = L_i(A, X)$, $\Sigma_A(X) = A \tilde{\otimes} X$.[1]

The definitions of the notions of the theory of topological vector spaces can be found in [2]. Any meaning of terms different from that assumed in [2] or introduced there only in the exercises is explained below. For definiteness, all linear spaces are assumed to be complex.

§1. Banach balls

Let A be a bounded absolutely convex[2] set in a separable linear topological space E. The linear span of A in E equipped with the norm

$$\|x\|_A = \inf\{\rho > 0 : x \in \rho A\}.$$

will be denoted by E_A. Sometimes it is convenient to assume that the norm $\|x\|_A$ is defined for all $x \in E$ but (according to the convention that $\inf \emptyset = +\infty$) $\|x\|_A = +\infty$ for $x \in E \setminus E_A$. Obviously $\breve{A} \subset A \subset \widehat{A}$, where

$$\breve{A} = \{x \in E_A : \|x\|_A < 1\} = \bigcup \{\rho A : \rho < 1\}$$

is the open, and

$$\widehat{A} = \{x \in E_A : \|x\|_A \leqslant 1\} = \bigcap \{\rho A : \rho > 1\}$$

is the closed unit ball in E_A. Therefore A is used as the closed unit ball in E_A if it is linearly closed, i.e. if every straight line passing through 0 intersects A in a closed interval; in other words, if A and its linear closure $\widehat{A}$ coincide.

Since A is bounded, the identity imbedding

$$i_A : E_A \to E$$

is continuous.

Definition 1. Any bounded and absolutely convex set in a locally convex separable space E will be called a *ball*. Any ball A for which E_A is complete (that is, E_A is a Banach space) will be called a *Banach ball*.

It is known ([2], Russian pp. 165–166) that every complete ball (in particular every closed ball in a complete or in a boundedly complete[3] space

1) The fulfillment of the other axioms of a D-category is easy to verify.

2) Absolutely convex = stably convex. An absolutely convex set can also be defined as a set $M \subset E$ such that if $x, y \in M$ and $|\lambda| + |\mu| \leq 1$, then $\lambda x + \mu y \in M$.

3) A boundedly complete (quasi-complete) space is a separable locally convex space where any bounded and closed set is complete.

as well as every bicompact or even only weakly bicompact ball in any space) is a Banach ball. But in general a Banach ball A, even if closed in E_A, is not necessarily closed in E (example: the unit ball of the space $C(a, b)$ considered in $L^1(a, b)$). Obviously the linear closure of a Banach ball is also a Banach ball.

1.1. *The image of a Banach ball under a continuous linear mapping is a Banach ball.*

Let A be a Banach ball in E, let $u \in L(E, F)$, and let $B = u(A)$, so that B is a ball in F, $u(E_A) = F_B$ and the restriction of u to E_A considered as a mapping into F_B is continuous. Since

$$u(\breve{A}) = u(\bigcup_{\rho<1} \rho A) = \bigcup_{\rho<1} \rho B = \breve{B},$$

it is also an open mapping. But a separable image of a Banach space under a continuous linear mapping is a Banach space.

1.2. *The product S of any family of Banach balls $(S^\alpha)_{\alpha\in A}$ is a Banach ball (in the product E of corresponding spaces E^α).*

S is a ball, as a product of balls. Let (x_n), where $x_n = (x_n^\alpha)_{\alpha \in A}$, be a Cauchy sequence in E_S. Then (x_n^α) is a Cauchy sequence in $E^\alpha_{S^\alpha}$ for every $\alpha \in A$ and therefore (x_n^α) converges to an $x^\alpha \in E^\alpha_{S^\alpha}$. Let $x = (x^\alpha)_{\alpha\in A}$. For any $\epsilon > 0$ there exists an n_ϵ such that for any $m, n > n_\epsilon$,

$$x_n - x_m \in \varepsilon S, \qquad \text{i.e.} \qquad x_n^\alpha - x_m^\alpha \in \varepsilon S^\alpha$$

for all $\alpha \in A$. Passing to the limit as $m \to \infty$, we obtain that $x_n^\alpha - x^\alpha \in 2\epsilon S^\alpha$ for all $\alpha \in A$, i.e. $x_n - x \in 2\epsilon S$ for any $n > n_\epsilon$. But this means that $x_n - x$, and consequently x belong to E_S and $x_n \to x$ in E_S.

1.3. *The algebraic sum of two Banach balls is a Banach ball.*

Let A and B be Banach balls in E. Then (by 1.2) $A \times B$ is a Banach ball in $E \times E$. Since $A + B$ is the image of $A \times B$ under the continuous linear mapping $(x, y) \to x + y$ of the Cartesian product $E \times E$ into E, it remains to apply 1.1.

Definition 2. A *fundamental system of Banach balls* in a locally convex separable space E is any family of Banach balls in E such that every Banach ball in E is contained in a ball of this family.

1.4. *The fundamental system of Banach balls $(S^\alpha)_{\alpha\in A}$ in E is ordered by inclusion, is directed, and covers E.*

Let $\alpha, \beta \in A$. Since (by 1.3) $S^\alpha + S^\beta$ is a Banach ball, there exists a $\gamma \in A$ such that $S^\alpha + S^\beta \subset S^\gamma$. But then also $S^\alpha \cup S^\beta \subset S^\gamma$, which proves the first statement. The second statement follows from the fact that every $x \in E$ is contained in the Banach ball Λx, where $\Lambda = \{\lambda: |\lambda| \leq 1\}$.

1.5. *The intersection of any family of Banach balls is a Banach ball.*

Let $(S^\alpha)_{\alpha \in A}$ be a family of Banach balls in E, and let $S = \bigcap_{\alpha \in A} S^\alpha$. For every $x \in E$ we have

$$\begin{aligned} \|x\|_S &= \inf\{\rho > 0: x \in \rho S\} = \inf\{\rho > 0: x \in \rho S^\alpha \quad \text{for all} \quad \alpha \in A\} \\ &= \inf\{\rho > 0: \|x\|_{S^\alpha} \leqslant \rho \quad \text{for all} \quad \alpha \in A\} \\ &= \inf\{\rho > 0: \sup_{\alpha \in A} \|x\|_{S^\alpha} \leqslant \rho\} = \sup_{\alpha \in A} \|x\|_{S^\alpha}. \end{aligned}$$

Thus

$$E_S = \{x \in E : \sup_{\alpha \in A} \|x\|_{S^\alpha} < +\infty\}.$$

The Cauchy sequence (x_n) in E_S is a Cauchy sequence in every E_{S^α}, and consequently it converges in E_{S^α}, and therefore also in E, to some x^α. Inasmuch as E is separable, $x = x^\alpha$ does not depend on α. For every $\epsilon > 0$ there exists an n_ϵ such that

$$\|x_n - x_m\|_S = \sup_{\alpha \in A} \|x_n - x_m\|_{S^\alpha} \leqslant \varepsilon$$

for all $m, n > n_\epsilon$. Letting $m \to \infty$ we obtain

$$\|x_n - x\|_S = \sup_{\alpha \in A} \|x_n - x\|_{S^\alpha} \leqslant \varepsilon$$

for all $n > n_\epsilon$, so that $x \in E_S$ (together with $x_n - x$) and $x_n \to x$ in E_S.

1.6. *It follows from* 1.5 *that if a ball A in E is contained in a Banach ball, there exists a smallest Banach ball containing A. This ball will be denoted by $A^\sim$.*

In view of 1.3 and 1.5, *the set of all Banach balls in a locally convex separable space, ordered by inclusion, is a lattice. Every subset of this lattice has a greatest lower bound, and every subset which is bounded from above has a least upper bound.*

1.7. *A barrel T in a separable locally convex space E absorbs every Banach ball A.*

In fact, the continuity of i_A implies that $T \cap E_A = i_A^{-1}(T)$ is a barrel, and hence it is a neighborhood of zero in E_A.

Let X be a separable linear topological space. In the sequel the completion of X, i.e. a complete separable linear topological space containing X as a dense (topological and vector) subspace, will be denoted by $X^\wedge$. It is known that $X^\wedge$ always exists (and $X^\wedge$ is defined uniquely to within an isomorphism which is the identity on X). Then every mapping $\phi \in L(X, Y)$ can be extended uniquely and continuously to $\phi^\wedge \in L(X^\wedge, Y^\wedge)$, and if $\phi = \psi \circ \chi$, then $\phi^\wedge = \psi^\wedge \circ \chi^\wedge$.

Let A be a ball in E. The closed unit ball in $E_A^\wedge$ will be denoted by $A^\wedge$.

1.8. *$i_A^\wedge(E_A^\wedge) \subset E$ if and only if there exists a Banach ball containing A. Then* a) *$A^\sim$ (cf. 1.6) and the linear closure of the ball $i_A^\wedge(A^\wedge)$ coincide, and* b) *if $u \in L(E, F)$ and B is a linearly closed Banach ball in F such that $u(A) \subset B$, then $u(A^\sim) \subset B$.*

In fact, if $i_A^\wedge(E_A^\wedge) \subset E$, then, in view of 1.1, $i_A^\wedge(A^\wedge)$ is a Banach ball in E containing A. Conversely, let C be a Banach ball in E containing A. We have $i_A^\wedge = i_C^\wedge \circ i_{AC}^\wedge$, where i_{AC} is the identity imbedding of E_A into E_C. Since $E_C^\wedge = E_C$, $i_C^\wedge$ is the identity imbedding and

$$i_A^\wedge(E_A^\wedge) = i_{AC}^\wedge(E_A^\wedge) \subset E_C \subset E.$$

Moreover, if the ball C is linearly closed, then (since $A^\wedge$ is the closure of A in $E_A^\wedge$) $i_A^\wedge(A^\wedge) = i_{AC}^\wedge(A^\wedge) \subset [i_{AC}^\wedge(A)]^- = A^- \subset C$, where $^-$ denotes the closure in E_C. Then the linear closure of $i_A^\wedge(A^\wedge)$ is also contained in C. Since C is an arbitrary Banach ball containing A, the linear closures of $i_A^\wedge(A^\wedge)$ and $A^\sim$ coincide. Finally, $u \circ i_A = i_B \circ u_{AB}$, where u_{AB} is the restriction of u to E_A and is considered as a mapping into F_B. Since $u_{AB}^\wedge(A') \subset [u_{AB}^\wedge(A)]^- = [u(A)]^- \subset B$, where $^-$ denotes the closure in F_B, we have $u^\wedge(i_A^\wedge(A^\wedge)) = i_B^\wedge(u_{AB}^\wedge(A^\wedge)) = u_{AB}^\wedge(A^\wedge) \subset B$, and therefore also $u(A^\sim) \subset B$.

1.9. *$A^\wedge$ is the smallest linearly closed Banach ball in $E_A^\wedge$ containing A.*

In fact, if $E = E_A^\wedge$, then $i_A^\wedge$ is the identity mapping of E onto itself and statement 1.8 a) implies that $A^\wedge \subset A^\sim$; the inverse inclusion is obvious.

§2. Local completeness and completion

A sequence of elements of a locally convex separable space E is said to be locally convergent in the sense of Köthe [5] if it is contained in some E_A,[1] and if it is convergent in this E_A. Local convergence implies convergence in the topology of the space E. The converse is, in general, not true. However, in a space E of denumerable type[2] any convergent sequence converges locally (cf. [5], §28.3).

---

1) The ball A in [5] is required to be closed.

2) I.e. possessing a denumerable fundamental system of neighborhoods of zero (= metrizable).

Definition 3. We shall call a sequence of elements of the space E a *local Cauchy sequence*, if it is contained in an E_A and if it is a Cauchy sequence there. E will be called *locally complete* if every local Cauchy sequence in E converges.

2.1. *It is obvious that the image of a local Cauchy sequence under a continuous linear mapping is a local Cauchy sequence.*

2.2. *E is locally complete if and only if it is (b)-complete in the sense of Słowikowski* [9], *i.e. if $\phi\hat{}(G\hat{}) \subset E$ for any normed space G and for an arbitrary $\phi \in L(G, E)$.*

Indeed, every $z\hat{} \in G\hat{}$ is a limit of a local Cauchy sequence (z_n) from G. By 2.1, $\phi(z_n)$ is then a local Cauchy sequence in E. Consequently, if E is locally complete, $\phi(z_n)$ converges to some $x \in E$. But in such a case $\phi\hat{}(z\hat{}) = x$ and therefore $\phi\hat{}(G\hat{}) \subset E$; that is, E is (b)-complete. Conversely, let (x_n) be a local Cauchy sequence in E, i.e. a Cauchy sequence in some E_A. Then x_n converges to some $\hat{x}$ in $\hat{E}_A$, Consequently, if E is (b)-complete, $x_n = i_A\hat{}(x_n)$ converges to $i_A\hat{}(\hat{x}) \in E$ and therefore E is locally complete.

2.3. *It is obvious that sequential completeness of E (and the more so completeness or bounded completeness) implies local completeness.*

The converse is, in general, not true. Thus a Banach space is always locally complete in the weak topology but, as we know, not always sequentially complete (example: c_0). However,

2.4. *A locally complete space E of denumerable type is complete.*

Let $\hat{x} \in \hat{E}$. Since $\hat{E}$ is also of denumerable type, $\hat{x} = \lim x_n$, where $x_n \in E$, and, according to the remark of Köthe mentioned above, (x_n) converges to $\hat{x}$ locally, i.e. converges in some $\hat{E}_C$. But then (x_n) is a Cauchy sequence in E_A, where $A = C \cap E$. Since E is locally complete, we see that (x_n) converges to an x in E and therefore also in $\hat{E}$. Consequently $\hat{x} = x \in E$, and therefore $\hat{E} = E$; that is, E is complete.

2.5. *E is locally complete if and only if every bounded subset of it (or, equivalently, every ball) is contained in a Banach ball.*

Let A be a ball in E. If E is locally complete, then, by virtue of 2.2, $i_A\hat{}(\hat{E}_A) \subset E$. But then A is contained in a Banach ball $i_A\hat{}(\hat{A}) \subset E$. Conversely, let G be a normed space with unit ball S and let $\phi \in L(G, E)$, so that $\phi(S)$ is a ball in E. If any ball in E is contained in a Banach ball, then $\phi\hat{}(S) = \phi(S) \subset A$, where A is a linearly closed Banach ball in E. Then, by 1.8 b), $\phi\hat{}(S^\sim) \subset A$ also. But, in view of 1.9, $S^\sim$ is the closed unit ball of the space $G^{\wedge}$. Consequently $\phi^{\wedge}(G^{\wedge}) \subset E_A \subset E$, and therefore (by 2.2) E is locally complete.

2.6. *It follows immediately from 2.5 that local completeness of a space is preserved if one equips the space with any other topology with the same family of bounded sets.*

2.7. Theorem. *Every locally convex separable space E is a dense topological vector subspace of a locally complete space $E^{\sim}$ which satisfies the following condition: Every mapping $\phi \in L(E, F)$, where F is an arbitrary locally complete space, can be extended continuously to $E^{\sim}$, i.e. it can be written as $\phi = \bar{\phi} \circ i$, where i is the identity imbedding of E into $E^{\sim}$ and $\bar{\phi} \in L(E^{\sim}, F)$.*

Proof. For every subspace H of the space $E^{\wedge}$ we define $H^{(1)} = \bigcup A^{\sim}$, where A runs through the set of all balls contained in H ($A^{\sim}$ exists since $E^{\wedge}$ is complete; see 1.6). If $x, y \in H^{(1)}$ are such that $x \in A^{\sim}$ and $y \in B^{\sim}$, where A and B are balls contained in H, then (since $\lambda C^{\sim} = (\lambda C)^{\sim}$ and $(2A + 2B)^{\sim}$ are convex sets containing $(2A)^{\sim}$ and $(2B)^{\sim}$)

$$x + y \in A^{\sim} + B^{\sim} = \frac{1}{2}[(2A)^{\sim} + (2B)^{\sim}] \subset (2A + 2B)^{\sim};$$

thus $x + y \in H^{(1)}$. Since $\lambda x \in (\lambda A)^{\sim}$, we have $\lambda x \in H^{(1)}$ for all λ. Thus $H^{(1)}$ is a vector space. We shall assume that $H^{(1)}$ and all the other subspaces of $E^{\wedge}$ are equipped with the induced topology. Now we shall construct (by transfinite induction) an increasing transfinite sequence of subspaces $E^{(\delta)}$. a) Let $E^{(0)} = E$. b) If $E^{(\gamma)}$ has already been constructed, we set $E^{(\gamma+1)} = (E^{(\gamma)})^{(1)}$. c) If δ is a limit number and if we have already constructed an increasing family of subspaces $(E^{(\gamma)})_{0 \leq \gamma < \delta}$, we set $E^{(\delta)} = \bigcup_{\gamma < \delta} E^{(\gamma)}$. Since all $E^{(\gamma)}$ are contained in $E^{\wedge}$, the sequence can be stabilized so that there exists a smallest ϑ for which $E^{(\vartheta+1)} = E^{(\vartheta)}$. Then $E^{\sim} = E^{(\vartheta)}$ has the required properties. In fact, since $E \subset E^{\sim} \subset E^{\wedge}$, E is a dense subspace of $E^{\sim}$. If A is a ball in $E^{\sim}$, then $A \subset (E^{\sim})^{(1)} = E$ and, in view of 2.5, $E^{\sim}$ is locally complete. Let $\phi \in L(E, F)$, where F is a locally complete space. We shall show that $\phi^{\wedge}(E^{(\delta)}) \subset F$ for all δ. In fact, a) $\phi^{\wedge}(E^{(0)}) = \phi(E) \subset F$. b) Let $\phi^{\wedge}(E^{(\gamma)}) \subset F$ and let A be a ball in $E^{(\gamma)}$. Since F is locally complete, $\phi^{\wedge}(A) \subset B$, where B is a linearly closed Banach ball. Then, in view of 1.8 b), also $\phi^{\wedge}(A^{\sim}) \subset B \subset F$. On the basis of the definition of $E^{(\gamma+1)}$ we conclude that $\phi^{\wedge}(E^{(\gamma+1)}) \subset F$ also. c) If $\phi^{\wedge}(E^{(\gamma)}) \subset F$ for all $\gamma < \delta$, where δ is a limit number, then also

$$\varphi^{\wedge}(E^{(\delta)}) = \bigcup_{\gamma<\delta} \varphi^{\wedge}(E^{(\gamma)}) \subset F.$$

Thus $\phi^{\wedge}(E^{\sim})(=\phi^{\wedge}(E^{(\vartheta)}))\subset F$. Denote the restriction of $\phi^{\wedge}$ to $E^{\sim}$ (considered as a mapping into F) by $\bar{\phi}$. It is easy to see that $\bar{\phi}\in L(E^{\sim}, F)$ and $\phi=\bar{\phi}\circ i$.

Definition 4. The space $E^{\sim}$ constructed in the proof of Theorem 2.7 will be called the *local completion* of the space E.

It is easy to see that $E^{\sim}$ is the smallest locally complete subspace of $E^{\wedge}$ containing E. In particular, if E is locally complete, $E^{\sim}=E$ (besides, one can see this from the construction of $E^{\sim}$). Therefore the existence of noncomplete but locally complete spaces noted above shows that, in general, $E^{\sim}$ and $E^{\wedge}$ do not coincide. However,

2.8. *If E is a space of denumerable type, $E^{\sim}=E^{\wedge}$.*

In fact, $E^{\wedge}$ is a space of denumerable type, and therefore in view of Köthe's remark mentioned above, here already $E^{(1)}=E^{\wedge}$.

Definition 5. Let E be a vector space and $(E_\alpha)_{\alpha\in A}$ a family of its subspaces ordered by inclusion and covering E. Further, let every E_α be equipped with a locally convex topology, and, moreover, if $E_\alpha\subset E_\beta$, let the identity imbedding of E_α into E_β be continuous. Then E equipped with the strongest locally convex topology in which the identity imbedding of each E_α into E is continuous is called the *inductive limit of the spaces* E_α and is denoted by $\varinjlim E_\alpha$.

Definition 6. A locally convex separable space E representable as the inductive limit of normed spaces will be called a *bornological* space.

E is a bornological space if and only if all its absolutely convex subsets, which absorb every bounded subset, are neighborhoods of zero. These sets in any locally convex separable space E form a fundamental system of neighborhoods of zero for the weakest bornological topology majorizing that of the space E. The space obtained by furnishing E with this bornological topology is called *the bornological space associated with* E and will be denoted by E_{born}. It is easy to see that *a continuous linear mapping of a bornological space F into E is also continuous as a mapping of F into E_{born}.*

2.9. *The local completion $E^{\sim}$ of a bornological space E is a bornological space.*

Let i be the identity imbedding of E into $E^{\sim}$ and let j be the identity mapping of $E^{\sim}$ onto $E^{\sim}_{\text{born}}$. Since i is continuous and E is a bornological space, the preceding remark allows us to conclude that the mapping $j\circ i$ is continuous. But $E^{\sim}_{\text{born}}$ possesses the same family of bounded sets as $E^{\sim}$, and consequently, in view of 2.6, E and $\widetilde{E}$ are locally complete simultaneously. Therefore, in view of 2.7, $j \circ i = \overline{(j \circ i)} \circ i$ and $\overline{j \circ i}$ is continuous. But since i is a dense imbedding while j is an identity and j^{-1} is continuous, $j = \overline{j \circ i}$. Thus j is continuous. This and the continuity of j^{-1} mean that the topologies of the spaces $\widetilde{E}$ and $\widetilde{E}_{\text{born}}$ coincide, i.e. $\widetilde{E}$ is a bornological space.

Definition 7. Locally convex bornological spaces will be called *spaces of type* (α) or *α-spaces.*

§3. β-spaces

Definition 8. Every locally convex separable space representable as the inductive limit of Banach spaces is called a β-space. The topology of a β-space will be called a *β-topology.*

3.1. *Every β-space is a bornological space.*

As Grothendieck, who introduced the notion of a β-space (cf. [4], Introduction), noted, every space of type (F) (i.e. every complete locally convex separable space of denumerable type) is a β-space. The union of any family of β-spaces, the quotient space of a β-space with respect to a closed subspace, and the inductive limit of β-spaces are β-spaces. It is easy to see (cf. §5) that also the inductive tensor product of β-spaces is a β-space. It is possible to prove that if the intersection of a family of β-spaces is bornological,[1)] it is a β-space. On the other hand, a closed subspace of a β-space and the projective limit of a sequence of β-spaces are not necessarily β-spaces.

A vector space E equipped with a locally convex separable topology t will be denoted by E_t.

3.2. *Absolutely convex sets in a locally convex separable space E_t absorbing all its Banach balls form a fundamental system of neighborhoods of zero for the weakest β-topology $\beta(t)$ in E majorizing t.*

Obviously, these sets form a fundamental system of neighborhoods of zero for the strongest locally convex topology i in E such that the identity imbeddings $i_A \colon E_A \to E_t$ are continuous for all Banach balls A from E_t. Since Banach balls are bounded sets, $i \geq t$, and hence, in particular, it follows that i is separable. Since, moreover, $E_i = \varinjlim E_A$ (cf. 1.4), i is a

1) Hence, in any case, for every family with the cardinality smaller than the first strongly inaccessible aleph (Mackey-Ulam Theorem [5]).

β-topology. Let i' be an arbitrary β-topology in E majorizing t. This means that $E_{i'} = \varinjlim B_\alpha$, where B_α are Banach spaces. Thus absolutely convex subsets of E absorbing the unit balls S_α of all the spaces B_α are absolutely convex neighborhoods of zero in $E_{i'}$. But then S_α are Banach balls in $E_{i'}$ and consequently (since $i' \geq t$) also in E_t. Therefore every neighborhood of zero for i is a neighborhood of zero for i'; that is, $i' \geq i$. Consequently i is the weakest β-topology $\beta(t)$ which majorizes t.

We shall call $\beta(t)$ the β-topology associated with t.

3.2$'$. *Obviously, $E_{\beta(t)}$ has the same family of Banach balls as E_t, and $\beta(t)$ is the strongent locally convex topology in E having this property.*

It follows immediately from 3.2 that

3.3. *E_t is a Banach space if and only if every absolutely convex set absorbing all Banach balls is a neighborhood of zero in E_t.*

3.4. *A locally convex separable space E_t is a β-space if and only if for any locally convex space F every $\phi \in \mathfrak{L}(E, F)$ bounded on all Banach balls in E_t is continuous.*

In fact, if the last condition is satisfied, then, in view of 3.2$'$, the identity mapping of E_t onto $E_{\beta(t)}$ is continuous, i.e. $\beta(t) \leq t$. Since, on the other hand, always $t \leq \beta(t)$, it follows that $t = \beta(t)$, i.e. E_t is a β-space. Conversely, let E_t be a β-space and let $\phi \in \mathfrak{L}(E, F)$. Suppose that $\phi(A)$ can be absorbed by any absolutely convex neighborhood of zero V in F for every Banach ball A in E_t. Then $\phi^{-1}(V)$ is an absolutely convex set in E absorbing all Banach balls, i.e. (by 3.3) $\phi^{-1}(V)$ is a neighborhood of zero in E_t. By the same token ϕ is continuous.

3.5. *Let E and F be β-spaces, $(S_\alpha)_{\alpha \in A}$ a fundamental system of Banach balls in E and $\phi \in \mathfrak{L}(E, F)$. If ϕ is a one-to-one mapping and $(\phi(S_\alpha))_{\alpha \in A}$ is a fundamental system of Banach balls in F, then ϕ is a topological and a vector isomorphism of E onto F.*

In view of 1.4, $\phi(E) = F$. Since, moreover, ϕ is bounded on all Banach balls in E and ϕ^{-1} is bounded on all Banach balls in F, it remains to apply 3.4.

3.6. *Every α-space E is a β-space.*

Every absolutely convex set in E absorbing all Banach balls absorbs all bounded sets (since, in view of 2.5, every bounded set in E is contained in a Banach ball). This means that every such absolutely convex set is a neighborhood of zero (since E is a bornological space), and it remains to apply 3.3.

3.6′. *It follows from* 3.1 *and* 3.6 *that the class of* α*-spaces and the class of locally complete* β*-spaces coincide.*

Definition 9. A locally convex separable space representable as the inductive limit of a sequence of Banach spaces (being by the same token a β-space) is called a *space of type* (LB).

Remark. Unlike in [5], we do not require that this sequence be strictly increasing. Its spaces can even all coincide, so that all Banach spaces belong to the class of spaces of type (LB).

Considering spaces of type (LB) we shall use the following special cases of Theorems A and B of the Introduction to Grothendieck's memoir [4].

3.A. *Let* E *be a space of type* (F) *and let* F *be a locally convex separable space being the union of a sequence* (F_n) *of Banach spaces such that their identity imbeddings into* F *are continuous. Then for every* $u \in L(E, F)$ *there exists an* n *such that* $u(E) \subset F_n$ *and* u *considered as a mapping into* F_n *is continuous.*

3.B. *A continuous linear mapping of a space of type* (LB) *onto a* β*-space is an open mapping.*

Note the following two properties of the spaces of type (LB) implied by these theorems:

3.7. *Any space* E *of type* (LB) *possesses an increasing fundamental sequence of Banach balls.*

In fact, $E = \varinjlim E_n$, where E_n are Banach spaces. Since the identity imbeddings $E_n \to E_{n+1}$ are continuous, we may assume without loss of generality that the unit balls S_n of the spaces E_n satisfy the condition $2S_n \subset S_{n+1}$ $(n = 1, 2, \cdots)$. Let S be an arbitrary Banach ball in E. Since E_S is a space of type (F) and the identity imbedding of E_S into E is continuous, Theorem 3.A implies that there exists an n such that $E_S \subset E_n$ and the identity imbedding of E_S into E_n is continuous. The latter means that there exists a $\rho > 0$ such that $S \subset \rho S_n$. Choosing k such that $2^k > \rho$, we obtain $S \subset 2^k S_n \subset S_{n+k}$.

3.8. *A locally convex separable space* F_t *possesses a fundamental sequence of Banach balls* (S_n) *if and only if* $F_{\beta(t)}$ *is a space of type* (LB), *and in this case* (S_n) *plays the role of a fundamental sequence of Banach balls for an arbitrary locally convex separable topology* t' *in* F *such that* $t' \leq t$.

First of all, it follows from 1.4 that (S_n) contains an increasing subsequence that is also a fundamental sequence of Banach balls in F_t.

Therefore, without loss of generality, we may assume that the sequence (S_n) is increasing. Since, by 1.4, (S_n) covers F, 3.2 yields $F_{\beta(t)} = \varinjlim F_{S_n}$, i.e. $F_{\beta(t)}$ is a space of type (LB). Conversely, let $F_{\beta(t)}$ be a space of type (LB) and let t' be a locally convex separable topology in F majorizing the topology t. According to 3.7, there exists a fundamental sequence of Banach balls (S_n) in $F_{\beta(t)}$. Since $\beta(t)$ is a β-topology and $\beta(t) \geq t'$, we see that $\beta(t) \geq \beta(t')$ and the identity mapping of $F_{\beta(t)}$ onto $F_{\beta(t')}$ is continuous. But then, in view of Theorem 3.B, $\beta(t) = \beta(t')$, and consequently, in view of 3.2′, (S_n) is a fundamental sequence of Banach balls in $F_{t'}$ also.

Remark. Not all the spaces of type (LB) are locally complete.

In fact, B. M. Markov [7] constructed a separable inductive limit E of a sequence of Banach spaces E_n such that there exists a bounded set A which is not contained in any E_n. As we have seen in the proof of 3.7, every Banach ball of the space E is contained in an E_n. It follows from a theorem of Grothendieck ([3], Theorem 9) that A is contained in the closure of a certain Banach ball. By the same token E is also an example of a space with a Banach ball whose closure is not a Banach ball.

§4. Inductive topology in the space of continuous linear mappings

Definition 10. Let E and F be locally convex separable spaces. The β-topology $i = \beta(s)$ associated with the topology of simple convergence will be called the *inductive topology* in $L(E, F)$.

Thus, according to 3.2 and 3.2′, absolutely convex neighborhoods of zero in $L_i(E, F)$ are absolutely convex sets in $L(E, F)$ absorbing every Banach ball of the space $L_s(E, F)$. $L_i(E, F)$ has the same Banach balls as $L_s(E, F)$.

4.1. *The Banach balls of the space $L_s(E, F)$ are bounded in $L_b(E, F)$, and consequently they are Banach balls also in $L_b(E, F)$.*

In fact, the sets $T(A, V) = \{u \in L(E, F): u(A) \subset V\}$, where A are bounded sets in E and V are absolutely convex closed neighborhoods of zero in F, form a fundamental system of neighborhoods of zero in $L_b(E, F)$. But these sets are barrels in $L_s(E, F)$ and, by 1.7, they absorb all Banach balls.

4.2. *It follows from 4.1 and 3.2 that the inductive topology in $L(E, F)$ majorizes the topology b of uniform convergence on bounded sets (called here also the topology of bounded convergence).*

Now we shall describe the inductive topology in $L(E, F)$ under special assumptions on E and F.

Definition 11. A locally convex separable space in which every absolutely convex closed set absorbing all bounded sets is a neighborhood of zero is called a *quasi-barreled space.*

Obviously, bornological spaces are quasi-barreled spaces.

4.3. *The local completion* $E^{\sim}$ *of a quasi-barreled space* E *is a quasi-barreled space.*

If V is a barrel in $E^{\sim}$ absorbing all bounded sets, then $U = V \cap E$ is a barrel in E with the same property, and consequently U is a neighborhood of zero in E. But then V, containing the closure of U in $E^{\sim}$, is a neighborhood of zero in $E^{\sim}$.

Definition 12. Any family of absolutely convex closed neighborhoods of zero in a locally convex space E such that every neighborhood of zero absorbs a neighborhood of this family will be called a *quasi-fundamental system of neighborhoods of zero.*

The unit ball of a normed space forms a quasi-fundamental system of neighborhoods of zero.

4.4. **Theorem.** *Let* E *be a quasi-barreled space,* F *a locally complete space,* $(U_\alpha)_{\alpha \in \mathrm{A}}$ *a fundamental system of neighborhoods of zero in* E, *and* $(V_\beta)_{\beta \in \mathrm{B}}$ *q quasi-fundamental system of neighborhoods of zero in* F. *For every* $\phi\colon \mathrm{B} \to \mathrm{A}$ *we set*

$$T_\varphi = \{u \in \mathcal{L}(E, F)\colon u(U_{\varphi(\beta)}) \subset V_\beta \quad \text{for all} \quad \beta \in \mathrm{B}\}. \tag{3}$$

Then $(T_\phi)_{\phi \in \mathrm{A}^\mathrm{B}}$ *is a fundamental system of Banach balls in* $L_i(E, F)$, *and* $L_i(E, F)$ *is an* α*-space.*

Proof. It is easy to see that $T_\phi \subset L(E, F)$. Let $x \in E$. For every $\beta \in \mathrm{B}$ there exists a λ_β such that $x \in \lambda_\beta U_{\phi(\beta)}$. Then it follows from (3) that

$$T_\varphi x \subset \bigcap_{\beta \in \mathrm{B}} \lambda_\beta V_\beta = B. \tag{4}$$

But B is absorbed by every neighborhood of zero, i.e. B is bounded in F. Consequently T_ϕ is bounded in $L_s(E, F)$. The absolute convexity of T_ϕ is implied by the absolute convexity of all V_β. Thus T_ϕ is a ball in $L_s(E, F)$. We set $L(E, F) = G$ and let (u_n) be a Cauchy sequence in $G_\phi = G_{T_\phi}$. For every $\epsilon > 0$ there exists an n_ϵ such that

$$\frac{u_n - u_m}{\varepsilon} \in T_\varphi \quad \text{for all} \quad m, n > n_\varepsilon, \tag{5}$$

i.e.

$$\frac{u_n - u_m}{\varepsilon}(U_{\varphi(\beta)}) \subset V_\beta \quad \text{for all } m, n > n_\epsilon \text{ and } \beta \in \mathrm{B}. \tag{6}$$

But since F is locally complete, 2.5 implies that $B \subset S$, where S is a Banach ball in F. Then it follows from (4) and (5) that

$$\frac{u_n(x) - u_m(x)}{\varepsilon} \in S \quad \text{for all} \quad m, n > n_\varepsilon.$$

Thus $(u_n(x))$ is a Cauchy sequence in F_S, and therefore $(u_n(x))$ converges in F_S (and consequently also in F) to a certain $u(x)$. Obviously $u \in \mathfrak{L}(E, F)$. Now, taking $m \to \infty$ in (6) and taking into account that all V_β are closed, we obtain, after passing to the limit,

$$\frac{u_n - u}{\varepsilon}(U_{\varphi(\beta)}) \subset V_\beta \quad \text{for all } m, n > n_\epsilon \text{ and } \beta \in \mathrm{B}.$$

Hence it follows that $u_n - u \in \epsilon T_\phi$ for all $n > n_\epsilon$. But this means that $u_n - u$, and therefore also u, belongs to G_ϕ, and $u_n \to u$ in G_ϕ, so that G_ϕ is complete. Thus T_ϕ is a Banach ball in $L_s(E, F)$, and consequently also in $L_i(E, F)$.

Now let T be an arbitrary bounded set in $L_i(E, F)$; in particular, an arbitrary Banach ball. In view of 4.2, T is bounded in $L_b(E, F)$, i.e. $T(A)$ is bounded in F for every bounded A from E, and therefore $U^\beta = \{x \in E \colon Tx \subset V_\beta\}$ absorbs all bounded sets in E. But U^β is closed and absolutely convex. Since E is a quasi-barreled space, we conclude that U^β is a neighborhood of zero in E. Therefore with every $\beta \in \mathrm{B}$ one can associate $\phi(\beta) \in \mathrm{A}$ so that $U_{\phi(\beta)} \subset U^\beta$. But then $T \subset T_\phi$, and therefore $L_i(E, F)$ is locally complete.

Remark. Since in the last part of the proof we made use only of the boundedness of T in $L_b(E, F)$, and since T_ϕ are Banach balls in $L_b(E, F)$, we have proved simultaneously that *if E is a quasi-barreled space and if F is locally complete, then $L_b(E, F)$ is locally complete.*

4.5. *If E is a quasi-barreled space, then $E'_i = \varinjlim E'_{U^0}$, where U runs through a decreasing fundamental system of neighborhoods of zero in F; in other words, absolutely convex neighborhoods of zero for the inductive topology in the conjugate space of a quasi-barreled space are absolutely convex sets of mappings absorbing all equicontinuous mappings.*

In fact, $E' = L(E, \mathbf{C}^1)$, where $\mathbf{C}$ is the field of complex numbers, and in (3) one may take the system formed by the unit ball $\Lambda = \{\lambda \in \mathbf{C} \colon |\lambda| \leq 1\}$ as $(V_\beta)_{\beta \in \mathrm{B}}$. Then $T_\phi = T_\alpha = U^0_\alpha$.

Such a form of inductive topologization of E' (without any assumptions on E) has been considered by many authors.

Remark. Even in the case of the conjugate space the inductive topology i can be stronger than the "strong" topology b. It will be so, for instance, when E is not a "regular" space of denumerable type ([3], Theorem 7).

4.6. *If E is a quasi-barreled space and F is a locally complete space, then the mapping Φ associating with every $\bar{u} \in L(E^{\sim}, F)$ its restriction u to E is a topological and a vector isomorphism of $L_i(E^{\sim}, F)$ onto $L_i(E, F)$.*

In view of Theorem 2.7, Φ is an isomorphism of $L(E^{\sim}, F)$ onto $L(E, F)$. Let $M^{\sim}$ denote the closure of the set $M \subset E$ in $E^{\sim}$ and let i be the identity imbedding of E into $E^{\sim}$. Taking into account the fact that the V_β are closed and using the notation of Theorem 4.4, we obtain

$$\begin{aligned}\Phi^{-1}(T_\varphi) &= \{\bar{u} \in L(E^{\sim}, F): \ \bar{u}(i(U_{\varphi(\beta)})) \subset V_\beta \quad \text{for all} \quad \beta \in B\} \\ &= \{\bar{u} \in L(E^{\sim}, F): \ \bar{u}(U_{\varphi(\beta)}^{\sim}) \subset V_\beta \quad \text{for all} \quad \beta \in B\}.\end{aligned}$$

But since $(U_\alpha)_{\alpha \in A}$ is a fundamental system of neighborhoods of zero in E and E is dense in $E^{\sim}$, $(U_\alpha^{\sim})_{\alpha \in A}$ is a fundamental system of neighborhoods of zero in $E^{\sim}$. In view of 4.3 and 4.2, $(\Phi^{-1}(T_\phi))_{\phi \in A^B}$ is a fundamental system of Banach balls in $L_i(E^{\sim}, F)$, and it remains to apply 3.5.

Definition 13. Any family (A_α) of bounded sets covering a bornological space E and such that every absolutely convex set absorbing all A_α is a neighborhood of zero in E is called a *determining system of bounded sets.*

Every fundamental system of Banach balls of a β-space is a determining system of its bounded sets.

4.7. *It is obvious that if (A_α) is a determining system of bounded sets in a bornological space E, F is a locally convex space and $u \in \mathfrak{L}(E, F)$, then the boundedness of u on all A_α is a sufficient condition for u to be continuous.*

4.8. **Theorem.** *Let E be a bornological space, F a locally complete space, $(A_\alpha)_{\alpha \in A}$ a determining system of bounded sets in E and $(B_\beta)_{\beta \in B}$ a fundamental system of linearly closed Banach balls in F. For every $\psi: A \to B$ we get*

$$T^{\psi} = \{u \in \mathcal{L}(E, F): u(A_\alpha) \subset B_{\psi(\alpha)} \quad \textit{for all} \quad \alpha \in A\}. \tag{7}$$

Then $(T^{\psi})_{\psi \in B^A}$ is a fundamental system of Banach balls in $L_i(E, F)$, and $L_i(E, F)$ is an α-space.

Proof. In view of 4.7, $T^\psi \subset L(E, F)$. Since $(A_\alpha)_{\alpha \in \mathrm{A}}$ covers E, (7) implies the boundedness of T^ψ in $L_s(E, F)$. The absolute convexity of T^ψ is implied by the absolute convexity of all B_β. Thus T^ψ is a ball in $L_s(E, F)$. Let us set $L(E, F) = G$ and let (u_n) be a Cauchy sequence in $G^\psi = G_{T^\psi}$. For every $\epsilon > 0$ there exists an n_ϵ such that $u_n - u_m \in \epsilon T$ for all $m, n > n_\epsilon$, i.e.

$$\frac{u_n - u_m}{\varepsilon}(A_\alpha) \subset B_{\psi(\alpha)} \text{ for all } m, n > n_\epsilon \text{ and } \alpha \in \mathrm{A}. \tag{8}$$

Let $x \in E$. Then $x \in A_\alpha$ for a certain $\alpha \in \mathrm{A}$, and consequently

$$\frac{u_n(x) - u_m(x)}{\varepsilon} \in B_{\psi(\alpha)} \quad \text{for all} \quad m, n > n_\varepsilon.$$

By the same token $(u_n(x))$ is a Cauchy sequence in $F^\alpha = F_{B_{\psi(\alpha)}}$. Therefore $(u_n(x))$ converges in F^α, and consequently also in F, to a certain $u(x)$. Since F is separable, $u(x)$ is the same for all α for which $x \in A_\alpha$. Obviously, $u \in \mathcal{L}(E, F)$. Taking into account the fact that every B_β is closed in F_{B_β}, we obtain after passing to the limit

$$\frac{u_n - u}{\varepsilon}(A_\alpha) \subset B_{\psi(\alpha)} \text{ for all } n > n_\epsilon \text{ and } \alpha \in \mathrm{A},$$

i.e. $u_n - u \in \epsilon T^\psi$ for all $n > n_\epsilon$. But this means that $u_n - u$, and therefore also u, belongs to G^ψ and $u_n \to u$ in G^ψ, so that G^ψ is complete. Thus T^ψ is a Banach ball in $L_s(E, F)$, and therefore also in $L_i(E, F)$.

Now let T be an arbitrary bounded set in $L_i(E, F)$. In view of 4.2, T is bounded in $L_b(E, F)$, i.e. $T(A)$ is bounded in F for every bounded subset A of E. Since F is locally complete, one can associate $\phi(\alpha) \in B$ with every $\alpha \in \mathrm{A}$ so that $T(A_\alpha) \subset B_{\psi(\alpha)}$. But this means that $T \subset T^\psi$.

4.9. *If E is a space of type (DF) (cf. [3]) and F is a space of type (F), the inductive topology in $L(E, F)$ and the topology of uniform convergence on bounded sets coincide, Moreover, $L_i(E, F)$ $(= L_b(E, F))$ is a space of type (F).*

E possesses a fundamental system of bounded sets (A_n) and F possesses a fundamental system of neighborhoods of zero (V_n). Therefore $T(A_n, V_n)$ (cf. 4.1) form a countable fundamental system of neighborhoods of zero in $L_b(E, F)$. Further, since F is complete, $L_b(E, F)$ is complete (cf. [3], Corollary 2 of Theorem 3). Consequently $L_b(E, F)$ is a space of type (F). But then b is a β-topology. Since $b \geq s$, we conclude that $b \geq \beta(s)$. But (by 4.2) $\beta(s) \geq b$. Thus $b = \beta(s) = i$.

4.10. *If E is a space of type (F) with a decreasing fundamental sequence of absolutely convex closed neighborhoods of zero (U_n) and if F is a locally convex separable space possessing a fundamental sequence of linearly closed Banach balls (S_n), then $L_i(E, F)$ is a space of type (LB) with the fundamental sequence of Banach balls*

$$T_n = \{u \in \mathscr{L}(E, F) \colon u(U_n) \subset S_n\}. \tag{9}$$

Similarly as in the proof of 3.8, we may assume that the sequence (S_n) is increasing. Let $E_n = E_{U_n}$[1] and $F_n = F_{S_n}$, let π_n be the canonical mapping of E onto E_n, and let ω^n be the identity imbedding of F_n into F. It is easy to see that $u_n \longrightarrow \omega^n u_n \pi_n$ is a continuous linear mapping of $L_s(E_n, F_n)$, and consequently also of $L_b(E_n, F_n)$, into $L_s(E, F)$, and that T_n is the image of the unit ball

$$R_n = \{u_n \in L(E_n, F_n) \colon u_n(\pi_n(U_n)) \subset S_n\}$$

of the Banach space $L_b(E_n, F_n)$ under this mapping. Consequently T_n is a Banach ball in $L_s(E, F)$. Let $u \in L(E, F)$. In view of 3.A and 1.4 there exists a p such that $u = \omega^p \bar{u}_p$, where $\bar{u}_p \in L(E, F_p)$. Further, since $\bar{u}_p^{-1}(S_p)$ is a neighborhood of zero in E, there exists a q such that $\bar{u}_p(U_q) \subset S_p$. Let $n = \max(p, q)$, so that $U_n \subset U_q$ and $S_p \subset S_n$. Denoting the identity imbedding of F_p into F_n by $\omega^p{}_n$, we have $\omega^p{}_n \bar{u}_p(U_n) \subset S_n$. Hence it follows easily that $\omega^p{}_n \bar{u}_p = u_n \pi_n$, where $u_n \in \mathfrak{L}(E_n, F_n)$. Since, moreover, $u_n(\pi_n(U_n)) \subset S_n$ and $\pi_n(U_n)$ is the unit ball of the space E_n, we see that $u_n \in R_n$. But then, according to that which was said above, $u = \omega^p \bar{u}_p = \omega^n \omega^p{}_n \bar{u}_p = \omega^n u_n \pi_n \in T_n$. Thus (T_n) is an increasing sequence of Banach balls covering $L_s(E, F)$. Let $L_{i'}(E, F) = \varinjlim (L(E, F))_{T_n}$. Every absolutely convex neighborhood of zero in $L_i(E, F)$ absorbing all Banach balls of $L_s(E, F)$ absorbs, in particular, all T_n, and consequently it is a neighborhood of zero in $L_{i'}(E, F)$. By the same token the mapping $\epsilon \colon L_{i'}(E, F) \longrightarrow L_i(E, F)$ is continuous. But $L_{i'}(E, F)$ is a space of type (LB) and $L_i(E, F)$ is a β-space. Therefore, on the basis of Theorem 3.B, we conclude that ϵ is also an open mapping. By the same token $i' = i$, so that $L_i(E, F)$ is a space of type (LB). Finally, let T be an arbitrary Banach ball in $L_i(E, F)$. Similarly as in the proof of Proposition 3.7 one can establish the existence of $\rho > 0$ and n such that $T \subset \rho T_n$. But since ρS_n is a Banach

---

1) E_U, where U is an absolutely convex neighborhood of zero in E, denotes the quotient space of E by the largest of its subspaces N_U contained in U equipped with the norm $n(\pi_U(x)) = \inf\{\rho > 0 \colon x \in \rho U\}$, where π_U is the canonical mapping of E onto E_U.

ball, there exists a k such that $\rho S_n \subset S_{n+k}$. Then it follows from (9) that also $\rho T_n \subset T_{n+k}$. Consequently $T \subset T_{n+k}$, and the proposition is proved.

§5. Tensor α-products of β-spaces

Definition 14. The *projective tensor product* $E \otimes_\pi F$ of the locally convex spaces E and F is the tensor product $E \otimes F$ equipped with the locally convex topology π (*projective tensor topology*) defined by the condition: for any locally convex space G the linear mapping of $E \otimes_\pi F$ into G is continuous if and only if the bilinear mapping of $E \times F$ into G inducing it is continuous. The completion of $E \otimes_\pi F$ will be denoted by $E \hat{\otimes} F$.

5.1 ([4], **Chapter I, Proposition 2).** *When U and V run over quasi-fundamental systems of neighborhoods of zero in E and F, then $\Gamma(U \otimes V)$*[1] *runs through a quasi-fundamental system of neighborhoods of zero in $E \otimes_\pi F$. In particular, if E and F are normed spaces with unit balls A and B, $E \otimes_\pi F$ is normable and $\Gamma(A \times B)$ can be taken for the unit ball in $E \otimes_\pi F$. Then*

$$\|x \otimes y\| = \|x\| \|y\|. \tag{10}$$

Definition 15. The *inductive tensor product* $E \otimes_i F$ of the locally convex spaces E and F is the tensor product $E \otimes F$ equipped with the locally convex topology i (the *inductive tensor topology*) determined by the following condition: for any locally convex space G the linear mapping of $E \otimes_i F$ into G is continuous if and only if the bilinear mapping of $E \times F$ into G inducing it is separately continuous (i.e. continuous with respect to each argument for any fixed value of the other one).

5.2. ([4], **Chapter I, Proposition 14).** *If $E = \varinjlim E_\alpha$ and $F = \varinjlim F_\beta$, then $E \otimes_i F = \varinjlim E_\alpha \otimes_i F_\beta$.*

Obviously, $i \geq \pi$. If E and F are spaces of type (F) (in particular, Banach spaces), $i = \pi$ because every separately continuous bilinear mapping of $E \times F$ into any G is then a continuous mapping.

5.3. *The inductive tensor product $E \otimes_i F$ of β-spaces E and F is a β-space.*

In view of 5.2, the proof is reduced to the case when E and F are Banach spaces. Let S and T be the unit balls of these spaces. In view of (10), $E \to E \otimes y$ and $F \to x \otimes F$ are topological and vector isomorphisms of E and F into $E \otimes_\pi F$ for arbitrary nonzero $y \in F$ and $x \in E$. Therefore $S \otimes y$ and $x \otimes T$ are Banach balls in $E \otimes_\pi F$. Consequently, if $\tilde{\phi}$ is a linear mapping of $E \otimes_\pi F$ into G bounded on every Banach ball, and if ϕ

1) ΓM is the absolutely convex hull of M (= the smallest absolutely convex set containing M).

is the corresponding bilinear mapping of $E \times F$ into G, then $\phi(S \times y)$ $(=\tilde{\phi}(S \otimes y))$ and $\phi(x \times T)$ $(=\tilde{\phi}(x \otimes T))$ are bounded for every $y \in F$ and $x \in E$, i.e. ϕ is separately continuous. But then $\tilde{\phi} \in L(E \otimes_\pi F, G)$ and, according to 3.4, $E \otimes_i F = E \otimes_\pi F$ is a β-space.

Definition 16. The local completion $E \tilde{\otimes} F$ of the inductive tensor product $E \otimes_i F$ will be called the tensor α-product of the β-spaces E and F.

In view of 5.3, 2.9 and 3.1, $E \tilde{\otimes} F$ is an a-space.

5.4. *If E and F are spaces of type* (*F*), *then* $E \tilde{\otimes} F = E \hat{\otimes} F$ *and hence* $E \tilde{\otimes} F$ *is also a space of type* (*F*).

In fact, in view of 5.1, $E \otimes_i F = E \otimes_\pi F$ is a space of denumerable type, and it remains to apply 2.8.

5.5. *Let E and F be β-spaces with the fundamental systems of Banach balls* $(A_\alpha)_{\alpha \in \mathrm{A}}$ *and* $(B_\beta)_{\beta \in \mathrm{B}}$. *Then* a) $\{\Gamma(A_\alpha \otimes B_\beta)\}_{\alpha \in \mathrm{A}, \beta \in \mathrm{B}}$ *is a determining system of bounded sets in* $E \otimes_i F$, *and* b) *if G is a locally complete space, then any* $u \in \mathfrak{L}(E \otimes F, G)$ *bounded on all* $A_\alpha \otimes B_\beta$ *can be extended to a continuous linear mapping of* $E \tilde{\otimes} F$ *into* G.

Since (by 5.1) $\Gamma(A_\alpha \otimes B_\beta)$ are the unit balls in $E_\alpha \otimes_i F_\beta = E_\alpha \otimes_\pi F_\beta$ (where $E_\alpha = E_{A_\alpha}$ and $F_\beta = F_{B_\beta}$), and since $\Gamma(A_\alpha \otimes B_\beta)$ cover $E \otimes F$, the validity of statement a) follows from 5.2. Since the boundedness of u on $A_\alpha \otimes B_\beta$ implies the boundedness on $\Gamma(A_\alpha \otimes B_\beta)$, we have $u \in L(E \otimes_i F, G)$ (in view of 4.7), and the validity of statement b) follows from 4.6.

5.6. *Let X be a quasi-barreled space, Y a locally complete quasi-barreled space, and Z a locally complete space. The mapping*

$$\omega \colon L(X, Y) \otimes L(Y, Z) \to L(X, Z),$$

defined by the formula $\omega(u \otimes v) = v \circ u$ *can be extended to a continuous linear mapping*

$$\tilde{\omega} \colon L_i(X, Y) \tilde{\otimes} L_i(Y, Z) \to L_i(X, Z).$$

Let $E = L_i(X, Y)$, $F = L_i(Y, Z)$ and $G = L_i(X, Z)$. Let $(U_\alpha)_{\alpha \in \mathrm{A}}$, $(V_\beta)_{\beta \in \mathrm{B}}$ and $(W_\gamma)_{\gamma \in \Gamma}$ be fundamental systems of absolutely convex closed neighborhoods of zero in X, Y and Z. Let, further, $\phi \in \mathrm{A}^{\mathrm{B}}$, $\psi \in \mathrm{B}^{\Gamma}$, $\chi \in \mathrm{A}^{\Gamma}$ and

$$S_\varphi = \{u \in E : u(U_{\varphi(\beta)}) \subset V_\beta \quad \text{for all} \quad \beta \in \mathrm{B}\},$$
$$T_\psi = \{v \in F : v(V_{\psi(\gamma)}) \subset W_\gamma \quad \text{for all} \quad \gamma \in \Gamma\},$$
$$R_\chi = \{w \in G : w(U_{\chi(\gamma)}) \subset W_\gamma \quad \text{for all} \quad \gamma \in \Gamma\}.$$

According to 4.4, E, F and G are locally complete spaces with the fundamental systems of Banach balls (and bounded sets) $(S_\phi)_{\phi\in A^B}$, $(T_\psi)_{\psi\in B^\Gamma}$, and $(R_\chi)_{\chi\in A^\Gamma}$. But

$$\omega(S_\varphi \otimes T_\psi) = \{v \circ u : u \in S_\varphi, v \in T_\psi\} \subset R_{\varphi\circ\psi}.$$

By the same token ω is a bounded mapping on all $S_\phi \otimes T_\psi$, and it remains to apply 5.5.

5.7. *Let E, F and G be locally convex separable spaces, U and V neighborhoods of zero in E and F, and S a linearly closed Banach ball in G. Any $u \in \mathfrak{L}(E \otimes F, G)$ satisfying the condition $u(U \otimes V) \subset S$ can be extended to $u^\wedge \in L(E \hat{\otimes} F, G)$ so that $u^\wedge(\overline{\Gamma(U \otimes V)}) \subset S$.*

In fact, since S is absolutely convex, $u(\Gamma(U \otimes V)) \subset S$. Therefore $u = \omega \circ u'$, where $u' \in L(E \otimes_\pi F, G_S)$ and ω is the identity imbedding of G_S into G. Since G_S is a Banach space with the closed unit ball S, u' can be extended to $u'^\wedge \in L(E \hat{\otimes} F, G_S)$ so that $u'^\wedge(\overline{\Gamma(U \otimes V)}) \subset S$. Then $u^\wedge = \omega \circ u'^\wedge$ possesses all the properties stated above.

§6. Exponential law

6.1. **Theorem.** *Let E and F be β-spaces and let G be a locally complete space. The mapping*

$$\Omega : L(E \otimes_i F, G) \to \mathcal{L}(E, \mathcal{L}(F, G)),$$

defined by the formula

$$\Omega(u) = \tilde{u}, \quad \text{where } \tilde{u}_x(y) = u(x \otimes y) \quad \text{for all} \quad x \in E, y \in F, \tag{11}$$

is a topological and a vector isomorphism of $L_i(E \otimes_i F, G)$ onto $L_i(E, L_i(F, G))$. Moreover, $\Omega \circ \Phi$, where Φ associates with every $\bar{u} \in L(E \tilde{\otimes} F, G)$ its restriction to $E \otimes F$, is a topological and a vector isomorphism of $L_i(E \hat{\otimes} F, G)$ onto $L_i(E, L_i(F, G))$.

Proof. Let $(A_\alpha)_{\alpha\in A}$, $(B_\beta)_{\beta\in B}$ and $(C_\gamma)_{\gamma\in\Gamma}$ be fundamental systems of linearly closed Banach balls in E, F and G. For every $\omega \in \Gamma^{A\times B}$ we set

$$V^\omega = \{u \in \mathcal{L}(E \otimes F, G) : u(A_\alpha \otimes B_\beta) \subset C_{\omega(\alpha,\beta)} \quad \text{for all} \quad \alpha \in A, \beta \in B\} \tag{12}$$

and

$$W^\omega = \{w \in \mathcal{L}(E, \mathcal{L}(F, G)) : w_x(y) \in C_{\omega(\alpha,\beta)} \quad \text{for all} \quad x \in A_\alpha, y \in B_\beta, \alpha \in A, \beta \in B\}. \tag{13}$$

Since $C_{\omega(\alpha,\beta)}$ is absolutely convex,

$$V^{\omega} = \{u \in \mathcal{L}(E \otimes F, G) : u(\Gamma(A_\alpha \otimes B_\beta)) \subset C_{\omega(\alpha,\beta)} \quad \text{for all} \quad \alpha \in \mathrm{A}, \beta \in \mathrm{B}\}.$$

Taking 5.5 a) and 4.8 into account, we conclude that $(V^{\omega})_{\omega \in \Gamma^{\mathrm{A}\times\mathrm{B}}}$ is a fundamental system of Banach balls in $L_i(E \otimes_i F, G)$. Further, according to 4.8, the sets

$$T^{\chi} = \{v \in \mathcal{L}(F, G) : v(B_\beta) \subset C_{\chi(\beta)} \quad \text{for all} \quad \beta \in \mathrm{B}\}$$

with all the possible $\chi \in \Gamma^{\mathrm{B}}$ form a fundamental system of Banach balls in $L_i(F, G)$. Therefore, again according to 4.8, the sets

$$\widetilde{W}^{\widetilde{\omega}} = \{w \in \mathcal{L}(E, L(F, G)) : w(A_\alpha) \subset T^{\widetilde{\omega}(\alpha)} \quad \text{for all} \quad \alpha \in \mathrm{A}\}$$
$$= \{w \in \mathcal{L}(E, \mathcal{L}(F, G)) : w(A_\alpha) \subset T^{\widetilde{\omega}(\alpha)} \quad \text{for all} \quad \alpha \in \mathrm{A}\}$$

with all the possible $\widetilde{\omega} \in (\Gamma^{\mathrm{B}})^{\mathrm{A}}$ form a fundamental system of Banach balls in $L_i(E, L_i(F, G))$. But $\omega \to \widetilde{\omega}$, where $\widetilde{\omega}$ is defined by the formula $\widetilde{\omega}(\alpha)(\beta) = \omega(\alpha, \beta)$ for all $\alpha \in \mathrm{A}$ and $\beta \in \mathrm{B}$, is a one-to-one mapping of $\Gamma^{\mathrm{A}\times\mathrm{B}}$ onto $(\Gamma^{\mathrm{B}})^{\mathrm{A}}$. Since, moreover, $W^{\omega} = \widetilde{W}^{\widetilde{\omega}}$, it follows that $(W^{\omega})_{\omega \in \Gamma^{\mathrm{A}\times\mathrm{B}}}$ is a fundamental system of Banach balls in $L_i(E, L_i(F, G))$.

If now $u \in V^{\omega}$, comparing relations (12) and (13) yields $\widetilde{u} \in W^{\omega}$. Conversely, let $w \in W^{\omega}$ and let u be the corresponding element of $\mathcal{L}(E \otimes F, G)$ defined by the condition

$$u(x \otimes y) = w_x(y) \quad \text{for all} \quad x \in E, \ y \in F. \tag{14}$$

We have $u(A_\alpha \otimes B_\beta) \subset C_{\omega(\alpha,\beta)}$ for all $\alpha \in \mathrm{A}$ and $\beta \in \mathrm{B}$, since $u \in L(E \otimes_i F, G)$. Then (14) means that $w = \Omega(u)$. Thus $W^{\omega} = \Omega(V^{\omega})$. But, as was shown, $(V^{\omega})_{\omega \in \Gamma^{\mathrm{A}\times\mathrm{B}}}$ and $(W^{\omega})_{\omega \in \Gamma^{\mathrm{A}\times\mathrm{B}}}$ are fundamental systems of Banach balls in $L_i(E \otimes_i F, G)$ and $L_i(E, L_i(F, G))$. Since the latter are β-spaces we conclude, by 3.5, that Ω is an isomorphism of $L_i(E \otimes_i F, G)$ onto $L_i(E, L_i(F, G))$. According to 4.6, $\Omega \circ \Phi$ is an isomorphism of $L_i(E \widetilde{\otimes}_i F, G)$ onto $L_i(E, L_i(F, G))$.

6.2. *If E and F are spaces of type (F), then $(E \hat{\otimes} F)'_i$ and $L_i(E, F'_i)$ are isomorphic.*

In fact, it suffices to take $G = \mathbf{C}^1$ in 6.1 ($\mathbf{C}$ is the field of complex numbers) and then to take 5.4 into account.

It is not known whether isomorphism is preserved when one replaces i by b.

6.3. *Let E and F be β-spaces and let G be a locally complete space. The mapping*

$$\Omega_0 : L(E, L_i(F, G)) \to \mathcal{L}(F, \mathcal{L}(E, G)),$$

defined by the formula

$$\Omega_0(\widetilde{u}) = \widetilde{u}', \text{ where } \widetilde{u}'_y(x) = \widetilde{u}_x(y) \text{ for all } x \in E, \ y \in F, \tag{11_0}$$

is a topological and a vector isomorphism of $L_i(E, L_i(F, G))$ onto $L_i(F, L_i(E, G))$.

In fact, let $\Omega_0 = \Omega' \circ \Omega^{-1}$, where Ω is defined by (11), and Ω' by the analogous formula

$$\Omega'(u) = \widetilde{u}', \text{ where } \widetilde{u}'_y(x) = u(x \otimes y) \quad \text{for all} \quad x \in E, y \in F.$$

Since the tensor product is symmetric, it remains to apply 6.1.

6.4. *Let E and F be spaces of type (F) and let G be a locally convex separable space with a fundamental system of linearly closed Banach balls (S_n). The mapping*

$$\Omega: L(E \hat{\otimes} F, G) \to \mathcal{L}(E, \mathcal{L}(F, G)),$$

defined by (11) *is a topological and a vector isomorphism of $L_i(E \hat{\otimes} F, G)$ onto $L_i(E, L_i(F, G))$.*

As in the proof of proposition 3.8, the sequence (S_n) may be assumed to be an increasing sequence. Let (U_n) and (V_n) be decreasing fundamental sequences of neighborhoods of zero in E and F. We set

$$P_n = \{u^\wedge \in L(E \hat{\otimes} F, G) : u^\wedge(U_n \otimes V_n) \subset S_n\}$$

and

$$Q_n = \{w \in \mathcal{L}(E, \mathcal{L}(F, G)) : w_x(y) \in S_n \quad \text{for all} \quad x \in U_n, \ y \in V_n\}.$$

Since $E \otimes F$ is dense in $E \hat{\otimes} F$, $u^\wedge$ is determined uniquely by its restriction u to $E \otimes F$. Then, in view of 5.7,

$$P_n = \{u^\wedge \in L(E \hat{\otimes} F, G) : u^\wedge(\overline{\Gamma(U_n \otimes V_n)}) \subset S_n\}.$$

But $E \hat{\otimes} F$ is a space of type (F) with the decreasing fundamental sequence of neighborhoods of zero $(\overline{\Gamma(U_n \otimes V_n)})$. In view of 4.10 we conclude that (P_n) is a fundamental sequence of Banach balls in $L_i(E \hat{\otimes} F, G)$. Further,

$$Q_n = \{w \in \mathcal{L}(E, \mathcal{L}(F, G)) : w(U_n) \subset T_n\},$$

$$\text{where } T_n = \{v \in \mathcal{L}(F, G) : v(V_n) \subset S_n\}.$$

But, in view of 4.10, (T_n) is a fundamental sequence of Banach balls in $L_i(F, G)$. Therefore (again by 4.10) (Q_n) is a fundamental sequence of Banach balls in $L_i(E, L_i(F, G))$. Obviously, $\Omega(P_n) \subset Q_n$. Conversely, let $w \in Q_n$ and let u be the corresponding element of $\mathfrak{L}(E \otimes F, G)$ defined by condition (14). We have $u(U_n \otimes V_n) \subset S_n$, so that, in view of 5.7, u can be extended to $\hat{u} \in L(E \hat{\otimes} F, G)$. Then (14) means that $w = \Omega(\hat{u})$. Thus $\Omega(P_n) = Q_n$. But, as we have shown, (P_n) and (Q_n) are fundamental systems of Banach balls in $L_i(E \hat{\otimes} F, G)$ and $L_i(E, L_i(F, G))$. Since the latter are β-spaces, we conclude by 3.5 that Ω is an isomorphism of $L_i(E \hat{\otimes} F, G)$ onto $L_i(E, L_i(F, G))$.

Remark. Proposition 6.4 is not a special case of 6.1, since G can fail to be a locally complete space (cf. 3.7 and the Remark at the end of §3).

In the same way as Theorem 6.1 implies proposition 6.3, proposition 6.4 implies:

6.5. *If E and F are spaces of type (F) and G is a locally convex separable space with a fundamental sequence of Banach balls, then Ω_0 defined by (11_0) is a topological and a vector isomorphism of $L_i(E, L_i(F, G))$ onto $L_i(F, L_i(E, G))$.*

Finally, we shall analyze Ω defined by formula (11) as a mapping of $L_b(E \hat{\otimes} F, G)$ into $L_b(E, L_b(F, G))$, and we shall mention the case when Ω establishes an isomorphism of these spaces.

6.6. *Let E, F and G be locally convex separable spaces. The mapping $\Omega: L(E \hat{\otimes} F, G) \to \mathfrak{L}(E, \mathfrak{L}(F, G))$ defined by (11) is a continuous linear mapping of $L_b(E \hat{\otimes} F, G)$ into $L_b(E, L_b(F, G))$. If for every bounded subset C of $E \hat{\otimes} F$ there exist bounded sets $A \subset E$ and $B \subset F$ such that $C \subset \overline{\Gamma(A \otimes B)}$, then Ω is an open mapping.*

We set $X = L(E \hat{\otimes} F, G)$ and $Y = L(E, L_b(F, G))$, and let $\mathfrak{A}$, $\mathfrak{B}$ and $\mathfrak{C}$ be the families of all the bounded sets in E, F and $E \hat{\otimes} F$. Let $\mathfrak{W}$ be the system of all the absolutely convex closed neighborhoods of zero in G. If $u \in X$, then for every $W \in \mathfrak{W}$ there exist neighborhoods of zero U in E and V in F such that $u(U \otimes V) \subset W$. Furthermore, for each $B \in \mathfrak{B}$ there exists a $\rho > 0$ such that $\rho B \subset V$. But then $u(\rho U \otimes B) \subset W$, i.e. $\tilde{u}(\rho U) \subset T(B, W)$ (cf. 4.1), where $\tilde{u} = \Omega(u)$. Since $\{T(B, W)\}_{B \in \mathfrak{B}, W \in \mathfrak{W}}$ is a fundamental system of neighborhoods of zero in $L_b(F, G)$, it follows that $\tilde{u} \in Y$, i.e. $\Omega(X) \subset Y$. Let $A \in \mathfrak{A}$. Since the relations $u(A \otimes B) \subset W$ and $\tilde{u}(A) \subset T(B, W)$ are equivalent,

$$\Omega(T(A \otimes B, W)) = T(A, T(B, W)) \cap \Omega(X). \tag{15}$$

But $\{T(A, T(B, W))\}_{A \in \mathfrak{A}, B \in \mathfrak{B}, W \in \mathfrak{W}}$ is a fundamental system of neighborhoods of zero in Y_b and $T(A \otimes B, W)$ is a neighborhood of zero in X_b. Consequently $\Omega: X_b \to Y_b$ is a continuous mapping. Finally, suppose that for every $C \in \mathfrak{C}$ there exists an $A \in \mathfrak{A}$ and $B \in \mathfrak{B}$ such that $C \subset \overline{\Gamma(A \otimes B)}$. The last inclusion implies that $T(C, W) \supset T(\overline{\Gamma(A \otimes B)}, W)$. But since W is an absolutely convex closed set, $T(\overline{\Gamma(A \otimes B)}, W) = T(A \otimes B, W)$. Therefore taking (15) into account, we obtain

$$\Omega(T(C, W)) \supset T(A, T(B, W)) \cap \Omega(X).$$

Since $\{T(C, W)\}_{C \in \mathfrak{C}, W \in \mathfrak{W}}$ is a fundamental system of neighborhoods of zero in X_b, we conclude that, under our assumptions, Ω is an open mapping of X_b onto the space $\Omega(X)$ equipped with the topology induced by that of Y_b.

Definition 17. A locally convex separable space will be called a *countably centered space* if for any sequence of neighborhoods of zero (U_n) in this space there exists a neighborhood of zero U which is absorbed by all U_n.

Remark. Every space of type (DF) is a countably centered space ([3], Lemma 2). There exist also other countably centered spaces (cf. [8], p. 185).

6.7. *If E is a countably centered space, F a quasi-barreled space of type (DF), and G a space of type (F), then (in the notation of the proof of 6.6) $\Omega(X) = Y$.*

Let (B_m) be a fundamental sequence of bounded sets in F and (W_n) a fundamental system of absolutely convex closed neighborhoods of zero in G. If $w \in Y$, then for any m and n there exists a neighborhood of zero U_{mn} in E such that $w(U_{mn}) \subset T(B_m, W_n)$. Since E is a countably centered space, there exists a neighborhood of zero U in E absorbed by all U_{mn} so that for any m and n we have $\rho_{mn} U \subset U_{mn}$ for a certain $\rho_{mn} > 0$. Hence $w(U) \subset T(\rho_{mn} B_m, W_n)$, Thus $w_x(\rho_{mn} B_m) \subset W_n$ for all $x \in U$. Since W_n is absolutely convex and closed, we conclude that $w_x(\overline{\Gamma \bigcup_m \rho_{mn} B_m}) \subset W_n$ for all $x \in U$. But $V_n = \overline{\Gamma \bigcup_m \rho_{mn} B_m}$ is an absolutely convex closed set in F absorbing all bounded sets, and consequently (since F is a quasi-barreled space) V_n is a neighborhood of zero. By the same token we have established that there exist a neighborhood of zero U in E and (for any n) a neighborhood of zero V_n in F such that $u(U \otimes V_n) \subset W_n$, where u is the element of $\mathfrak{L}(E \otimes F, G)$ corresponding to w and defined by condition (14). This means that $u \in L(E \otimes_\pi F, G)$, and, since G is complete, u can be extended to $\hat{u} \in X$. But then (14) means that $w = \Omega(\hat{u})$.

6.8. *If E is a space of type (DF), F a quasi-barreled space of type (DF) and G a space of type (F), then Ω is an isomorphism of $L_b(E\hat{\otimes}F, G)$ onto $L_b(E, L_b(F, G))$.*

Since E and F are spaces of type (DF), for every bounded subset C of $E\hat{\otimes}F$ there exist bounded sets A in E and B in F such that $C\subset \overline{\Gamma(A\otimes B)}$ ([4], Chapter I, Proposition 5.2). In such a case the validity of the proposition follows immediately from 6.6, 6.7 and the remark on Definition 17.

Similarly to 6.3 and 6.5 we have:

6.9. *If E and F are quasi-barreled spaces of type (DF) and G is a space of type (F), then Ω_0 defined by (11_0) is a topological and a vector isomorphism of $L_b(E, L_b(F, G))$ onto $L_b(F, L_b(E, G))$.*

BIBLIOGRAPHY

[1] N. Bourbaki, *Topologie générale*, Chap. 10, 2ième éd., Hermann, Paris, 1961.

[2] ———, *Espaces vectoriels topologiques*, Actualités Sci. Indust., nos. 1189, 1229, 1230, Hermann, Paris, 1953, 1955, Russian transl., IL, Moscow, 1959. MR 14, 880; 17, 1109.

[3] A. Grothendieck, *Sur les espaces (F) et (DF)*, Summa Brasil. Math. 3 (1954), 57–123; Russian transl., Matematika 2 (1958), no. 3, 81–127. MR 17, 765.

[4] ———, *Produits tensoriels topologiques et espaces nucléaires*, Mem. Amer. Math. Soc. No. 16 (1955). MR 17, 763.

[5] G. Köthe, *Topologische lineare Räume*. I, Die Grundlehren der math. Wissenschaften, Band 107, Springer-Verlag, Berlin, 1960. MR 24 #A411.

[6] B. Maissen, *Über Topologien im Endomorphismenraum eines topologischen Vektorraumes*, Math. Ann. 151 (1963), 283–285. MR 27 #4049.

[7] B. M. Makarov, *Some pathological properties of inductive limits of B-spaces*, Uspehi Mat. Nauk 18 (1963), no. 3 (111), 171–178. (Russian) MR 27 #2839.

[8] D. A. Raĭkov, *Certain properties of completely bounded linear operators*, Moskov. Gos. Ped. Inst. Učen. Zap. 188 (1962), 171–191. (Russian)

[9] W. Słowikowski, *On continuity of inverse operators*, Bull. Amer. Math. Soc. 67 (1961), 467–470. MR 24 #A1590.

[10] E. Spanier, *Quasi-topologies*, Duke Math. J. 30 (1963), 1–14. MR 26 #1847.

[11] A. S. Švarc, *Duality of functors*, Dokl. Akad. Nauk SSSR 148 (1963), 288–291 = Soviet Math. Dokl. 4 (1963), 89–92. MR 29 #5887.

Translated by H. Swiatak

Since $\|\Psi(\lambda)\| \leq |\Psi(\lambda)| \leq 1$, the equality obtained can be extended to arbitrary $f, g \in H$ through a limiting process. This, however, implies that (1) holds. The lemma is proved.

2°. Let A be some selfadjoint operator in H, and E_λ its resolution of the identity. The lemma cannot be immediately applied to the operator measure $E(\Delta)$,[1)] since $\mathrm{Tr}(E(\Delta))$ is in general infinite. However, it is possible to proceed in the following manner. We select in H some unbounded operator D, with a dense domain of definition $\mathfrak{D}(D)$, for which there exists a bounded operator D^{-1} with finite H-norm, such that $D^{-1}(Df) = f(f \in \mathfrak{D}(D))$. We now consider the operator measure $\Theta(\Delta) = D^{-1*}E(\Delta)D^{-1}$ for which $\rho(\Delta) = \mathrm{Tr}(\Theta(\Delta)) \leq \|E(\Delta)\||D^{-1}|^2 \leq |D^{-1}|^2 < \infty$. Applying the lemma to $\Theta(\Delta)$, for arbitrary $f, g \in \mathfrak{D}(D)$ we can write

$$(E(\Delta)f, g) = (E(\Delta)D^{-1}Df, D^{-1}Dg) = (\Theta(\Delta)Df, Dg)$$

$$= \int_\Delta (\Psi(\lambda)Df, Dg)\, d\rho(\lambda). \tag{3}$$

This implies the following Parseval-type equality:

$$(f, g) = \int_{-\infty}^{\infty} d(E_\lambda f, g) = \int_{-\infty}^{\infty} (\Psi(\lambda)Df, Dg)\, d\rho(\lambda) \quad (f, g \in \mathfrak{D}(D)). \tag{4}$$

If $\Psi(\lambda)$ were constructed for $\Omega(\Delta) = E(\Delta)$, then it would be the projection operator on the eigensubspace corresponding to the eigenvalue λ. In our case, the role of the projection operator must be played, generally speaking, by the meaningless expression $D^*\Psi(\lambda)D$. Nevertheless, everything in this situation can be given precise meaning. Namely, it can be shown that for any denumerable set U of vectors $u \in \mathfrak{D}(A) \cap \mathfrak{D}(D)$ for which also $Au \in \mathfrak{D}(D)$, the following identity holds for almost all λ:

$$\Psi(\lambda)D(A - \lambda E)u = 0. \tag{5}$$

Relation (5) can be written in equivalent form: for any $f \in H$ and $u \in U$

$$(\Psi(\lambda)f, D(A - \lambda E)u) = (f, \Psi(\lambda)D(A - \lambda E)u) = 0. \tag{6}$$

If the operators D and $A - \lambda E$ commute, and if the elements u form a dense set in H, then it follows from (6) that $(A - \lambda E)D^*\Psi(\lambda)f = 0$, i.e. $D^*\Psi(\lambda)f$ is an eigenvector. In the general case, instead of this fact, it is necessary to restrict ourselves to equality (5) or (6).

---

1) Everywhere in what follows, $f(\Delta)$ shall denote the increment of f_λ on Δ.

So let us prove (5). For $u \in U$ and f in some denumerable dense set R in H, we have

$$(\Psi(\lambda) D(A - \lambda E) u, f) = \lim_{\nu \to \infty} \left(\frac{\Theta(\Delta_\nu)}{\rho(\Delta_\nu)} D(A - \lambda E) u, f \right)$$

$$= \lim_{\nu \to \infty} \frac{1}{\rho(\Delta_\nu)} (D^{-1*} E(\Delta_\nu)(A - \lambda E) u, f) = \lim_{\nu \to \infty} \frac{1}{\rho(\Delta_\nu)} (E(\Delta_\nu)(A - \lambda E) u, D^{-1} f)$$

$$= \lim_{\nu \to \infty} \frac{1}{\rho(\Delta_\nu)} \int_{\Delta_\nu} (\mu - \lambda) \, d(E_\mu u, D^{-1} f). \tag{7}$$

The measure $(E(\Delta)u, D^{-1}f)$ is absolutely continuous with respect to $\rho(\Delta)$. In fact, if $\rho(\Delta) = 0$, then also $\Theta(\Delta) = 0$ (cf. page 228), and thus, for $g, h \in \mathfrak{D}(D)$, we have $(E(\Delta) g, h) = (\Theta(\Delta) Dg, Dh) = 0$. Since $\mathfrak{D}(D)$ is dense in H, it follows that $E(\Delta) = 0$, and the assertion is proved.

From the absolute continuity of this measure it follows that the limit on the right side of (7) on a set $\Lambda_{u,f}$ of full measure equals zero. But then on a set of full measure $\Lambda = \bigcap_{u \in U,\, f \in R} \Lambda_{u,f}$ it also equals zero for any $u \in U$ and $f \in R$. By the denseness of R in H, we conclude from this that (5) holds for $\lambda \in \Lambda$.

Thus we arrive at the following theorem.

Theorem. *Let $E(\Delta)$ be a resolution of the identity corresponding to a selfadjoint operator A acting on a separable Hilbert space H, and let D be an arbitrary unbounded operator with a dense domain of definition $\mathfrak{D}(D)$, for which there exists the operator D^{-1} with finite H-norm, such that $D^{-1}(Df) = f$, $f \in \mathfrak{D}(D)$. Then for $f, g \in \mathfrak{D}(D)$, and for any Δ, the Parseval equality*

$$(E(\Delta) f, g) = \int_{\Delta} (\Psi(\lambda) Df, Dg) \, d\rho(\lambda) \tag{8}$$

holds, where $\Psi(\lambda)$ is a nonnegative operator depending on λ with a uniformly bounded H-norm for almost all λ, and $d\rho(\lambda)$ is some nonnegative finite measure. Roughly speaking, the operator $D^* \Psi(\lambda) D$ is the projector on the eigensubspace corresponding to the eigenvalue λ. More precisely, this implies the following: for every denumerable family of vectors $u \in \mathfrak{D}(A) \cap \mathfrak{D}(D)$ such that Au is also in $\mathfrak{D}(D)$, the identity

$$\Psi(\lambda) D(A - \lambda E) u = 0 \tag{9}$$

holds for almost all λ (in the sense of $d\rho(\lambda)$).

Parseval's equality (8) can easily be written (nonuniquely) in an equivalent form, containing the eigenvectors. To this end, we denote by

$\phi_\alpha(\lambda)$ $(\alpha = 1, \cdots, N_\lambda \leq \infty)$ the full set of orthonormal eigenvectors of the completely continuous nonnegative operator $\Psi(\lambda)$ (λ fixed), corresponding to the eigenvalues $\nu_\alpha(\lambda) > 0$, and we set $\Psi_\alpha(\lambda) = \sqrt{\nu_\alpha(\lambda)}\, \phi_\alpha(\lambda)$. Writing in (8) the expansion $(\Psi(\lambda) Df, Dg)$ in a bilinear series of the eigenvectors $\Psi_\alpha(\lambda)$, we obtain that *for any $f, g \in \mathfrak{D}(D)$ and Δ, Parseval's equality holds in the form*

$$(E(\Delta) f, g) = \int_\Delta \sum_{\alpha=1}^{N_\lambda} \overline{(\Psi_\alpha(\lambda), Df)}\, (\Psi_\alpha(\lambda), Dg)\, d\rho(\lambda) \quad (N_\lambda \leqslant \infty). \tag{10}$$

Here $\Psi_\alpha(\lambda)$ are the orthogonal vectors in H for which $D^* \Psi_\alpha(\lambda)$ are the eigenelements of the operator A corresponding to the eigenvalue λ in the sense that for each u appearing in the theorem

$$(\Psi_\alpha(\lambda), D(A - \lambda E) u) = 0 \quad (\alpha = 1, \ldots, N_\lambda). \tag{11}$$

Note that the last equality follows from (6), if we set $f = \Psi_\alpha(\lambda)$ in it.

3°. If the operator D is such that D^{-1} has a finite H-norm, then it is suitable for the construction of eigenfunction expansions of any selfadjoint operator A. However, for a fixed A the choice of D is freer: Observing the proof of §2°, we see that it is sufficient to require only that $\operatorname{Tr}(D^{-1*} E(\Delta) D^{-1}) < \infty$.

In this connection, the following remark is useful: Let R_z be the resolvent of A; if D is only such that $\mathfrak{D}(D)$ is dense and $|R_z^l D^{-1}| < \infty$ for some regular z and some sufficiently high power l, then the theorem holds for such A and D.

In fact, for any finite Δ there exists $\epsilon_\Delta > 0$ such that

$$E(\Delta) \leqslant \varepsilon_\Delta \int_{-\infty}^{\infty} \frac{1}{|z - \lambda|^{2l}}\, dE_\lambda = \varepsilon_\Delta (R_z^l)^* R_z^l.$$

From this follows the inequality

$$\operatorname{Tr}\,(D^{-1*} E(\Delta) D^{-1}) \leqslant \varepsilon_\Delta \operatorname{Tr}\,(D^{-1*} (R_z^l)^* R_z^l D^{-1}) = \varepsilon_\Delta\, |R_z^l D^{-1}|^2 < \infty,$$

which proves the assertion.

More generally: it is possible to require only that there exist a positive function $\gamma(\lambda)$ in the spectrum of A such that $|\gamma(A) D^{-1}| < \infty$. We observe that the subsequent construction of this section can also be carried out, replacing the resolvent by the operator $\gamma(A)$.

It follows from what we have said that if a power of the resolvent itself has finite H-norm, then we can set $D = E$ and simply differentiate $E_\cdot(\Delta)$. In this case $\Psi(\lambda)$ is the projection operator onto the eigensubspace corresponding to λ.

Let us consider a second special case of the theorem, for which we can draw some conclusions on the behavior of the eigenfunctions.[1)] Let $H = L_2$, where L_2 is constructed on the locally compact space R with absolute additive measure dp, finite on compact sets. Let us assume that the selfadjoint operator A in question is such that for some regular z, some power R_z^l of its resolvent R_z is an integral operator with kernel $K(p, q)$ $(p, q \in R)$, for which, for almost all q,

$$\int |K(p,q)|^2 dp < \infty. \tag{12}$$

We construct a function $\delta(p) \geq 1$ such that

$$\iint |K(p,q)|^2 \frac{1}{\delta^2(q)} dpdq < \infty. \tag{13}$$

If we define the operator D as the multiplication operator on L_2 by the function $\delta(p)$, then the inverse operator D^{-1} will be the multiplication operator by $1/\delta(p)$, and condition (13) will imply that $\| R_z^l D^{-1} \| < \infty$. Thus, for the operator A in question, D may be taken to be just the multiplication operator by $\delta(p)$. The operator $\Psi(\lambda)$ constructed in the theorem is an operator of finite H-norm in L_2, so that it is a Hilbert-Schmidt operator, i.e. it is generated by a kernel $\Psi(p, q; \lambda)$ for which, for almost all λ,

$$\iint |\Psi(p,q;\lambda)|^2 dpdq = \| \Psi(\lambda) \|^2 \leqslant C^2 < \infty. \tag{14}$$

Thus Parseval's equality (8) can be written in the form

$$\begin{aligned}(E(\Delta) f, g) &= \int_\Delta (\Psi(\lambda) Df, Dg)\, d\rho(\lambda) \\ &= \int_\Delta \left\{ \iint \Psi(p,q;\lambda)\, \delta(q) f(q)\, \delta(p) \overline{g(p)}\, dpdq \right\} d\rho(\lambda) \\ &= \int_\Delta \left\{ \iint \psi(p,q;\lambda) f(q)\, \overline{g(p)}\, dpdq \right\} d\rho(\lambda).\end{aligned} \tag{15}$$

Here we have set $\psi(p, q; \lambda) = \Psi(p, q; \lambda)\, \delta(p)\, \delta(q)$; in (15), $f, g \in \mathfrak{D}(D)$, i.e. we must take functions f and g from L_2 such that δf and δg are also in L_2. As already explained, $D^* \Psi(\lambda) D$ in the sense defined serves as the projection operator on the eigensubspace of λ, so that $\psi(p, q; \lambda)$ is the kernel of this operator, and for each variable p and q it is in the corresponding sense an eigenfunction of the operator A. We shall not make this definition more

1) These results were obtained jointly by G. I. Kac, A. G. Kostjučenko, and the author.

precise, since it immediately follows from the discussion on pp. 229–231. We shall only remark that in the case when A is generated from an elliptic differential expression, $\psi(p, q; \lambda)$ is the ordinary spectral function of the problem.

Inequality (14) now shows that for almost all λ

$$\iint \frac{|\psi(p, q; \lambda)|^2}{\delta^2(p)\,\delta^2(q)}\, dp\,dq \leqslant C^2 < \infty. \tag{16}$$

Thus we arrive at the following statement: If for some regular z, some power R_z^l of the resolvent is an integral operator with kernel $K(p, q)$ for which (12) holds for almost all q, then for every function $\delta(p) \geq 1$ satisfying (13) the estimate (16) holds for almost all λ. In other words, since $\psi(p, q; \lambda)$ is an eigenfunction of A for each p and q, (16) is an estimate for almost all the eigenfunctions.

We now assume in addition that the integral

$$\int |K(p, q)|^2\, \delta^2(q)\, dq = \chi(p) \tag{17}$$

exists for each p and is a function bounded in each compactum. We first establish the equality

$$\iint K(p, \xi)\, \overline{K(q, \eta)}\, \psi(\xi, \eta; \lambda)\, d\xi\, d\eta = \frac{1}{|z-\lambda|^{2l}}\, \psi(p, q; \lambda).^{1)} \tag{18}$$

Indeed, the double integral on the left of (18) exists, since, by Bunjakovskiĭ's inequality,

$$\iint |K(p, \xi)\, \overline{K(q, \eta)}\, \psi(\xi, \eta; \lambda)|\, d\xi\, d\eta$$
$$\leqslant \left(\iint |K(p, \xi)\, \overline{K(q, \eta)}|^2 \delta^2(\xi)\, \delta^2(\eta)\, d\xi\, d\eta\right)^{\frac{1}{2}} \left(\iint \frac{|\psi(\xi, \eta; \lambda)|^2}{\delta^2(\xi)\, \delta^2(\eta)}\, d\xi\, d\eta\right)^{\frac{1}{2}}$$
$$\leqslant C \sqrt{\chi(p)\, \chi(q)}. \tag{19}$$

From (17) it follows that $R_z^{l*} u$, $R_z^{l*} v \in \mathfrak{D}(D)$ for finite (i.e. vanishing outside some compactum) bounded functions u and v. Thus with the help of (15), and remembering (19), we can write

1) This equality essentially expresses the fact that an eigenfunction of an operator is also an eigenfunction of its resolvent.

$$\int_\Delta\int\int\left\{\int\int K(p,\xi)\,\overline{K(q,\eta)}\,\psi(\xi,\eta;\lambda)\,d\xi d\eta\right\}\overline{u(p)}\,v(q)\,dpdq\,d\rho(\lambda)$$
$$=\int_\Delta\left\{\int\int\psi(\xi,\eta;\lambda)\,(R_z^{l*}v)(\eta)\,\overline{(R_z^{l*}u)(\xi)}\,d\xi d\eta\right\}d\rho(\lambda)=(E(\Delta)R_z^{l*}v,R_z^{l*}u)$$
$$=(R_z^l E(\Delta)R_z^{l*}v,u). \tag{20}$$

Due to the fact that the operator $R_zR_z^{l*}$ is an integral operator with the kernel

$$\int K(p,\xi)\,\overline{K(q,\xi)}\,d\xi,$$

we can readily show that the operator $R_z^lE(\Delta)R_z^{l*}$ is also an integral operator with the kernel

$$\int_\Delta\frac{\psi(p,q;\lambda)}{|z-\lambda|^{2l}}\,d\rho(\lambda)$$

(see the detailed discussion in [4], pp. 114–115). Thus (20) may be written as

$$\int_\Delta\int\int\left\{\int\int K(p,\xi)\,\overline{K(q,\eta)}\,\psi(\xi,\eta;\lambda)\,d\xi d\eta\right\}\overline{u(p)}\,v(q)\,dpdqd\rho(\lambda)$$
$$=\int_\Delta\int\int\frac{\psi(p,q;\lambda)}{|z-\lambda|^{2l}}\,\overline{u(p)}\,v(q)\,dpdqd\rho(\lambda),$$

From this, by the arbitrariness of u, v and Δ, (18) follows.

From (18) and (19) it follows that the estimate

$$|\psi(p,q;\lambda)|\leqslant|z-\lambda|^{2l}C\sqrt{\chi(p)\chi(q)} \tag{21}$$

holds for almost all λ.

Thus, if in addition the kernel $K(p, q)$ and the function $\delta(q)$ are such that the integral (17) exists for each p and is a bounded function on each compactum, then the estimate (21) holds for almost all λ and almost all p and q. This estimate, as well as estimate (16), is an estimate of growth for almost all the eigenfunctions of the operator A.

4°. We make some additional nontrivial observations. Let $A_1, \cdots, A_n$ be a finite number of selfadjoint operators, acting on Hilbert space H and commuting in the sense that the respective resolutions of the identity $E_{\lambda_1}^{(1)}, \cdots, E_{\lambda_n}^{(n)}$ commute for arbitrary $\lambda_1, \cdots, \lambda_n$. Then, as is known [10], it is possible to construct an operator measure $E(\Delta)$ where Δ is a Borel set in the n-dimensional space of the points $\lambda = (\lambda_1, \cdots, \lambda_n)$, having the properties analogous to the properties of ordinary spectral measures. In particular, for each $f, g \in H$ and finite Δ, the equality

$$(f, g) = \int_{-\infty}^{\infty} \cdots \int d_\lambda (E_\lambda f, g), \quad (A_j E(\Delta) f, g) = \int_{\Delta} \cdots \int \lambda_j d_\lambda (E_\lambda f, g)$$

holds.

All the previous constructions can be carried over without change to expansions related to the n-dimensional spectral measure $E(\Delta)$. We shall not formulate these results in detail.

We also remark that in all the previous discussion we could have considered generalized resolutions of the identity (i.e. resolutions obtained by extensions to more general spaces). All the results hold mutatis mutandis in this case.

5°. We pause briefly on some available applications of these results.

For $H = L_2$ with ordinary Lebesgue measure in a finite domain in n-dimensional space, it is convenient to take D to be a differential operator having a bounded left inverse, for example $\partial^n/\partial x_1 \cdots \partial x_n$ or $(\partial^2/\partial x_1^2 + \cdots + \partial^2/\partial x_n^2)^m$ [3,4]. For unbounded domains, such an operator is conveniently provided by a multiplier increasing to ∞, which guarantees the boundedness in some sense of the operator D^{-1} [8,9]. In the cases considered (as generally in function spaces), it is useful to adopt the terminology and techniques of generalized functions [1,2].

In the case of operators (in L_2) arising from differential expressions $\mathfrak{S}$, some detailed general results are given [4]. In particular, in this case equality (6) implies, for example, that $D^*\Psi(\lambda) f$ is a generalized solution of the equation $\mathfrak{S}[u] = \lambda u$. If $\mathfrak{S}$ is elliptic, then the generalized solution will be an ordinary solution, and we arrive at the case of expansion by ordinary eigenfunctions [1,2]. From this viewpoint the theory of eigenfunction expansions for elliptic equations was given in [4]. A similar picture occurs for arbitrary differential operators in which generalized solutions are ordinary solutions, in particular, for ordinary differential operators (for the latter the theory of expansions was outlined in [4], and carried out in [11]). Finally, we remark that the approach of §4° to the natural cases permits the construction of expansions in eigenfunctions of differential operators of the type of multiple Fourier integral expansions.

In the case of a Hilbert space H, constructed in the usual manner with a positive definite kernel $K(p, q)$, it is convenient to take D in the same manner as in L_2. Expansion in eigenfunctions of differential operators in this space reduces to the generalized theorem of Bochner on positive definite functions, in the case where $e^{i\lambda x}$ is replaced by the eigenfunctions of these operators. These results, which contain previous investigations [12,13] for ordinary differential equations, are presented in [14,15]. Their extension to the case of general kernels $K(p, q)$ is contained in [7]. We note, in addition, that the expansion of a positive definite kernel of the type in the ordinary n-dimensional theorem of Bochner (with n-fold integrals over λ) can be obtained on the basis of the remarks of §4°.

Eigenfunction expansions of operators acting on a Hilbert space with a scalar product of the type of a Dirichlet integral were carried out in [1,2], and also in [5].

We now show that it is possible to obtain estimates for eigenfunctions of differential operators, as from the results of §2°. Consider, for example, the case of Schrödinger's equation $\Delta u + c(p)u = \lambda u$ in three dimensions. If $c(p)$ is bounded below (i.e. $c(p) \geq c_0 > -\infty$), then it can be shown that there exists a z for which the kernel $K(p, q)$ of the resolvents R_z satisfies the inequality $|K(p, q)| \leq e^{-|p-q|}/|p - q|$. (In [16], A. G. Kostjučenko gave broad conditions on differential operators for which detailed estimates hold.) From this estimate it follows that inequality (13) will be satisfied if we set $l = 1$ and $\delta^2(p) = 1 + |p|^{3+\epsilon}$ $(\epsilon > 0)$. In this case

$$\chi(p) = \int |K(p, q)|^2 \delta^2(q)\, dq \leqslant \int \frac{e^{-2|p-q|}}{|p-q|^2}(1 + |q|^{3+\varepsilon})\, dq$$

$$= \int \frac{e^{-2|s|}}{|s|^2}(1 + |p - s|^{3+\varepsilon})\, ds \leqslant A(1 + |p|^{3+\varepsilon}).$$

Since $\psi(p, q; \lambda)$ regarded as a function of p for fixed λ and arbitrarily varying parameter q runs through a basis in the eigensubspace corresponding to λ, so, by considering the above inequality and the estimate (21), we readily conclude that almost all the eigenfunctions of the equation increase to ∞ no faster than $|p|^{3/2+\epsilon}$, $\epsilon > 0$.

We remark that without restrictions on the coefficients of the differential operator (elliptic, hyperbolic, or arbitrary), it is always possible to conclude [3,4,9] that n-fold integrals of eigenfunctions increase to ∞ like polynomials; this is related to the finite H-norms of the operator D^{-1}.

BIBLIOGRAPHY

[1] I. M. Gel'fand and A. G. Kostjučenko, *Expansion in eigenfunctions of differential and other operators*, Dokl. Akad. Nauk SSSR **103** (1955), 349–352. (Russian) MR **17**, 388.

[2] I. M. Gel'fand and G. E. Šilov, *Generalized functions*. Vol. 3: *Theory of differential equations*, Fizmatgiz, Moscow, 1958; English transl., Academic Press, New York, 1967. MR **21** #5142b; **36** #506.

[3] Ju. M. Berezanskiĭ, *On expansion according to eigenfunctions of general self-adjoint differential operators*, Dokl. Akad. Nauk SSSR **108** (1956), 379–382. (Russian) MR **18**, 323.

[4] ———, *Eigenfunction expansions of self-adjoint operators*, Mat. Sb. **43** (85) (1957), 75–126. (Russian) MR **21** #3646.

[5] F. E. Browder, *Eigenfunction expansions for formally self-adjoint partial differential operators*. I. Proc. Nat. Acad. Sci. U.S.A. **42** (1956), 769–771. MR **19**, 1061.

[6] L. Gårding, *Eigenfunction expansions*, Proc. Summer Seminar Partial Differential Equations (Boulder, Col., 1957), Interscience, New York, 1964, pp. 301–325. MR **29** #2525.

[7] K. Maurin, *Entwicklung positiv definiter Kerne nach Eigendistributionen. Differenzierbarkeit der Spektralfunktion eines hypoelliptischen Operators*, Bull. Acad. Polon. Sci. Sér. Sci. Math. Astronom. Phys. **6** (1958), 149–155. MR **20** #4738.

[8] G. I. Kac, *Expansion in characteristic functions of self-adjoint operators*, Dokl. Akad. Nauk SSSR **119** (1958), 19–22. (Russian) MR **21** #3647.

[9] ———, a) *Generalized elements of a Hilbert space*, Ukrain. Mat. Ž. **12** (1960), 13–24. (Russian) MR **24** #A1029. b) *Spectral decompositions of self-adjoint operators in terms of generalized elements of a Hilbert space*, Ukrain. Mat. Ž. **13** (1961), no. 4, 13–33. (Russian) MR **26** #1764.

[10] A. I. Plesner and V. A. Rohlin, *Spectral theory of linear operators*. II, Uspehi Mat. Nauk **1** (1946), no. 1 (11), 71–191; English transl., Amer. Math. Soc. Transl. (2) **62** (1967), 29–175. MR **9**, 43.

[11] L. B. Rabinovič, *Eigenfunction expansions for ordinary differential operators*, Diploma thesis, Kiev Univ., 1958. (Russian)

[12] A. Ja. Povzner, *On differential equations of Sturm-Liouville type on a half-axis*, Mat. Sb. **23** (65) (1948), 3–52; English transl., Amer. Math. Soc. Transl. (1) **4** (1962), 24–101. MR **10**, 299.

[13] M. G. Kreĭn, *On a general method of decomposing Hermite-positive nuclei into elementary products*, C. R. (Dokl.) Acad. Sci. URSS 53 (1946), 3–6. MR 8, 277.

[14] Ju. M. Berezanskiĭ, *Generalization of Bochner's theorem to expansions according to eigenfunctions of partial differential equations*, Dokl. Akad. Nauk. USSR 110 (1956), 893–896. (Russian) MR 19, 1061.

[15] ———, *Representation of positive definite kernels by eigenfunctions of differential equations*, Mat. Sb. 47 (89) (1959), 145–176. (Russian) MR 21 #2774.

[16] A. G. Kostjučenko, *On the spectral properties of elliptic operators*, Dokl. Akad. Nauk SSSR 115 (1957), 34–37. (Russian) MR 20 #1068.

Translated by:
C. K. Chu

Amer. Math. Soc. Transl.
(2) Vol. 93, 1970

THEOREMS ON POSITIVE SOLUTIONS OF EQUATIONS OF THE SECOND KIND WITH NONLINEAR OPERATORS*

UDC 517.948

V. Ja. STECENKO AND A. R. ESAJAN

In this article we investigate equations of the second kind

$$\lambda x = Ax + f,$$

where A is a nonlinear operator operating in Banach space E with cone K and leaving this cone invariant.[1] We determine the ranges of values of the parameter λ for which this equation has a solution belonging to K. The theorems proved below represent a further development of the results of M. A. Krasnosel'skiĭ [1] relative to positive nonlinear operators and can be applied to the proof of existence theorems for solutions of nonlinear integral equations, and to boundary value problems in the theory of differential equations.

1. **Definition.** We will say that the operator A is *continuous from the left* (*from the right*) *at the point* x^* if for every sequence x_n:

$$x_n \leqslant x^* \ (x_n \geqslant x^*) \ (n = 1, 2, \ldots),$$

the condition $\|x_n - x^*\| \to 0$ implies that

$$\|Ax_n - Ax^*\| \to 0 \ (n \to \infty).$$

If A is continuous from the right and from the left at x^*, we will say that A is *continuous over the cone* K *at* x^*.

We have

Theorem 1.[2] *Let* A *be a positive linear operator relative to the normal cone* K (*cf.* [1], [3]). *Then* A *is continuous over* K *at any point* x^* *of the space* E.

Proof. First note that since A is linear, left continuity at $x^* \in E$ implies right continuity at the same point. Hence it suffices to show that A is continuous from the left at each point $x^* \in E$.

Let

$$\|x_n - x^*\| \to 0 \text{ and } x_n \leqslant x^* \ (n = 1, 2, \ldots).$$

---

* Translation of Mat. Sb. 68(110) (1965), 473–486.

1) We use the terminology of [1].

2) This theorem supplements the corresponding result in [2].

It is necessary to show that $\|Ax_n - Ax^*\| \to 0$. Assume the contrary. Then there is a positive number β and a subsequence $\{x_{n_k}\}$ of $\{x_n\}$ such that

$$\| Ax_{n_k} - Ax^* \| > \beta \quad (k = 1, 2, \ldots). \tag{1}$$

Without loss of generality we can assume that

$$\| x_{n_k} - x^* \| < \frac{1}{k^3} \quad (k = 1, 2, \ldots), \tag{2}$$

because if this were not the case, we could use another subsequence for which (2) would be true. Let $x^* - x_{n_k} = u_k \in K$; we now consider the series $\sum_{k=1}^{\infty} ku_k$. By (2) this series converges strongly. Denote its sum by u_0. It is clear that $u_0 \in K$ and $\theta \leq ku_k \leq u_0$ $(k = 1, 2, \cdots)$; hence $\theta \leq kAu_k \leq Au_0$ since A is positive. Therefore

$$\| kAu_k \| \leqslant M \| Au_0 \|,$$

where M is the constant of normality of K. This means that

$$\| Au_k \| \to 0 \quad (k \to \infty).$$

On the other hand, $\|Au_k\| = \|A(x^* - x_{n_k})\| > \beta$. So we have a contradiction, and the theorem is proved.

Corollary 1. *Let the cone K be normal, and let linear operator B satisfy the condition*

$$B \leqslant A \quad (\text{or} \quad -A \leqslant B),$$

where A is a positive $(AK \subset K)$ operator. Then B is continuous over K.

Proof. The operators A and $A - B$ are positive on K. Thus by Theorem 1 they are continuous over the cone. Suppose we have a sequence x_n:

$$x_n \geqslant x_0, \ x_n \to x_0.$$

Then

$$(A - B)\, x_n \to (A - B)\, x_0, \qquad -Ax_n \to -Ax_0,$$

whence it follows that $-Bx_n \to -Bx_0$, i.e. $Bx_n \to Bx_0$, and consequently that B is continuous from the right at x_0. Therefore B is continuous over the cone at x_0.

Corollary 2. *Let the cone K be normal, let the linear operator A be positive, and let the nonlinear operator B be such that*

$$\theta \leqslant Py - Bx \leqslant Ay - Ax \quad (x, y \in K, \ y \geqslant x).$$

Then B is continuous over K at each point $x^* \in K$.

Furthermore, it is easy to see that if K is reproducing and the linear positive operator A is continuous over K, then A is continuous in E.

In fact, let it follow from the conditions

$$x_n \in K, \quad x_n \to \theta$$

that $Ax_n \to \theta$. Pick an arbitrary sequence $h_n \in E$, $h_n \to \theta$ $(n = 1, 2, \cdots)$.

By virtue of the nonoblateness of reproducing cones we have

$$h_n = x_n - x_n' \quad (x_n, x_n' \in K), \quad \| x_n \| = C \| h_n \|, \quad \| x_n' \| \leqslant C \|h_n\|,$$

whence it is clear that x_n, x_n', Ax_n, $Ax_n' \to \theta$. Therefore

$$Ah_n = Ax_n - Ax_n' \to \theta.$$

So A is continuous at θ and consequently in all of E.

From this argument and Theorem 1 we have in particular the well-known result [2]

A linear operator A which is positive on a normal reproducing cone K is continuous.

Definition. An operator B is *sublinear* [4] relative to the linear operator A if the following relations are satisfied:

$$B\theta = \theta, \tag{3}$$

$$By - Bx \leqslant Ay - Ax \quad (y \geqslant x \geqslant \theta). \tag{4}$$

Note that if B is sublinear on K relative to A, then $Bx \leq Ax$ $(x \in K)$.

If for each x from some set $M \subset K$ there exist a natural number p and a positive number α such that

$$A^p x \leqslant \alpha u_0 \quad (p = p(x);\ u_0 \in M),$$

then we say that A is u_0*-bounded from above* over K on the set M.

Theorem 2. *Let the operator B be monotone on a normal cone K and sublinear relative to the positive operator A. Let*

$$BK_g \subset K \ (g \in K),^{1)} \tag{5}$$

let A be u_0-bounded from above on the set K_g, and let

$$A^m u_0 \leqslant \lambda_0 u_0. \tag{6}$$

Then the equation

$$\lambda x = Bx + f \tag{7}$$

has for all $f \in K_{(\varepsilon + \sqrt[m]{\lambda_0})g}$ $(\epsilon > 0)$ at least one solution $x^* \in K$ for all λ:

---

1) K_g is the set of elements of the form $g + x$ $(x \in K)$.

$$\sqrt[m]{\lambda_0} < \lambda \leqslant \Lambda, \tag{8}$$

where

$$\Lambda = \sup_{\lambda} \left\{ \frac{f}{\lambda} \geqslant g \right\}.$$

Here there can be no solutions congruent over K in K_g.

Proof. Along with equation (7) we investigate the auxiliary equation

$$\lambda y = Ay + f, \quad f \in K_{(\varepsilon + \sqrt[m]{\lambda_0})g}.$$

We construct successive approximations for both equations:

$$x_0 = y_0 = \frac{1}{\lambda} f,$$

$$x_{n+1} = \frac{1}{\lambda}(Bx_n + f), \quad y_{n+1} = \frac{1}{\lambda}(Ay_n + f) \quad (n=0, 1, 2, \ldots). \tag{9}$$

From (8) it follows that $x_0 = y_0 \in K_g$, because $f/\lambda \geq f/\Lambda \geq g$. Note, moreover, that $\Lambda > \sqrt[m]{\lambda_0}$, since for $f \in K_{(\varepsilon + \sqrt[m]{\lambda_0})g}$ we have the inequality $f/(\epsilon + \sqrt[m]{\lambda_0}) \geq g$.

We will show that

$$1)\ x_n \leqslant y_n, \quad 2)\ x_n \leqslant x_{n+1}, \quad 3)\ y_n \leqslant y_{n+1} \quad (n=0, 1, 2, \ldots).$$

We will prove first inequality 2). We have $x_1 = (Bx_0 + f)/\lambda \geq x_0$. We assume that $x_s \geq x_{s-1}$. Then $x_{s-1}, x_s \in K_g$, and therefore

$$x_s = \frac{1}{\lambda}(Bx_{s-1} + f) \leqslant \frac{1}{\lambda}(Bx_s + f) = x_{s+1}.$$

Inequality 3) is proved analogously. We will now prove inequality 1). Since $x_0 = y_0$,

$$x_1 = \frac{1}{\lambda}(Bx_0 + f) \leqslant \frac{1}{\lambda}(Ay_0 + f) = y_1.$$

Since $x_2 \geq x_1$ we have $x_2 \in K_g$, and consequently

$$x_2 = \frac{1}{\lambda}(Bx_1 + f) \leqslant \frac{1}{\lambda}(Ax_1 + f) \leqslant \frac{1}{\lambda}(Ay_1 + f) = y_2.$$

By induction we will have that $x_n \leq y_n$ $(n = 0, 1, 2, \cdots)$. Obviously

$$x_1 - x_0 = \frac{1}{\lambda} Bx_0 \leqslant \frac{1}{\lambda} Ax_0 = y_1 - y_0,$$

and so forth; we will get

$$x_n - x_{n-1} \leqslant y_n - y_{n-1} \quad (n = 1, 2, \ldots). \tag{10}$$

From (10) it will follow that

$$0 \leqslant x_{n+p} - x_n \leqslant \sum_{k=n+1}^{n+p} \frac{A^k f}{\lambda^{k+1}}. \tag{11}$$

The functional series

$$R_\lambda f = \sum_{k=0}^{\infty} \frac{A^k f}{\lambda^{k+1}} \tag{12}$$

converges strongly for any $f \in K_g$, and $|\lambda| > \sqrt[m]{\lambda_0}$. In fact, since A is u_0-bounded on K_g, there exist p and α such that

$$A^p f \leqslant \alpha u_0 \quad (\alpha > 0,\ f \in K_g);$$

whence by virtue of the positiveness of A we get

$$\theta \leqslant A^{p+km} f \leqslant \alpha A^{km} u_0 \quad (k = 1, 2, \ldots).$$

But

$$A^{km} u_0 \leqslant \lambda_0^k u_0 \quad (k = 1, 2, \ldots).$$

Further, we have

$$\sum_{k=0}^{\infty} \frac{\| A^k f \|}{|\lambda|^{k+1}} = \sum_{k=0}^{m+p-1} \frac{\| A^k f \|}{|\lambda|^{k+1}} + \sum_{n=1}^{\infty} \sum_{s=0}^{m-1} \frac{\| A^{p+nm+s} f \|}{|\lambda|^{p+nm+s+1}}.$$

Moreover,

$$A^s (A^{p+km} f) \leqslant A^s (\alpha A^{km} u_0) \leqslant \alpha \lambda_0^k A^s u_0 \quad (s = 0, 1, \ldots, m-1).$$

Therefore

$$\sum_{k=0}^{\infty} \frac{\| A^k f \|}{|\lambda|^{k+1}} \leqslant \sum_{k=0}^{m+p-1} \frac{\| A^k f \|}{|\lambda|^{k+1}} + \alpha M \sum_{s=0}^{m-1} \frac{\| A^s u_0 \|}{|\lambda|^{p+s+1}} \sum_{n=1}^{\infty} \left(\frac{\lambda_0}{|\lambda|^m} \right)^n,$$

from which the strong convergence of (12) follows for $|\lambda| > \sqrt[m]{\lambda_0}$. By the normality of K and inequality (11) we have

$$\| x_{n+p} - x_n \| \leqslant M \cdot \varepsilon = \varepsilon_1.$$

Thus $\{x_n\}$ converges, i.e.

$$x_n \to x^*.$$

By Corollary 2 of Theorem 1 the operator B is continuous over K; in particular it is continuous from the left. Since the sequence $\{x_n\}$ is monotone increasing, we can go to the limit in the first relation in (9). We get

$$\lambda x^* = Bx^* + f.$$

In order to finish the proof it remains to show that in K_g there are no congruent solutions. Assume the contrary; moreover, let

$$\lambda x_1 \equiv Bx_1 + f, \quad \lambda x_2 \equiv Bx_2 + f$$

and $x_1 \geq x_2 \geq g$ $(x_1 \neq x_2)$. The equation

$$\lambda y = Ay + f \tag{13}$$

has a solution y^* for any $f \in K_g$ and for $|\lambda| > \sqrt[m]{\lambda_0}$. We will construct for (13) successive approximations in two different ways. Let

$$1)\ y_0^{(1)} = x_1, \quad y_n^{(1)} = \frac{1}{\lambda}(Ay_{n-1}^{(1)} + f)$$

$$(n = 1, 2, \ldots)$$

$$2)\ y_0^{(2)} = x_2, \quad y_n^{(2)} = \frac{1}{\lambda}(Ay_{n-1}^{(2)} + f).$$

We will show that sequences $\{y_n^{(1)}\}$ and $\{y_n^{(2)}\}$ converge to a solution of (13). Let

$$y_n^{(1)} \to y_1, \quad y_n^{(2)} \to y_2.$$

We have

$$y_0^{(1)} = x_1,$$

$$y_1^{(1)} = \frac{1}{\lambda}(Ay_0^{(1)} + f) = \frac{1}{\lambda}(Ax_1 + f) \geqslant \frac{1}{\lambda}Bx_1 + \frac{1}{\lambda}f =$$

$$= \frac{1}{\lambda}(\lambda x_1 - f) + \frac{1}{\lambda}f = x_1 = y_0^{(1)},$$

$$y_2^{(1)} = \frac{1}{\lambda}(Ay_1^{(1)} + f) \geqslant \frac{1}{\lambda}(Ay_0^{(1)} + f) = y_1^{(1)};$$

and by induction we get

$$y_n^{(1)} \geqslant y_{n-1}^{(1)} \quad (n = 1, 2, \ldots). \tag{14}$$

Analogously we prove that

$$y_n^{(2)} \geqslant y_{n-1}^{(2)} \quad (n = 1, 2, \ldots).$$

It is easy to see that

$$y_n^{(s)} = \frac{1}{\lambda^n}A^n x_s + \sum_{k=0}^{n-1} \frac{A^k f}{\lambda^{k+1}} \quad (s = 1, 2;\ n = 1, 2, \ldots). \tag{15}$$

In fact,

$$y_n^{(s)} = \frac{1}{\lambda}(Ay_{n-1}^{(s)} + f) = \frac{1}{\lambda}\left[A\,\frac{1}{\lambda}(Ay_{n-2}^{(s)} + f) + f\right]$$

$$= \frac{1}{\lambda^2}A^2 y_{n-2}^{(s)} + \frac{1}{\lambda^2}Af + \frac{1}{\lambda}f = \ldots = \frac{1}{\lambda^n}A^n x_s + \sum_{k=0}^{n-1}\frac{A^k f}{\lambda^{k+1}}.$$

Since $x_s \in K_g$ $(s = 1, 2)$, there exist $p(s)$ and $\alpha(s)$ such that

$$A^p x_s \leqslant \alpha u_0.$$

By the positiveness of A we will have

$$A^{nm+p}x_s \leqslant \alpha A^{nm} u_0 \leqslant \alpha \lambda_0^n u_0.$$

Let

$$u_n(s) = \frac{1}{\lambda^n}A^n x_s.$$

Then

$$\theta \leqslant u_{nm+p}(s) \leqslant \frac{\alpha}{\lambda^p}\left(\frac{\lambda_0}{\lambda^m}\right)^n u_0 \quad (n = 1, 2, \ldots).$$

Since K is normal and $\lambda > \sqrt[m]{\lambda_0}$,

$$u_{nm+p} \to \theta \quad (n \to \infty).$$

If we go to the limit in

$$y_{nm+p}^{(s)} = \frac{1}{\lambda^{nm+p}}A^{nm+p}x_s + \sum_{k=0}^{nm+p-1}\frac{A^k f}{\lambda^{k+1}}$$

we get that the sequence $\{y_{nm+p}^{(s)}\}$ converges to $R_\lambda f$:

$$y_{nm+p}^{(s)} \to R_\lambda f \quad (n \to \infty).$$

Since the sequence $\{y_n^{(s)}\}$ $(n = 1, 2, \cdots)$ is monotone,

$$y_n^{(s)} \to R_\lambda f \quad (s = 1, 2),$$

that is, $y_1 = y_2 = R_\lambda f$. Further,

$$x_1 - x_2 = \frac{1}{\lambda}(Bx_1 - Bx_2) \leqslant \frac{1}{\lambda}(Ax_1 - Ax_2) \leqslant \frac{1}{\lambda}(Ay_0^{(1)} - Ay_0^{(2)}) = y_1^{(1)} - y_1^{(2)},$$

$$x_1 - x_2 \leqslant \frac{1}{\lambda}(Ax_1 - Ax_2) \leqslant \frac{1}{\lambda}(Ay_1^{(1)} - Ay_1^{(2)}) = y_2^{(1)} - y_2^{(2)};$$

and by induction we prove that

$$\theta \leqslant x_1 - x_2 \leqslant y_n^{(1)} - y_n^{(2)} \quad (n = 1, 2, \ldots).$$

Going to the limit in this inequality as $n \to \infty$, we get $x_1 = x_2$, which is a contradiction. So there are no solutions in K_g which are conjugate among themselves. The theorem is proved.

Remark. If $g \in E_f$, then as Λ it is possible to take any number satisfying the inequality

$$\Lambda \leqslant \frac{1}{\|g\|_f}.$$

In fact, in this case

$$\frac{f}{\Lambda} \geqslant \frac{f}{\dfrac{1}{\|g\|_f}} = \|g\|_f \cdot f \geqslant g.$$

For $g = \theta$ Theorem 2 can be formulated in the following form:

Theorem 3. *Let a positive operator B, which is monotone on the normal cone K, be sublinear relative to the positive operator A which is u_0-bounded from above; moreover, let*

$$A^m u_0 \leqslant \lambda_0 u_0.$$

Then equation (7) *has at least one solution in K for $\lambda > \sqrt[m]{\lambda_0}$ and any $f \in K$.*

2. In the proof that there are no solutions in K congruent to each other, it is necessary, as in Theorem 2, in order to find a solution of (13) to construct successive approximations

$$x_0 = g, \quad x_n = \frac{1}{\lambda}(Ax_{n-1} + f) \quad (n = 1, 2, \ldots), \tag{16}$$

starting from different elements $g \in K$. Here it is necessary to prove that the successive approximations $x_n(g)$ converge (and moreover converge to a solution of (13)). We also establish some facts relative to the convergence of successive approximations (16) to a solution of

$$\lambda x = Ax + f \tag{13'}$$

with linear operator A. Note that in the following theorems we do not assume that A is continuous.

Theorem 4. *Let the cone K be normal and the operator A be positive, and let*

$$A^m u_0 \leqslant \lambda_0 u_0 \quad (u_0 \in K,\ u_0 \neq \theta).$$

Then for $\lambda > \sqrt[m]{\lambda_0}$ and any $f \in E_{u_0}$ the successive approximations (16) *converge to a solution x^* of equation* (13) *beginning from any null approximation*

$g \in E_{u_0}$ *for which*

$$A^{p_0}(\lambda g - Ag - f) \in K \tag{17}$$

for some p_0 $(p_0 = 0, 1, \cdots)$.

Proof. If g is an arbitrary element of E_{u_0}, then

$$-\|g\|_{u_0} \cdot u_0 \leqslant g \leqslant \|g\|_{u_0} \cdot u_0.$$

We have

$$x_n(g) = \frac{1}{\lambda^n} A^n g + \sum_{k=0}^{n-1} \frac{A^k f}{\lambda^{k+1}}.$$

Further,

$$-\|g\|_{u_0} \cdot \left(\frac{\lambda_0}{\lambda^m}\right)^n u_0 \leqslant \frac{1}{\lambda^{nm}} A^{nm} g \leqslant \|g\|_{u_0} \cdot \left(\frac{\lambda_0}{\lambda^m}\right)^n u_0.$$

Since

$$\|g\|_{u_0} \left(\frac{\lambda_0}{\lambda^m}\right)^n u_0 \to \theta \quad (n \to \infty),$$

for $\lambda > \sqrt[m]{\lambda_0}$, we have

$$u_n(g) \equiv \frac{1}{\lambda^{nm}} A^{nm} g \to \theta \quad (n \to \infty),$$

by virtue of which

$$\lim_{n \to \infty} \| x_n(g) - R_\lambda f \| = 0. \tag{18}$$

From (17) it follows that

$$\frac{1}{\lambda_0^p} A^{p_0} g - \frac{1}{\lambda^{p_0+1}} A^{p_0+1} g \leqslant \frac{1}{\lambda^{p_0+1}} A^{p_0} f.$$

But

$$\frac{1}{\lambda^{p_0+1}} A^{p_0} f = \sum_{k=0}^{p_0} \frac{A^k f}{\lambda^{k+1}} - \sum_{k=0}^{p_0-1} \frac{A^k f}{\lambda^{k+1}},$$

i.e.

$$\frac{1}{\lambda^{p_0}} A^{p_0} g + \sum_{k=0}^{p_0-1} \frac{A^k f}{\lambda^{k+1}} \leqslant \frac{}{\lambda^{p_0+1}} A^{p_0+1} g + \sum_{k=0}^{p_0} \frac{A^k f}{\lambda^{k+1}},$$

from which we get that $x_{p_0} \leq x_{p_0+1}$. By virtue of the positiveness of A

$$x_{p_0+1} = \frac{1}{\lambda}(Ax_{p_0} + f) \leqslant \frac{1}{\lambda}(Ax_{p_0+1} + f) = x_{p_0+2},$$

and by induction we will have

$$x_{p_0+k-1} \leqslant x_{p_0+k} \qquad (k = 1, 2, \ldots).$$

Thus the sequence of approximations (16) beginning from the number p_0 is monotone. A is continuous over K because K is normal. Taking (18) into consideration and passing to the limit in (16), we have

$$\lambda R_\lambda f = A R_\lambda f + f,$$

i.e.

$$R_\lambda f \equiv \sum_{k=0}^{\infty} \frac{A^k f}{\lambda^{k+1}}$$

is a solution (for $\lambda > \sqrt[m]{\lambda_0}$) of equation (13).

Remark 1. In Theorem 4, instead of condition (17) we could require that

$$-A^{p_0}(\lambda g - Ag - f) \in K.$$

The proof in this case is the same, except that the successive approximations x_n $(n = 0, 1, 2, \cdots)$ form, beginning from some n, a monotone decreasing sequence.

Remark 2. The solution

$$R_\lambda f = \sum_{k=0}^{\infty} \frac{A^k f}{\lambda^{k+1}}$$

of equation (13) is unique in E_{u_0}.

In fact, let $x^* \in E_{u_0}$ and $\lambda x^* = Ax^* + f$, i.e.

$$A^{q_0}(\lambda x^* - Ax^* - f) \in K$$

for any q_0. Then, constructing successive approximations (16), beginning with the element x^*, we will have

$$x_0 = x^*,\ x_1 = \frac{1}{\lambda}(Ax^* + f) = x^*, \ldots, x_n = x^*, \ldots,$$

i.e. the sequence $\{x_n\}$ is stationary. On the other hand, by Theorem 4 $x_n \to R_\lambda f$.

Theorem 5. *Let the cone K be normal, let the positive operator A be u_0-bounded from above, and let*

$$A^m u_0 \leqslant \lambda_0 u_0 \quad (u_0 \in K,\ u_0 \neq \theta).$$

Then for $\lambda > \sqrt[m]{\lambda_0}$ and for any $f \in L(K)$ successive approximations (16) *converge to solution $R_\lambda f$ of equation* (13), *beginning from any null approximation $g \in L(K)$ satisfying one of the conditions*

$$A^{q_0}(\lambda g - Ag - f) \in K, \quad A^{q_0}(\lambda g - Ag - f) \in -K, \tag{19}$$

where q_0 *is an arbitrary natural number* ($q_0 = 0, 1, 2, \cdots$).

Proof. First let $g \in K$. We construct successive approximations (16). For g we will find p and α such that the inequality $A^p g \leq \alpha u_0$ is satisfied. Then $A^{nm+p} g \leq \alpha \lambda_0^n u_0$ ($n = 0, 1, \cdots$) and therefore

$$\theta \leqslant \frac{1}{\lambda^{nm+p}} A^{nm+p} g \leqslant \frac{\alpha}{\lambda^p} \left(\frac{\lambda_0}{\lambda^m}\right)^n u_0 \quad (n = 0, 1, \ldots),$$

i.e.

$$x_{nm+p}(g) \to R_\lambda f \quad (n \to \infty) \tag{20}$$

for $\lambda > \sqrt[m]{\lambda_0}$. Let the first of the relations in (19) be satisfied. Then, as in the proof of the preceding theorem, we get

$$x_{q_0} \leqslant x_{q_0+1} \leqslant \ldots \leqslant x_{q_0+n} \leqslant \ldots \quad (n = 1, 2, \ldots).$$

From this condition and conditions (20) we have

$$x_n(g) \to R_\lambda f \quad (n \to \infty).$$

Since K is normal and consequently A is continuous from the left, we can pass to the limit in (16). We will have

$$AR_\lambda f = \lambda R_\lambda f - f,$$

i.e. $R_\lambda f$ is a solution of equation (13).

Now let $g \in L(K)$; then $g = g_1 - g_2$ ($g_1, g_2 \in K$). Moreover, let $A^\gamma g_1 \leq \beta u_0$. We have

$$x_n(g) = \frac{1}{\lambda^n} A^n (g_1 - g_2) + \sum_{k=0}^{n-1} \frac{A^k f}{\lambda^{k+1}}.$$

Further

$$\frac{1}{\lambda^{\gamma+nm}} A^{\gamma+nm}(g_1 - g_2) \leqslant \frac{1}{\lambda^{\gamma+nm}} A^{\gamma+nm} g_1 \leqslant \frac{\beta}{\lambda^\gamma}\left(\frac{\lambda_0}{\lambda^m}\right)^n u_0,$$

i.e. $x_{\gamma+nm}(g) \to R_\lambda f$ ($n \to \infty$). Since beginning from some q_0 the sequence $\{x_n(g)\}$ is monotone, $x_n(g) \to R_\lambda f$. By going to the limit in (16) we get the assertion of the theorem.

Remark. The solution $R_\lambda f$ of (13) is unique in $L(K)$.

If the conditions of Theorem 5 are satisfied, if the operator B is sublinear relative to the operator A on the cone K, and if $y^* \in K$ is a solution of the equation

$$\lambda y = By + f \quad (\lambda > \sqrt[m]{\lambda_0}),$$

then the successive approximations (16) converge to a solution x^* of the equation

$$\lambda x = Ax + f,$$

if we take y^* as the null approximation. Using this fact, it is possible to substantially simplify the proof of the second part of Theorem 2.

3. If it is known that the operator A is continuous (bounded), then the following versions of Theorems 2 and 3 are valid.

Theorem 2′. *Let the operator B, which is monotone on the normal cone K, be sublinear relative to the positive bounded operator A, and let $BK_g \subset K$ $(g \in K)$.*

Then equation (7) *has at least one solution in K for all $f \in K_{[\epsilon+r(A)]g}$ and for $r(A) < \lambda \leq \Lambda$.*[1] *There are no congruent solutions in K_g. Here as in Theorem 2 we write*

$$\Lambda = \sup_{\lambda}\left\{\frac{f}{\lambda} \geqslant g\right\}.$$

For $g = \theta$ this theorem can be stated as follows:

Theorem 3′. *Let the positive operator B which is monotone on the normal cone K be sublinear relative to the bounded operator A. Then* (7) *has at least one solution in K for all $f \in K$ and for $\lambda > r(A)$. There are no congruent solutions in K.*

The proofs of Theorems 2′ and 3′ are completely analogous to the proofs of Theorems 2 and 3 respectively, and are therefore omitted.

4. We will show that if the operator B does not have the property of sublinearity but K has some additional properties, then the assertions of the preceding theorems remain valid.

Theorem 6. *Let the linear operator A be u_0-bounded from above, and let $A^m u_0 \leq \lambda_0 u_0$, $B \leq A$, where the operator B is monotone and positive. Let one of the two following conditions be fulfilled:*

a) *K is a regular cone, and the operator B is continuous over it.*

b) *K is a normal cone, and the operator B is completely continuous.*

Then (7) *has at least one solution in K for any $f \in K$ and for $\lambda > \sqrt[m]{\lambda_0}$.*

Proof. Considering the equations

---

[1] $r(A)$ is the spectral radius of A.

$$\lambda x = Bx + f, \qquad \lambda y = Ay + f \ (\lambda > 0)$$

and constructing for each of these equations successive approximations

$$x_0 = y_0 = \frac{1}{\lambda} f, \ x_{n+1} = \frac{1}{\lambda}(Bx_n + f),$$

$$y_{n+1} = \frac{1}{\lambda}(Ay_n + f) \ (n = 0, 1, \ldots) \tag{21}$$

we get, as in Theorem 2, that

$$x_n \leqslant x_{n+1}, \ y_n \leqslant y_{n+1}, \ x_n \leqslant y_n \ (n = 1, 2, \ldots).$$

The sequence $\{y_n(f)\}$ converges strongly for $|\lambda| > \sqrt[m]{\lambda_0}$ and for any $f \in K$, because

$$y_n(f) = \frac{1}{\lambda}(Ay_{n-1} + f) = \sum_{k=0}^{n} \frac{A^k f}{\lambda^{k+1}}$$

and $\lim_{n\to\infty} y_n = R_\lambda f$, and $y_n \leq R_\lambda f$ $(n = 1, 2, \cdots)$.

First let the first condition of the theorem be fulfilled. Then

$$x_1 \leqslant x_2 \leqslant \ldots \leqslant x_n \leqslant \ldots \leqslant R_\lambda f. \tag{22}$$

The sequence $\{x_n\}$ converges because K is regular. We now introduce the notation

$$\lim_{n\to\infty} x_n = x^*.$$

Since B is continuous over the cone, it is possible to pass to the limit in the first of the relations in (21) taking into consideration that $\{x_n\}$ is monotone. We get

$$x^* = \frac{1}{\lambda}(Bx^* + f) \qquad (\lambda > \sqrt[m]{\lambda_0}),$$

i.e. that x^* is a solution of (7).

Now let the second condition of the theorem be fulfilled. From (22) it is clear that $\|x_n\| \leq M \|R_\lambda f\|$ $(n = 1, 2, \cdots)$. From the complete continuity of B we get the compactness of the sequence $\{Bx_n\}$ $(n = 1, 2, \cdots)$, i.e. the compactness of the sequence $\{\lambda x_{n+1} + f\}$ $(n = 1, 2, \cdots)$. Therefore $\{x_n\}$ $(n = 1, 2, \cdots)$ is itself compact. But every compact sequence monotone over the cone converges: $x_n \to x^*$ $(n \to \infty)$. The remainder of the proof is clear.

Remark 1. Passing to the limit as $n \to \infty$ in (22), we will have

$$x^* \leq R_\lambda f.$$

Remark 2. In Theorem 6 the requirement that B be continuous over the cone can be replaced by the weaker requirement that it be continuous over monotone sequences. More precisely, it is possible to require that

$$\| x_n - x^* \| \to 0 \ (n \to \infty;\ x_0 \leqslant x_1 \leqslant \ldots \leqslant x_n \leqslant \ldots)$$

imply

$$\| Bx_n - Bx^* \| \to 0$$

and correspondingly to assert that B is continuous over monotone sequences and maps every bounded monotone sequence into a compact one instead of asserting that B is completely continuous.

Remark 3. If B satisfies the condition

$$B(f_1 + f_2) \leqslant Bf_1 + Bf_2 \quad (f_1, f_2 \in K), \tag{23}$$

then it is superfluous to require that B be continuous over the cone, since in this case the other conditions of the theorem imply that B is continuous over the cone.

We will show this. Let $h_n \in K$ and $\|h_n\| \to 0$. Then

$$\theta \leqslant Bh_n - B\theta = Bh_n \leqslant Ah_n,$$

and, because the cone is normal and operator A is continuous over the cone,

$$\| Bh_n \| \leqslant M \| Ah_n \| \to 0 \ (n \to \infty).$$

We also have that B is continuous over the cone at the point $x = \theta$. Now let x_0 be an arbitrary element of K. By (23), for $h_n \in K$

$$\theta \leqslant B(x_0 + h_n) - Bx_0 \leqslant Bh_n$$

and consequently

$$\| B(x_0 + h_n) - Bx_0 \| \leqslant M \| Bh_n \| \to 0.$$

The proof of the following theorem is like that of Theorem 6.

Theorem 7. *Let the linear operator A be positive, let B be positive and monotone, and let $B \leq A$. Moreover, let one of the conditions* a), b) *of Theorem* 6 *be fulfilled.*

Then (7) *has at least one solution $x^* \in K$ for $\lambda > r(A)$ and for any $f \in K$. This solution can be obtained by the method of successive approximations* (*cf.* (21)).

The following theorems are a generalization of Theorems 6 and 7.

Theorem 8. *Let the linear operator A be u_0-bounded from above,*

$$A^m u_0 \leqslant \lambda_0 u_0, \ B \leqslant A, \ BK_g \subset K,$$

where g *is a fixed element of the cone, and one of the conditions* a) *and* b) *of Theorem* 6 *is fulfilled. Then* (7) *has at least one solution* $x^* \in K$ *for any* $f \in K_{(\varepsilon + \sqrt[m]{\lambda_0})g}$ *and for* $\sqrt[m]{\lambda_0} < \lambda \leq \Lambda$.

Theorem 9. *Let the linear positive operator* A *be bounded,*

$$B \leqslant A, \quad BK_g \subset K \quad (g \in K),$$

and let one of the two conditions a) *and* b) *of Theorem* 6 *be fulfilled.*

Then (7) *has at least one solution* $x^* \in K$ *for any*

$$f \in K_{(\varepsilon + r(A))g}, \quad r(A) < \lambda \leqslant \Lambda.$$

In Theorems 8 and 9

$$\Lambda = \sup_{\lambda} \left\{ \frac{f}{\lambda} \geqslant g \right\},$$

and the solution x^* can be obtained by the method of successive approximations (cf. (21)) in which we can take $x_0 = f/\lambda$ as the initial approximation. Note that in Theorem 8 it is sufficient to require that $AK \subset K$ and that the operator A be u_0-bounded on K_g, instead of requiring that A be u_0-bounded on the cone K.

BIBLIOGRAPHY

[1] M. A. Krasnosel'skiĭ, *Positive solutions of operator equations*, Fizmatgiz, Moscow, 1962; English transl., Noordhoff, Groningen, 1964. MR 26 #2862.

[2] I. A. Bahtin, M. A. Krasnosel'skiĭ and V. Ja. Stecenko, *On the continuity of positive linear operators*, Sibirsk. Mat. Ž. 3 (1962), 156–160. (Russian) MR 25 #2451.

[3] M. G. Kreĭn and M. L. Rutman, *Linear operators leaving invariant a cone in a Banach space*, Usephi Mat. Nauk 3 (1948), No. 1 (23), 3–95; English transl., Amer. Math. Soc. Transl. (1) 10 (1962), 199–325. MR 10, 256; MR 12, 341.

[4] V. Ja. Stecenko and A. R. Esajan, *On the problem of solvability of equations of the second kind*, Izv. Akad. Nauk Tadžik. SSR Otdel. Fiz.-Tehn. i Him. Nauk 1964, no. 2 (15), 13–35. (Russian)

Translated by:
H. N. Parker